D1174642

Response of Plants
to Multiple Stresses

Physiological Ecology
A Series of Monographs, Texts, and Treatises

Series Editor
Harold A. Mooney
Stanford University, Stanford, California

Editorial Board
Fakhri Bazzaz   F. Stuart Chapin   James R. Ehleringer
Robert W. Pearcy   Martyn M. Caldwell

List continues at the end of the volume.

Response of Plants to Multiple Stresses

Edited by

Harold A. Mooney
Department of Biological Sciences
Stanford University
Stanford, California

William E. Winner
Department of General Sciences
Oregon State University
Corvallis, Oregon

Eva J. Pell
Department of Plant Pathology
Pennsylvania State University
University Park, Pennsylvania

Ellen Chu
Developmental Editor

Academic Press, Inc.
Harcourt Brace Jovanovich, Publishers

San Diego   New York   Boston
London   Sydney   Tokyo   Toronto

Front cover photo: Abundant development of "knees" in dense stand of baldcypress growing in a periodically flooded soil. Courtesy of U.S. Forest Service.

Academic Press, Inc.
San Diego, California 92101

United Kingdom Edition published by
Academic Press Limited
24–28 Oval Road, London NW1 7DX

Library of Congress Cataloging-in-Publication Data

Response of plants to multiple stresses / [edited by] Harold A.
   Mooney, William E. Winner, Eva J. Pell.
      p.   cm. – (Physiological ecology series)
   Includes index.
   ISBN 0-12-505355-X
   1. Plants, Effect of stress on.   2. Plant physiological ecology.
I. Mooney, Harold A.   II. Winner, William E.   III. Pell, Eva J.
IV. Series.
QK754.R47   1991
581.2–dc20   90-23925
                                                                   CIP

PRINTED IN THE UNITED STATES OF AMERICA
91 92 93 94   9 8 7 6 5 4 3 2 1

Contents

Part II
Biotic Interactions

Part III
Plant Growth Forms

16. Shrub Life-Forms
Philip W. Rundel

17. Responses of Evergreen Trees to Multiple Stresses
R. H. Waring

18. Effects of Environmental Stresses on Deciduous Trees
T. T. Kozlowski

Contributors

Numbers in parentheses indicate the pages on which the authors' contributions begin.

P. G. Ayres (227), Department of Biological Sciences, University of Lancaster, Lancaster LA1 4YQ, United Kingdom

F. A. Bazzaz (283), Department of Organismic and Evolutionary Biology, Harvard University, Cambridge, Massachusetts 02138

B. D. Campbell (143), Unit of Comparative Plant Ecology (NERC), Department of Plant Sciences, The University, Sheffield S10 2TN, England

F. Stuart Chapin III (67), Department of Integrative Biology, University of California, Berkeley, Berkeley, California 94720

N. R. Chiariello (162), Department of Biological Sciences, Stanford University, Stanford, California 94305

James S. Coleman[1] (249), Department of Biology, Stanford University, Stanford, California 94305

Michael S. Dann (189), Department of Plant Pathology and Environmental Resources Research Institute, The Pennsylvania State University, University Park, Pennsylvania 16802

R. E. Dickson (4), Forestry Sciences Laboratory, North Central Forest Experiment Station, USDA Forest Service, Rhinelander, Wisconsin 54501

Christopher B. Field (35), Department of Plant Biology, Carnegie Institution of Washington, Stanford University, Stanford, California 94305

Donald R. Geiger (104), Department of Biology, University of Dayton, Dayton, Ohio 45469

J. P. Grime (143), Unit of Comparative Plant Ecology (NERC), Department of Plant Sciences, The University, Sheffield S10 2TN, England

S. L. Gulmon (162), Department of Biological Sciences, Stanford University, Stanford, California 94305

J. G. Isebrands (4), Forestry Sciences Laboratory, North Central Forest

---

[1] Current address: Department of Biology, Biological Research Laboratories, Syracuse University, Syracuse, New York 13244.

Experiment Station, USDA Forest Service, Rhinelander, Wisconsin 54501

Clive G. Jones (249), Institute of Ecosystem Studies, Mary Flagler Cary Arboretum, The New York Botanical Garden, Millbrook, New York 12545

Jon E. Keeley (329), Department of Biology, Occidental College, Los Angeles, California 90041

T. T. Kozlowski (391), Environmental Studies Program and Department of Biological Studies, University of California, Santa Barbara, Santa Barbara, California 93106

S. J. McNaughton (307), Department of Biology, Syracuse University, Syracuse, New York 13244

H. A. Mooney (129), Department of Biological Sciences, Stanford University, Stanford, California 94305

S. R. Morse (283), Department of Organismic and Evolutionary Biology, Harvard University, Cambridge, Massachusetts 02138

Eva J. Pell (189), Department of Plant Pathology and Environmental Resources Research Institute, The Pennsylvania State University, University Park, Pennsylvania 16802

Philip W. Rundel (345), Laboratory of Biomedical and Environmental Sciences, Department of Biology, University of California, Los Angeles, Los Angeles, California 90024

Edward J. Rykiel, Jr. (206), Center for Biosystems Modelling, Industrial Engineering Department, Texas A&M University, College Station, Texas 77843

E.-D. Schulze (89), Lehrstuhl Pflanzenökologie, Universität Bayreuth, D-8580 Bayreuth, Germany

Jerome C. Servaites (104), Research Institute, University of Dayton, Dayton, Ohio 45469

Peter J. H. Sharpe (206), Center for Biosystems Modelling, Industrial Engineering Department, Texas A&M University, College Station, Texas 77843

R. H. Waring (371), College of Forestry, Oregon State University, Corvallis, Oregon 97331

W. E. Winner (129), Department of General Science, Oregon State University, Corvallis, Oregon 97331

Preface

Plants in natural environments are subject to changing multiple stresses during their annual growth cycles. The structural and metabolic features characteristic of plants growing in diverse environments are thus, most likely, traits that have evolved to optimize growth and reproductive output under these changing conditions. Therefore, an important avenue toward understanding the responses of plants to multiple stresses is to compare traits of plants from diverse environmental regimes. Such comparisons will reveal the genetic potential and adaptive pathways for coping with environments characterized by different stress combinations. Evolutionary traits such as leaf longevity can be viewed as coarse adjustments to a specific environmental stress regime. Through short-term, plastic adjustments, plants can also adjust to local microsite conditions or conditions that change during their life span. The short-term adjustments generally interest the physiologist, whereas ecologists focus on the longer-term responses. In this book we bring together both evolutionary and physiological perspectives, with their differing time scales of concern, to develop a comprehensive view of how plants cope with stress in the environment.

Plants are integrated units. Stress responses cannot be fully evaluated except in the context of the whole plant. In virtually all natural environments, plants are subjected to conditions that reduce their potential growth: for instance, low light or temperature or limiting water or nutrient availability. As a stress is imposed, traits usually exhibit a cascade of responses occurring on different time scales and involving biochemical and morphological adjustments. These responses act in concert to compensate for environmental changes so that, it appears, maximum productivity for the new resource state is achieved. Such whole-plant integrated responses through time, involving many processes, have not been generally studied. Most studies of plant stress have focused on responses of a single process to a single stress factor over a short time. Extrapolating from such short-term results to effects on long-term yield may be misleading because of longer-term compensations.

In addition to evaluating stress responses in a whole-plant context, far more attention should be directed to examining the effects of interacting stresses. Such studies have been carried out for a number of air pollutants but not generally for natural stress factors. These studies should be designed to mimic the time course and intensity of these factors, as they occur in natural environments, so that the results from them are unambiguous.

It will be an enormous challenge to design and carry out studies in the future that are more integrated and span longer time scales and address interactions among multiple stresses. But without such research, we cannot develop a comprehensive understanding of plant stress responses. And without such an understanding, we cannot improve our management of crops and forests or predict the responses of plants to the new environmental regimes that global change will produce: altered atmospheric properties, including higher concentrations of CO^2 and often high concentrations of pollutants; more unpredictable rainfall; and warmer and perhaps more extreme temperatures than at present.

As a beginning toward understanding the evolutionary and physiological responses of whole plants to environmental stress, a group of ecologists and plant physiologists came together at Asilomar, California, to share their knowledge and viewpoints. This book is the result of the interactions that occurred at this meeting. We have organized the material into three sections reflecting the short-term responses of plants to multiple stresses, the immediate consequences of these responses to biotic interactions such as pathogens and herbivores, and the long-term evolutionary consequences of the adaptations of plants to specific environmental stress regimes. This third section reinforces the need to understand stress responses at the whole-plant level.

We hope this book will prove valuable to students of plant ecology and environmental physiology as well as to those involved in the management of plants and their environments. We thank the U.S. Department of Energy and the Electrical Power Research Institute for providing support for this meeting and the resulting synthesis.

I

Structure, Function, and Resources

1

Leaves as Regulators of Stress Response

R. E. Dickson   J. G. Isebrands

I. Introduction

The acclimation of plants to changing environmental factors or stresses involves both short-term physiological response and long-term physiological, structural, and morphological modifications. These changes help minimize stress in the plant and maximize the use of internal and external resources. Most stress responses are not fully expressed in isolated systems but occur as an integrated response of the whole plant. Some of these plant responses are very subtle, e.g., the activation of vascular invertase in the dark only if the leaf is attached to the plant (Eschrich and Eschrich, 1987). Other changes are more dramatic, e.g., shedding of leaves, flowers, and fruit in response to drought (Daie, 1988). Much information is available on specific physiological responses within single leaves (Geiger, 1987), but relatively little is known about whole-plant systems and whether the physiological mechanisms involved in response are initiated wholly within the leaf or result from whole-plant interactions. In addition, basic biological information on structural–functional interactions of whole plants is lacking for most plant species, and even less is known about how structural–functional interactions change with stress.

The whole plant can be divided into as many subsystems as desired. Leaves, stems, and roots are three major, but arbitrary, divisions. Various combinations of leaf, stem, or root may form integrated physiological units (Watson and Casper, 1984; Thomas and Watson, 1988). Such units may interact with and respond to environmental stress in relative isolation from the whole plant. The leaf, node, and subtending stem internode make up an integrated stem unit that is a basic building block for shoots (Kremer and Larson, 1982, 1983). The anatomical and physiological development of this stem unit is regulated by the leaf on the unit and by other developing leaves above. The growth of each component of a stem unit and of aggregates of stem units may be influenced by environmental stresses that significantly alter shoot development. Leaves, stems, and roots all receive environmental stimuli and respond in an integrated fashion to control the growth and development of the whole plant. It is, however, the metabolic responses of leaves to different environmental stresses that largely regulate the growth and development of both shoots and roots.

Many regulatory systems are present in leaves, and these systems change both as a function of leaf developmental stage (Dale and Milthorpe, 1983) and in response to different environmental factors. The major regulatory functions of leaves can be divided into two areas: control of anatomical and morphological development of the shoot and control of carbon allocation within the whole plant. We believe it is necessary to have a large pool of basic biological information concerning structural and functional development—knowledge derived from plants grown under more or less optimal conditions—to understand, interpret, and predict plant changes in response to stress. Such structural–functional information has been developed for cottonwood (*Populus deltoides* Bartr. Marsh.). This information could form the basis for similar studies on other plants and could also be used to predict morphological and physiological changes in shoots in response to environmental stresses. The objectives of this chapter are to review some of the anatomical and physiological information available from cottonwood and other plants and to show how this information might be used to predict the response of shoots (leaves, stems, branches) to multiple environmental stresses.

II. Leaf Regulation of Shoot Morphological Development

Shoot anatomical and morphological development is tightly controlled in both time and space. The development of leaves, stems, lateral branches, and other shoot components is the result of both genetic and environmental constraints operating through a variety of feedback systems. Leaves are major regulators of shoot development. Leaves control shoot development by regulating (1) leaf primordium initiation and development, (2) petiole and lamina vascularization of developing leaves, (3) lateral bud initiation and vascularization, and (4) stem primary and secondary vascular development. The vascular system, both xylem and phloem, connects essentially every organ and tissue of the plant and transports water, organic and mineral nutrients, hormones, and other essential products throughout the plant. It is essential to understand the anatomical development of different parts of the shoot and how leaves regulate this anatomical development before one can predict responses to stress.

A. Primordium Initiation and Development

Primordium initiation and the resulting phyllotactic patterns are important for shoot morphology because they determine leaf display. Over the years, several theories and hypotheses have been proposed to account for primordium initiation and the development of various phyllotactic patterns (Hicks, 1980; Larson, 1983; Chapman and Perry, 1987). Most of these hypotheses assume *de novo* creation of primordia on the apex, control of initiation by the apical meristem, and origin of the leaf vascular system within the primordia. Much anatomical evidence, however, indicates that the vascular system develops acropetally in continuum with the primary and secondary vascular system in the stem and does not originate within the primordia. The vascular system is highly organized and under strict genetic and physiological control (Larson, 1980, 1983).

The leaf trace concept of primordium initiation and subsequent leaf development provides a basis for studying and describing shoot development. Two major components of the leaf trace concept are the procambial system and the antecedent leaf. The procambial system consists of procambial stands, vascular traces that develop acropetally in the ground tissue of the apical meristem well before initiation of the primordia to which they will eventually connect, and the procambial leaf traces, vascular leaf traces that develop after primordium initiation and vascularization by the procambial strand. The antecedent leaf is an older leaf, or leaves, directly connected by vascular traces to other developing leaves and primordia. Both the anatomical and physiological events associated with the leaf trace concept have been worked out in great detail for cottonwood *(Populus deltoides)* (Dickson and Shive, 1982; Larson, 1980, 1982, 1983, 1984). This concept is particularly important because it provides an anatomical and physiological basis for modification of shoot morphology by environmental stimuli received and acted upon by the antecedent leaves.

The central vascular trace of a leaf is the perpetuating trace of the acropetally progressing primary vascular system. The procambial strand of the central trace develops acropetally in the residual meristem of the apex, then diverges to the point of primordium initiation (Fig. 1). Shortly after the initial vascularization of the primordium, the central trace branches

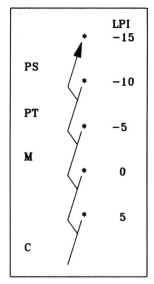

Figure 1 Central trace developmental sequence in cottonwood. The central trace is directly connected to every fifth leaf and increases in maturity for the procambial strand downward to the cambium, the area of secondary development. LPI (leaf plastochron index) 5 is the first functional antecedent leaf based on assimilate transport from the leaf. LPI −15 is an initiating primordium not yet in anatomically discernible contact with the procambial strand. PS, procambial strand; PT, procambial trace; M, metacambium; C, cambium. [Adapted from Larson (1980).]

and continues upward until it diverges again to another primordium one orthostichy above (in a plant with 2/5 phyllotaxy an orthostichy consists of five leaves). This acropetal development of the procambial system is not restricted to cottonwood. Similar patterns of vascular development have been found in a wide variety of plant species with quite different growth forms (Kuehnert and Larson, 1983; Larson, 1986a,b). In these plants, the primary vascular system also develops acropetally in continuum to form the highly organized and functionally integrated stem–node–leaf complex, even though leaves vary from simple to highly complex. The leaf trace system provides a template for all subsequent vascular development and a continuous system through which regulatory signals can be transported. Although there is little evidence of direct environmental influences on trace development, phyllotactic transitions in the vascular system may be controlled by physiological feedback mechanisms between developing leaves and the vascular system, thus providing a mechanism for plants to change leaf arrangements and leaf display in response to changing environmental conditions. Environmental conditions are very important, however, in initiating changes in both primordia and apical meristems. Changes from leaves to phyllodes or thorns and changes from a vegetative apex to flowers to dormant bud are all initiated by environmental changes (King, 1983). Perception of environmental change is primarily by antecedent leaves, and transmission of biochemical signals is through the vascular system.

A vital part of the leaf trace concept is the presence of the antecedent leaf or leaves. This leaf provides metabolites and perhaps hormones for primordium initiation and subsequent development. ^{14}C tracer studies have shown that recently mature leaves transport photosynthate only to developing leaves and primordia directly connected by vascular traces to the mature source leaf (Larson, 1977). Because the transport distances involved rule out diffusion, the first functional antecedent leaf must be capable of both CO_2 fixation and phloem transport (Dickson and Shive, 1982).

Antecedent leaves are the major receptors of environmental stimuli and thus exert considerable control over the development of the apex. An example of this control was found when cottonwood plants were transferred from long (18 h) days to short (8 h) days (Goffinet and Larson, 1981). These 12-leaf plants had about 14 young developing leaves and primordia in the bud (LPI 0 to LPI -14). After transfer to short days, all 14 of these preformed leaves and primordia expanded and matured as foliage leaves. The next two primordia that were initiated under short days (a new primordium was initiated about every two days in plants of this size) aborted, however, and the stipules expanded to form new bud scales. Subsequent primordia developed normally but remained dormant in the bud. Although it took about two weeks for the bud scales to expand and the dormant bud to form, the signal to initiate this developmental change probably arrived at the apex within one or two days after the antecedent leaves received the short-day stimuli.

B. Petiole and Leaf Vascularization

Initial vascular development in the petiole and lamina also proceeds acropetally in a precisely controlled sequence, which is similar in all plants studied to date (Larson, 1984). For example, a cottonwood leaf is served by three traces: a central trace and left and right lateral traces. During primordium initiation, a procambial strand diverges from the central trace of the sympodium (see Fig. 1), exits the stem at the node, and forms the central trace (Fig. 2: C_1) of the developing leaf primordium (Isebrands and Larson, 1977, 1980). The central trace is followed shortly in sequence by the procambial strands of the left and right lateral traces (Fig. 2: L_1 and R_1). Within two to three plastochrons (time interval between initiation of new primordia), subsidiary bundles (Fig. 2: C_2, C_3, L_2, L_3, etc.) begin to develop from the leaf basal meristem (Larson, 1984) and progress acropetally in the petiole and lamina as the primordium expands. The original traces and related subsidiary bundles diverge to form the secondary veins of the lamina. Components of the central trace diverge into both right and left halves of the lamina and continue to the lamina tip as well. Components of the left lateral traces are confined to their respective halves of the lamina (Isebrands and Larson, 1980). During vascularization of the leaf, additional subsidiary bundles develop from the leaf basal meristem downward

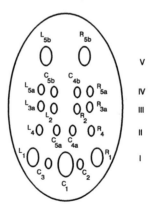

Figure 2 Schematic diagram of the arrangement of procambial leaf traces and their subsidiary bundles in a petiole transection 1 cm below the juncture of the petiole and lamina in a developing cottonwood leaf. Capital letters denote the leaf trace and related subsidiary bundles: central (C), right (R), or left (L). Numerical subscripts indicate the number of the subsidiary bundle, and small letters denote temporal origin of the bundle. Roman numerals (I–V) denote the temporal origin of the tiers of subsidiary bundles. [Adapted from Isebrands and Larson (1977).]

in the stem and merge with existing stem vascular traces to integrate the leaf and stem vascular system (Larson, 1984). Thus the veins of the leaf are intimately connected with the vasculature of the stem throughout development. The size and vigor of the original traces and the number of subsidiary bundles are probably controlled by metabolic processes in the antecedent leaves. This in turn determines the course of leaf development and length and width of the new leaf (Pieters and van den Noort, 1988).

C. Lateral Bud and Branch Vascularization

Despite the importance of lateral bud and branch development to crown architecture, relatively little detailed information is available concerning initial vascularization and subsequent development. In cottonwood, it appears that leaves also regulate the initiation and vascularization of lateral buds and branches. Lateral bud initiation begins very early during stem development. About three to four plastochrons after leaf primordium initiation, two bud traces diverge from the central trace of the primordium and develop upward to the area of bud initiation. These original bud traces branch and vascularize the first four leaves of the bud (Larson and Pizzolato, 1977; Pizzolato and Larson, 1977). A procambial strand then diverges from each of these central traces and vascularizes additional primordia in the bud. After bud break and elongation of the lateral branch, foliage leaf initiation, vascularization, and development are essentially the same as those on the terminal shoot (Richards and Larson, 1982). Subsequent primary and secondary vascular development of the branch is regulated by leaves of the branch and progress acropetally in the branch and basipetally into the main stem to form a functionally continuous transport system (Larson and Fisher, 1983). Thus continued lateral branch development requires a suitable environment for the branch and functional leaves on that branch (see Section IV,B).

D. Primary and Secondary Vascularization of the Stem

As the original procambial strands progress acropetally in the stem and as the related procambial traces and subsidiary bundles vascularize the leaf, additional subsidiary bundles develop basipetally from the leaf basal meristem and merge with the existing stem traces to integrate the leaf and stem vascular system (Larson, 1984). These vascular traces in the stem progress from primary tissue (protoxylem to metaxylem) to secondary tissue in an ordered developmental sequence associated with leaf ontogeny (Fig. 3).

Although associated initially with specific leaf traces, secondary development is not directly related to the relative age of the leaf connected to that trace but is more a function of physiological gradients within the maturing stem unit. Secondary vascular development is first recognized by the initiation of fibers with birefringent secondary walls (Fig. 3b). After initia-

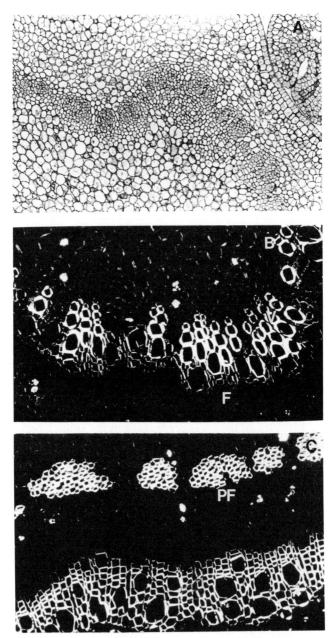

Figure 3 Micrographs of developing leaf traces in cottonwood stems. (A) Transverse section of primary tissue in the internode, showing developing procambial traces and subsidiary bundles. (B) Transverse section of internode showing the initiation of secondary tissue (i.e., fibers and vessels) beginning at the leaf traces. Polarized light microscopy. (C) Transverse section of internode below the secondary transition zone, showing the complete woody cylinder of secondary xylem and secondary wall development in the phloem fibers. Polarized light microscopy. [Adapted from Larson and Isebrands (1974).]

tion at the leaf traces, fiber differentiation progresses tangentially into the interfascicular regions while vessel initiation progresses radially in association with the existing leaf traces. After the leaf–node–internode stem unit matures, new secondary tissue forms a completely closed woody cylinder (Fig. 3c). Although the individual leaf traces eventually lose their identity in the closed cylinder of secondary tissue, they persist as relatively independent pathways from the leaf, through the stem and probably down through the roots (Kozlowski and Winget, 1963; Fisher *et al.*, 1983).

The initiation of secondary vascular tissue is associated with the cessation of internode elongation, which is in turn associated with leaf maturation. In cottonwood, the transition from primary to secondary tissue always occurs in the internode above the first fully expanded leaf (Larson and Isebrands, 1974). Thus, environmental factors regulating leaf size and maturation rate also regulate internode development and stem morphology (Goffinet and Larson, 1981).

III. Leaf Regulation of Carbon Allocation

One needs to know the basic patterns of carbon fixation and transport within both leaves and shoots to understand changes in these patterns in response to environmental stress (Chapter 5). Developing leaves may simultaneously transport photosynthate out of mature regions of the lamina to other parts of the shoot and import photosynthate to immature regions. Mature leaves normally export photosynthate only to developing leaves with direct vascular connections to that source leaf. The imposition of single or multiple stresses may change both the rates and patterns of these developmental processes. For example, loss of a single source leaf usually does not have a significant effect on growth of developing leaves because each sink leaf is connected by vascular traces to two or three other source leaves. Loss of several mature source leaves, however, may significantly influence new leaf development. Plants often compensate for such losses with an increase in new leaf growth and a decrease in lower stem and root growth (Bassman and Dickmann, 1982, 1985). Most shoot responses to stress depend on such compensation strategies.

A. Ontogenetic and Morphological Indices (PI and QMI)

To study and accurately relate structural development and physiological function within and between plants, it is necessary to quantify plant ontogeny with plant morphological indices. Although widely used, leaf or plant chronological age (e.g., time after planting, emergence, or bud break) is not adequately related to morphological or physiological development. Structural development and physiological functions are closely sequenced in time and space and form a continuum from leaf initiation to full expansion to senescence. The exact timing of this sequence (leaf initiation rate, leaf expansion rate, or number of leaves in the leaf-developing zone) is very similar within clonal material (Ceulemans *et al.*, 1988) or even within seedlings from a defined seed source (Larson and Isebrands, 1971). Even within clonal material growing under the same environmental conditions, however, different plants are at different ontogenetic stages or are slightly out of phase with one another. In plants with indeterminate shoot growth, leaf initiation and developmental rates are uniform for a particular plant and are related more to plant size than to small changes in environment (Pieters, 1983, 1986). In such plants, the plastochron index (PI), a numerical index of plant ontogeny, can be used to quantify developmental stage (Larson and Isebrands, 1971; Lamoreaux *et al.*, 1978; Ceulemans *et al.*, 1988). With this information, structural–functional studies can be conducted on different plants at different times but at the same developmental stage. With the plastochron index, lamina lengths are used to predict anatomical, biochemical, and physiological processes in different parts of the plant. Such processes can be predicted because they are related to the morphological events measured.

Unfortunately, the PI system cannot be applied to plants with determinate or semideterminate shoot growth. Studies on such plants require the development of a morphological index based on simple and reproducible stem and leaf measurements so that plants can be treated and sampled at specific ontogenetic stages. An example of this type of index was developed for northern red oak and named the Quercus Morphological Index (QMI) (Hanson *et al.*, 1986). Without such indices, it is difficult to define relationships among plant developmental stage, physiological processes, and environmental conditions and even more difficult to predict or interpret responses to environmental stresses.

B. Carbon Transport in Developing Leaves

Carbon transport is initially controlled within the leaf. Most simple leaves mature first at the apex. Photosynthate fixed by the apical region is then translocated out of the leaf through the midvein and petiole to other developing leaves above (Fig. 4a) (Dickson, 1986). Usually no photosynthate is translocated from the mature tip to developing basal lamina because of the spacial separation and lack of phloem bridges between leaf traces serving the tip and base. In compound leaves such as those of honeylocust (*Gleditsia triacanthos* L.) or green ash (*Fraxinus pennsylvanica* Marsh.), maturation progresses from the base to the tip (Larson and Dickson, 1986). At certain stages of leaf development, mature leaflets can translocate both to distal developing leaflets on the rachis and out of the leaf to other parts of the plant (Fig. 4b and c). To further complicate within-leaf transport patterns, plants such as northern red oak (*Quercus*

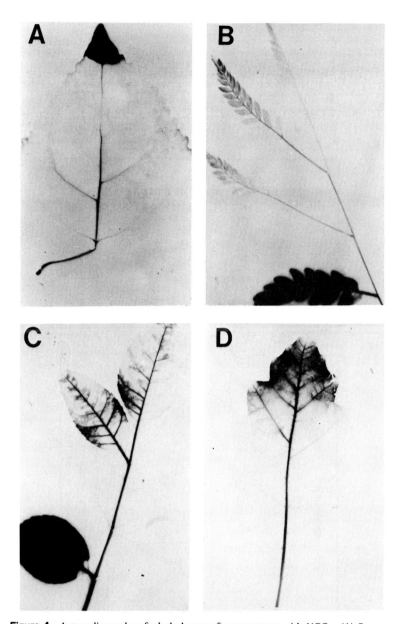

Figure 4 Autoradiographs of whole leaves after treatment with $^{14}CO_2$. (A) Spot treatment of lamina tip of cottonwood. (B, C) Whole lower leaflet treatment of compound leaves of honeylocust and green ash, respectively. (D) Expanding leaf of northern red oak imports ^{14}C assimilate to the developing lamina tip only. Note the direction of transport in each case: Cottonwood exports ^{14}C from the mature tip past the developing base and out of the leaf. Honeylocust and green ash leaves transport ^{14}C from the mature basal leaflets both acropetally to the developing leaflets and basipetally out of the leaf. Red oak imports ^{14}C only into the developing lamina tip, indicating relative basal maturity. [Adapted from Dickson (1986) and Larson and Dickson (1986).]

rubra L.) have a flushing growth habit in which all leaves of a flush expand and mature at about the same time. In northern red oak, leaves develop from base to tip. Therefore the tip of the lamina is slightly younger than the base and imports photosynthate after import to the base has stopped (Fig. 4d).

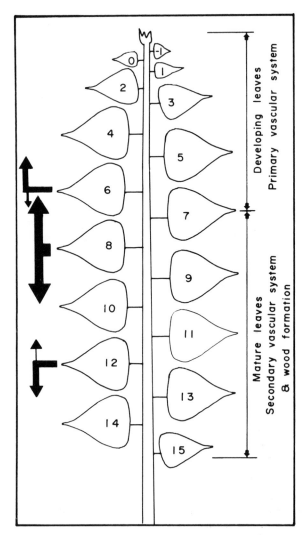

Figure 5   Diagram of a 16-leaf cottonwood plant showing the leaf plastochron index (LPI) numbering system, developing and mature leaf zones, areas of primary and secondary vascularization, and photosynthate transport. Arrows at left indicate direction of transport; arrow size indicates relative amounts. [From Dickson (1986).]

C. Carbon Transport within the Shoot

Transport from mature leaves depends on vascular connections between source and sink leaves and proximity to major sinks (Vogelmann *et al.*, 1982; Dickson, 1986; Geiger, 1987). For example, a recently mature source leaf on a 16-leaf cottonwood plant has direct vascular connections to sink leaves inserted three and five positions above the source leaf; thus a leaf at LPI 7 would transport primarily to sink leaves LPI 4 and LPI 2 (Fig. 5). Recently mature leaves (LPIs 7, 8, and 9) translocate upward to developing leaves and apex as well as downward to lower stem and roots. Older leaves (LPIs 12 to 15) transport primarily downward to lower stem and roots. Similar within-plant transport patterns hold for most plants with indeterminate growth during vegetative growth. For example, a mature honeylocust leaf eight leaves down from the apex transports about 30% to upper stem and leaves and 70% downward to lower stem and roots (Table I). The leaf at the eleventh position translocates essentially all photosynthate to lower stem and roots. Although the transport patterns described above hold for most plants with indeterminate growth, anomalous transport patterns have been found. For example, in tomato (*Lycopersicon esculentum* L.), the lower leaves translocate upward to developing leaves and apex, and the upper leaves translocate to lower stem and roots (Russell and Morris, 1983). Such patterns reflect the functioning of internal and exter-

Table I Transport of ^{14}C from Mature Honeylocust Leaves at Different Positions below the Apex[a]

| | Treated leaf type and position | | | |
| | Once-pinnate | | Bipinnate | |
Transport direction	8[b]	11	8	11
Upward				
Leaves	10.1[c]	1.0	18.4	0.7
Stem	20.5	1.0	17.4	4.0
Downward				
Leaves	0.2	0.2	0.1	1.0
Stem	53.0	39.9	42.0	76.0
Roots	16.2	57.9	22.1	18.3

[a] $^{14}CO_2$ was administered to either once-pinnate leaves on 20-leaf plants or bipinnate leaves on 26-leaf plants.

[b] Recently matured leaves were located at position 8 and fully matured leaves at position 11 below the apex.

[c] Values represent percentage of total ^{14}C translocated from source leaves and are the means of two replications. [From Larson and Dickson (1986).]

nal phloem and cannot be interpreted without detailed structural information.

Within-plant transport patterns are largely regulated by the sink strength of developing leaves, which compete with other sinks throughout the plant. Northern red oak seedlings provide an extreme example of such control. During a flushing episode when the stem and leaves of the second flush are rapidly expanding, more than 90% of the photosynthate translocated from first-flush leaves moves upward to the developing second flush, and about 5% moves to lower stem and roots (Fig. 6). During the lag phase, when second-flush leaves and stem are fully expanded, only about 5% of photosynthate from first-flush leaves is translocated upward, and 95% is translocated to lower stem and roots.

Plants respond to stress or environmental changes by both short- and long-term changes in allocation of carbon and other resources. Environmental changes are sensed primarily by leaves, which then respond metabolically, or with changes in growth rate, or both. This change in leaf growth rate changes sink strength and thus changes allocation within the

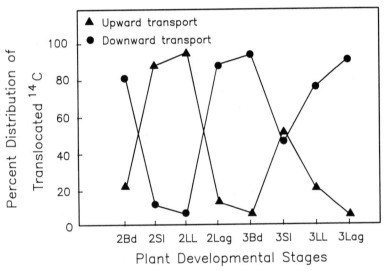

Figure 6   Distribution of translocated ^{14}C within northern red oak plants during two flushing episodes. Percentages are based on ^{14}C translocated from a first flush source leaf and recovered in plant parts 48 h after $^{14}CO_2$ treatment. Upward transport: sum of all ^{14}C recovered in developing stem and leaves above the first flush. Downward transport: sum of all ^{14}C recovered in first flush stem, tap root, and lateral roots. Developmental stages based on the *Quercus* morphological index (QMI; Hanson *et al.*, 1986), e.g., 2-bud stage: new developing flush 2 to 4 cm long; 2-leaf linear stage: stem elongation stopped and medium flush leaves in most rapid phase of expansion; 2-lag stage: second flush leaves have stopped elongation; 2-lag to 3-bud stage: interval between completion of one growth flush and the onset of the next flush (Dickson, unpubl. data).

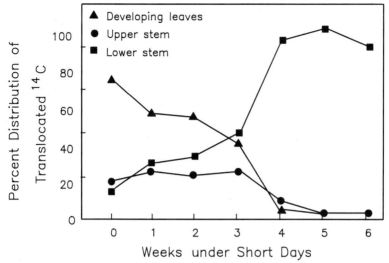

Figure 7 Distribution of translocated ^{14}C within cottonwood plants during dormancy induction. Percentages are based on total ^{14}C translocated from a source leaf at LPI 7 and recovered in plant parts 48 h after $^{14}CO_2$ treatment. Developing leaves: pooled LPI 0 to LPI 4. Upper stem: all stem tissue above the source leaf. Lower stem: all stem tissue below the source leaf to the cotyledons. [Adapted from Dickson and Nelson (1982).]

plant. For example, a short-term response occurs when leaves that are inhibited by water stress during the day expand more rapidly during the night. The rapid expansion increases the sink strength of developing leaves and causes carbon normally partitioned to roots at night to be partitioned instead to developing leaves (Bunce, 1978; Kerr *et al.,* 1985). However, the decrease in leaf sink strength during the day would increase transport to roots at that time.

An example of long-term response occurs when cottonwood plants are switched from long (18 h) to short (8 h) photoperiods. Under long days, cottonwood plants grow rapidly, and more than 80% of the photosynthate translocated from upper mature leaves (LPI 7 and 8) is partitioned to upper stems and developing leaves (Fig. 7). Under short days, leaf growth decreases rapidly, and carbon, normally allocated to leaf development, is reallocated to lower stem and roots. When leaf and upper stem growth stops, assimilates from mature leaves are directed primarily into storage compounds in lower stems and roots (Dickson and Nelson, 1982).

IV. Autonomy: The Relation of Plant Parts to the Whole System

Autonomy, or the relative inability of different parts of the shoot to exchange acquired resources, has a significant effect on the response of the whole shoot to environmental stresses. Plants are composed of modular units that increase in complexity from tiny intercellular organelles to whole shoot or root systems. These units combine into subsystems that tend to function semi-independently with respect to the whole plant. Such subsystems function autonomously, particularly with respect to the assimilation and use of carbon. The response of these subsystems to environmental stress largely determines the morphological and physiological adaptation of the whole plant. In this section we consider only three subsystems of increasing complexity: leaves, lateral branches, and ramets in clonal groups.

A. Leaves

As leaves expand and mature, they change from sinks to sources of assimilates. This maturation involves both structural and physiological changes. Mature source leaves fix carbon, use some for internal respiration and tissue maintenance (ca. 20–30%), and export the remainder (ca. 70–80%) (Geiger, 1987). Mature leaves under normal conditions do not import photosynthate via the phloem from other mature leaves, even if directly connected by vascular traces. It is possible for some photosynthate to leak from, or be transferred from, phloem to xylem and then move into mature leaves (Minchin and McNaughton, 1987). But such xylem-transported photosynthate, even if present in large quantities, is used within the vascular tissue or reloaded into the phloem and exported (Vogelmann *et al.*, 1985). Mature leaves can be induced to import small quantities of photosynthate, e.g., if severely stressed by aphid attack (Wu and Thrower, 1973) or by experimental manipulation of light and CO_2 levels (Thrower and Thrower, 1980; Fisher and Eschrich, 1985a,b; Eschrich and Eschrich, 1987). Aphid use of phloem contents can reverse the normal outward flow in sieve tubes because aphids are often sited over the major veins downstream from the phloem-loading and flow control mechanisms, which are located in minor veins. If mature leaves are induced to import assimilate by drastic treatments, the imported assimilate remains mostly in the veins. Mature leaves are usually incapable of using imported sugars because of active phloem reloading from the apoplast (outside of the plasmolemma in nonliving cell walls, etc), or blockage of symplastic (inside the plasmolemma or living protoplasts) transport pathways away from the phloem (Fisher, 1986; Turgeon, 1987). Thus, shading or other environmental stress usually leads to senescence and shedding of mature leaves that cannot maintain a favorable carbon balance.

B. Branches

Branches, as assemblages of leaves, are more complex subsystems. During early development, as the axillary bud elongates and foliage leaves expand from the preformed leaf primordia in the bud, the branch is a strong sink. Assimilate for early development of sylleptic branches (those developing from current-year buds) comes largely from the axillant leaf (Fisher *et al.*, 1983). Assimilate for the initial development of proleptic branches (those developing from dormant buds) comes from stem storage tissue in deciduous plants and from both stem and leaf storage in evergreens. If the axillant leaves are small or under stress, or if there is not enough assimilate available from storage, the lateral branch usually aborts during initial development. The sink strength of the branch decreases as more foliage leaves develop. When the new branch has enough mature leaves to supply developing leaves, it becomes photosynthetically independent of the main shoot (Dickson, 1986). Assimilate produced by branch leaves is distributed within the branch just as it is in the main shoot, i.e., upper mature leaves supply the developing leaves, and lower mature leaves supply the lower portions of the branch. Excess assimilate produced by mid- and lower-branch leaves is translocated to the main stem. Thus, branch assimilate is important for lower main stem and root growth, particularly after budset and maturation of the terminal branch leaves. This excess branch assimilate contributes little to height growth, however (Isebrands, 1982; Isebrands and Michael, 1986). Leaves of mid- and lower-crown branches export little assimilate to the current terminals or upper crown of the main stem or to other lower crown branches. Also, leaves on current terminals and upper branches do not export to the lower branches (Isebrands, 1982; Isebrands *et al.*, 1983). Thus, the leaves on any particular branch are responsible for production of assimilate for branch and leaf respiration, for cambial development, for vascular tissue development in that branch, and for development and maintenance of vascular connections between branch and main stem. If excess photosynthate is not produced by leaves on the branch because of shading or some other stress, cambial activity in the lower branch stops, vascular connections between the branch and newly developing xylem of the main stem are not maintained, and the branch is soon isolated from the water and mineral nutrient flow of the main stem (MaGuire and Hann, 1987). When such flow is disrupted, leaves on the branch rapidly senesce, and the whole branch dies.

C. Plant Clonal Groups

Individual plants, or ramets, within clonal groups behave much like the main stem and branch systems of single plants, although the potential for integration and sharing of assimilate and other resources among individual ramets is probably greater. In addition, the ability to share or transfer resources among ramets in a clone may be highly species specific (Magda *et al.*, 1988). Shading and other manipulative studies have shown that clonal plant groups are highly integrated systems. Shading of certain ramets, for example, increases photosynthesis in neighboring ramets, indicating increased sink strength (Hartnett and Bazzaz, 1983). Ramets detached from the genet and shaded during early development quickly die. In contrast, ramets connected to the genet and shaded continue early development, but begin to die when they approach maturity. This implies that assimilate is transported among the ramets during early development and that it stops when the shaded ramet matures. Similarly, ^{14}C transport studies have shown an increase in transport of current ^{14}C assimilate to the rhizome and roots of an adjacent shaded ramet but no ^{14}C transport to the shaded shoots (Flanagan and Moser, 1985). Such studies indicate that during early development, stored assimilate, and probably current photosynthate, is transported to the developing ramet. When that ramet matures to the stage that it should begin to transport basipetally to rhizome and roots, however, import is no longer possible. Thus, an individual ramet responds very much like a developing leaf or branch on a single plant. Crown architectural systems often seem very complex. They are, however, composed of numerous morphologically and physiologically autonomous subsystems that, when understood, decrease crown complexity greatly and, in turn, help explain response of the complete system to environmental stress.

V. Response to Environmental Stress

It is very difficult to describe a shoot response to stress. Shoots as a whole do not respond to stress in most cases. The individual parts of the shoot respond to stress in different ways and in different time frames. Moreover, predicting shoot response to a single stress is much easier than predicting responses to multiple stresses. Multiple stresses may be additive in their effects, or complex interactions may take place. Such interactions are common in response to multiple nutrient deficiencies (Reich and Schoettle, 1988) and to increased carbon dioxide (Larigauderie *et al.*, 1988) or ozone (Darrall, 1989) concentrations. With increased carbon dioxide concentration, the efficiencies of nitrogen or water use may increase, thus increasing shoot growth (Norby *et al.*, 1986; Larigauderie *et al.*, 1988). In contrast, with higher temperatures and higher carbon dioxide concentrations, shoot growth and biomass production may either increase, decrease, or stay the same, depending on the interactions of the carbon, nitrogen, and water cycles (Pastor and Post, 1988).

Plants respond to stress by changes in assimilation of carbon and in partitioning of carbon and other resources within the plant. Environmental changes are sensed primarily by leaves. Carbon partitioning and plant growth then change to compensate for the stress. Imposition of a stress usually initiates a series of changes that involve compensatory feedback among many parts of the plant. Such a cycle can be illustrated by response to nitrogen stress, i.e., a decrease in nitrogen supply. First, developing leaf growth slows, decreasing sink strength (Pieters and van den Noort, 1985; MacAdam *et al.*, 1989). Then, photosynthate builds up in source leaves, often shown by an increase in starch storage. As a result, transport is redirected to lower stem and roots (Constable and Rawson, 1982; Good and Williams, 1986). The increase in supply of current photosynthate to roots may increase root growth; increased root growth results in an increase in nitrogen uptake, and the cycle repeats. Although this is a simplified treatment of a very complex process with many feedback systems and areas for regulation, it illustrates how little we know about the structural–functional response to many environmental stresses. We clearly need research in this area, with careful review of existing information on plant responses to major environmental stresses.

Plant response to defoliation can be used as an example of whole-plant response to stress. Defoliation caused by plant herbivores can have dramatic effects on plant growth. Plants respond to defoliation with compensatory growth, increased assimilation of carbon, and production of antiherbivore defense compounds (Coley *et al.*, 1985). A partial or even complete defoliation of vigorously growing plants with adequate food reserves usually has little effect on overall growth. In vigorous nursery-grown *Populus* hybrids, 75–80% defoliation was required to decrease growth by 20% (Bassman *et al.*, 1982). Similar defoliation of plants that are already under other stresses, however, may cause death. Defoliated plants usually respond by decreasing lower stem and root growth, increasing new foliage and lateral branch growth, and increasing photosynthesis in both residual foliage and newly produced foliage (Bassman and Dickmann, 1982; Heichel and Turner, 1983; Tschaplinski and Blake, 1989a,b). Such changes in carbon partitioning and increases in carbon fixation rates may partially compensate for lost foliage. However, leaves are usually smaller and internodes shorter on refoliated trees relative to controls (Gregory and Wargo, 1986). Repeated defoliation of trees such as red maple and red oak can lead to severe decreases in number of buds and new leaves and eventually to branch dieback or death of the whole tree (Heichel and Turner, 1984).

Specific response to defoliation may be highly variable because of different genetic predisposition, different physiological states, and different amounts of stored food. The timing of defoliation can be very important. Defoliation in July and August is particularly damaging. Buds and new shoots formed after such late-seasonal defoliation are often killed during the winter, and carbohydrate depletion can limit regrowth the following year (Isebrands, 1982; Gregory and Wargo, 1986; Gregory *et al.*, 1986). New branch and foliage growth of deciduous plants in the spring depends entirely on stored carbohydrates, and new branches may abscise soon after initiation if carbohydrate is limiting. More important for tree growth, defoliation may initiate a cycle involving many stress factors. For example, low carbohydrate reserves in stems and roots increase susceptibility to cold winter temperatures, decrease foliage regrowth, decrease root growth, increase water stress, and increase susceptibility to root rots and other pathogens (Wargo and Montgomery, 1983; Gregory *et al.*, 1986). Such multiple stresses may cause top dieback, general progressive decline, and eventual death of the tree.

VI. The Shoot–Root Connection

The morphological, anatomical, and physiological differences between shoots and roots are so great that shoots and roots are often considered to be two entirely separate systems (Groff and Kaplan, 1988). Yet there is no question that relative shoot and root growth is closely controlled by multiple feedback systems, and understanding the complementary functions of shoots and roots is prerequisite to explaining whole-plant development. This functional equilibrium has been recognized for a long time (Crist and Stout, 1929) and extensively studied (Lyr and Hoffmann, 1967; Mooney, 1972; Brouwer, 1983; Lambers, 1983; Schulze, 1983); it can be viewed simply as changes in carbon allocation (de Wit and Penning de Vries, 1983) or as a much more complex system of carbon, mineral nutrients, and hormonal interactions (Fig. 8) (Carmi, 1986; Milligan and Dale, 1988; Zhang and Davies, 1989; Chapter 4). Even though complicated in detail, the functional equilibrium between shoots and roots can largely be explained by the production and partitioning of carbon associated with the uptake and use of nitrogen (Ågren and Ingestad, 1987; Chapter 6). These carbon–nitrogen interactions are then modified by the relative stress on the shoot or root systems (Hunt and Nicholls, 1986).

The changes in photosynthesis and carbon partitioning with nitrogen stress illustrate a simple functional response. Although photosynthesis is closely related to leaf nitrogen content (Ågren and Ingestad, 1987; Syvertsen, 1987; Chapter 2), under nitrogen stress the growth rates of developing leaves usually decrease before photosynthesis decreases (McDonald *et al.*, 1986). The resulting decrease in leaf sink strength shifts carbon allocation from developing leaves to lower stem and roots and provides more assimilate for root growth. Because the responses to nitrogen stress of leaf growth rate and photosynthetic rate take place in different age classes of leaves, it is important to determine leaf ontogenetic stage (see Section II,A) when studying effects of nitrogen stress (Tolley-Henry and Raper, 1986). In addition, plants under nitrogen stress may partition limited internal nitrogen to recently mature leaves so as to maximize photosynthesis and nitrogen use efficiency (Hunt *et al.*, 1985).

The above simplistic example of plant response to nitrogen stress becomes much more complicated when one considers the finer details of the

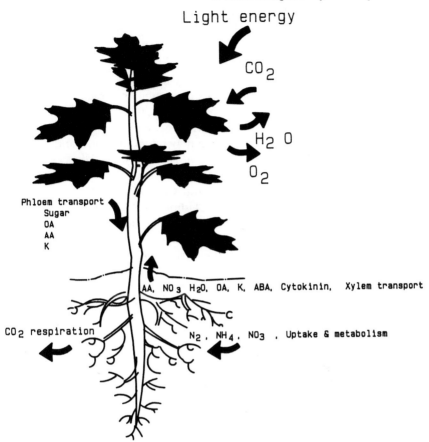

Figure 8   Diagram of a tree seedling showing major gas-exchange reactions, phloem and xylem transport compounds, and nitrogen uptake. All factors can be involved in complex uptake and metabolic interactions that are in turn modified by environmental stresses acting on both the shoot and root systems.

relative availability of ammonium or nitrate as influenced by site factors (Mladenoff, 1987); the uptake, transport, and metabolism of ammonium, nitrate, or their amino acid conversion products (Vogelmann *et al.*, 1985); the cycling of organic nitrogen compounds within plants (Lambers *et al.*, 1982; Peoples *et al.*, 1985); and how these interacting factors influence the functional equilibrium between shoots and roots.

Root growth is particularly responsive to changes in carbon allocation within shoots. In actively growing plants, roots are relatively weak sinks and tend to receive assimilate only from lower leaves or after developing leaf demand is met. In fact, assimilate stored in roots may be retranslocated to shoots in times of stress (Huber, 1983). Root growth may also depend primarily on current photosynthate (Van Den Driessche, 1987), although such dependence may be species specific (Philipson, 1988). The interaction between root growth and developing leaf demand for photosynthate can be seen in the frequently observed periodicity of root growth, common in many plants (Hoffmann and Lyr, 1973; Drew and Ledig, 1980), but particularly well documented in plants with strong episodic shoot growth (Borchert, 1978; Reich *et al.*, 1980; Sleigh *et al.*, 1984). Under mild stress, however, assimilate can build up in both shoots and roots. Under these conditions, root growth may be largely independent of shoots and depend more on the root environment (Kuhns *et al.*, 1985). Such excess assimilate is probably involved in the proliferation of roots in areas of high nutrient or water supply (Barta, 1976; Eissenstat and Caldwell, 1988). In addition, plants with limited assimilate because of shoot stress may also reallocate carbon to roots in areas favorable for growth. Studies on plants with split root systems have shown that nitrogen fertilization of part of the root system will increase ^{14}C transport to the fertilized roots (Singleton and Van Kessel, 1987).

In previous sections the regulatory role of leaves in the structural-functional integration of the shoot is emphasized. Leaves and roots are functionally integrated as well. Much evidence shows that the development of both depends on their complementary functions; carbon from leaves is necessary for root growth, and signals from roots may regulate leaf functions (Zhang and Davies, 1989). Evidence for structural control of root systems by leaves is weak however. Studies on water movement in plants have shown that single vessels or related groups of vessels may differentiate and extend from leaves down through the stems to roots, although water and dyes injected into these vessels in roots may diffuse laterally and spiral around the stem during ascent (Kozlowski and Winget, 1963). Additional studies with split root systems have shown that sectors of shoots and roots are often highly independent of the rest of the plant and may not transport laterally (Barta, 1976; Hansen and Dickson, 1979). Growth of the main roots follows changes in leaf growth. For example, cambium activation and the formation of vessels initiated by expanding leaves progress down the branches to the main stem and down throughout the main roots (Fayle, 1968; Denne and Atkinson, 1987). In contrast, growth and turnover of fine roots seem largely independent of shoot growth (Johnson-Flanagan and Owens, 1985; Eis, 1986); such fine root growth is more a function of root environment (Kuhns *et al.*, 1985; Eissenstat and Caldwell, 1988). Fine root growth, however, may be cyclic in plants with strong recurrent shoot flushes (Kummerow *et al.*, 1982). Nevertheless, fine roots are significant sinks for fixed carbon (Nambiar, 1987). Thus, environmental stresses that decrease carbon fixation may also decrease fine root growth.

Direct phloem connections between specific leaves and roots have not been demonstrated, although transport of ^{14}C assimilates to roots and bulb scales correspond to the phyllotactic patterns of the shoot (Tietema *et al.*, 1972). Perhaps the best evidence for leaf regulation of vascular development in roots is found in sugar beet (*Beta vulgaris* L.) (Stieber and Beringer, 1984). In developing sugar beet root tissue, each cambial ring is associated with a whorl of leaves. Each developing leaf is directly connected by vascular traces to a sector of the developing cambial ring and is apparently active in the induction and development of this vascular tissue in the outer layers of the root parenchyma. Lateral roots are also initiated at specific sections of the developing cambial ring and are probably directly connected by vascular traces to the developing leaves.

These findings suggest that there may be an intimate and direct connection between a leaf and a specific portion of the root system. If this is so, information in this area would be extremely helpful for understanding the structural–functional relations between shoots and roots and their response to environmental stresses.

VII. Whole-Plant Physiological-Process Models

As more fundamental information on the structural–functional relations of shoots and roots becomes available, increased opportunities will arise to develop whole-plant physiological-process models to aid in understanding the underlying mechanisms of plant growth. Such models can help us examine why a phenomenon occurs or how a stress affects plant growth (Landsberg, 1986; Leary, 1988). One approach to developing whole-plant process models is to base them on experimentally derived structural–functional relationships coupled with established scientific theory (Isebrands *et al.*, 1990). Key concepts for this approach come from a long series of anatomical and physiological experiments on trees conducted at our Wisconsin laboratory (Isebrands *et al.*, 1983; Larson, 1983; Dickson, 1986). They include recognition that: (1) individual leaves are the primary organizing units of the crown, (2) photosynthesis is the central physiological process, (3) carbon allocation is a major determinant of growth, (4) shoot–root interaction and feedback are essential growth components, (5) individual branches are independent physiological units, and (6) shoot growth is indirectly affected through direct stress effects on the physiological processes of the leaves.

An example of this approach is ECOPHYS, an interactive ecophysiological growth-process model of a poplar tree (Isebrands *et al.*, 1990; Rauscher *et al.*, 1988). The general structure of the model consists of a light interception and photosynthesis (i.e., photosynthate production) submodel linked to a photosynthate distribution and growth submodel (Fig. 9). The model simulates growth of a one-year-old poplar using the individual leaf as a primary biological unit, with data on clonal morphological and phenologi-

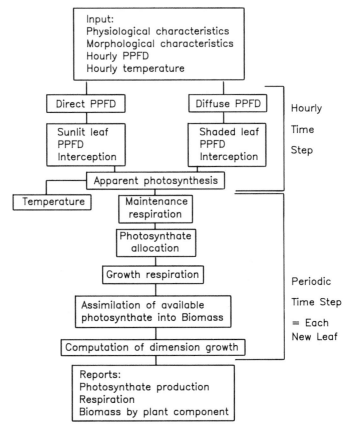

Figure 9   Schematic diagram of ECOPHYS, an ecophysiological growth-process model of poplar, showing general model structure. [Adapted from Isebrands *et al.* (1990).]

cal characteristics, seasonal environmental variables, and site variables as the major driving variables. Clonal traits include leaf size and shape, leaf orientation, and budset date. Environmental variables include hourly solar radiation and temperature.

The leaf intercepts light hourly over the day according to leaf and solar position, then produces assimilate for respiration and transport to growth centers throughout the plant. The model gives an hourly photosynthesis report that includes an output of total leaf area, leaf area in sun and shade, light intensity (i.e., photosynthetic photon flux density, or PPFD), and total photosynthesis per leaf and per tree. The assimilate is then partitioned to competing growth centers within the tree according to transport coefficients determined from ^{14}C tracer studies of poplars (Dickson, 1986). The individual tree components (i.e., leaves, stem, and roots) grow in response to available photosynthate after maintenance and growth respiration is subtracted. The model also gives a daily growth summary that includes height, diameter, number of leaves, biomass by component, and volume. In addition, there is an interactive graphical display of the tree so that one can see the leaves and their relation to the sun at any time of day and from any perspective.

For a test of interacting environmental factors, we examined the influence of light (i.e., PPFD) and temperature on the growth of first-year poplars at Rhinelander, Wisconsin (Rauscher *et al.*, 1988). In this test, we substituted individual high-sunlight days (i.e., 8 or 15 days) for low-sunlight days during the season and vice versa; low- and high-temperature days were also substituted (Table II). The days chosen for substitution had to be close to the replaced days so that long-term seasonal trends were not disturbed. The results show that light influences growth more than temperature does, and there is a positive interaction between light and temperature. The interaction of light and temperature can increase or decrease the biomass of individual trees by nearly 50% compared with that obtained under the actual environmental conditions present in 1979. If quantitative information were available on how other stresses affected photosynthetic rates and

Table II   Test Simulation of Different Light and Temperature Regimes for One-Year-Old Poplars over 52 Days from Late July to Early September in Rhinelander, Wisconsin

Test run[a]			Light sum[b] (E/m²)	Temperature sum[c] (°C)	Aboveground biomass (g)	Height (cm)	Diameter (cm)
No.	PPFD	Temp.					
1	−15	−15	686	448	25	80	1.14
2	−15	+15	682	728	27	83	1.15
3	−15	NC	682	589	27	83	1.16
4	−8	NC	773	589	31	90	1.21
5	NC	−15	914	448	37	101	1.25
6	NC	−8	914	508	39	105	1.26
7	*NC*	*NC*	*914*	*589*	*44*	*112*	*1.27*
8	NC	+8	914	681	44	114	1.27
9	NC	+15	914	728	45	114	1.27
10	+15	−15	1112	448	50	122	1.34
11	+8	NC	1031	589	51	124	1.33
12	+15	NC	1123	589	61	142	1.36
13	+15	+15	1123	728	66	151	1.36

[a] In the modified seasonal test runs, ±15 or 8 clear or cloudy days and ±15 or 8 high- or low-temperature days were substituted for contrasting days. The data for substitution came from our light and temperature weather traces.

[b] Light sum = intercepted PPFD summed over Julian days 204–256.

[c] Temperature sum = average daily temperature minus 5°C (e.g. 17.7°C − 5°C = 12.7°C for day 1) summed over Julian days 204–256. In test run No. 7, the summed light and temperature are used by the model to predict poplar growth from actual environmental conditions for 1979 in Rhinelander, WI. NC, no change from standard environmental conditions present in 1979 (from Rauscher *et al.*, 1988).

leaf growth (e.g., ozone; Reich, 1987), it could also be incorporated into the model and used to predict additional individual and interactive effects on plant growth.

The example given here illustrates how explanatory ecophysiological-process models could allow us to integrate, quantitatively and qualitatively, our scientific and heuristic knowledge about the processes governing tree growth. Probably the most important use of current process models is to help clarify critical knowledge gaps and uncertainties, to formulate testable hypotheses, and to provide a method for evaluating the significance and validity of research results (Rauscher *et al.*, 1988). But inadequate baseline structural–functional data and a poor understanding of underlying mechanisms of plant growth still limit widespread development and application of whole-plant process models.

From the viewer's perspective, plants are made up of (1) the anatomical, morphological, and physical topology that forms the visual structure and (2) the chemical, biochemical, and physiological functions on which stress operates to change the structure. It is useless to argue which came first or which is more important. Without a good understanding of both — and we believe an understanding developed first of structure then of function — it will be very difficult to understand or predict plant changes in response to stress. If the prediction of shoot response to a single stress is difficult because of deficiencies in basic structural–functional information, the prediction of shoot response to short-term or long-term multiple stresses may be nearly impossible. How can we provide the systematic understanding of the physiological feedback systems and the interactions that are present when stresses such as carbon dioxide, water, or nitrogen (both high or low levels) all influence carbon assimilation, allocation, and partitioning? In addition, short-term studies, which are much easier to conduct and control, cannot account for long-term responses to stress because of the ability of plants to acclimate to stress. More multidisciplinary research is needed in important plant species to examine leaf initiation, respiration, carbon transport, root development, and shoot–root feedback, as well as the effects of water, nutrient, and other environmental stresses on these processes.

We hope this paper will point out the importance of leaf development to shoot and plant development and stimulate research to explain more specific control mechanisms on structural–functional relations and how these mechanisms respond to environmental stress.

Acknowledgment

The authors would like to acknowledge the pioneering work of P. R. Larson, our mentor now retired, in the study of structural–functional relations in plants. Without his guidance and detailed research approach, much of what we know about the fundamental mechanisms of tree growth (cited here) would not be available.

References

Ågren, G. I., and Ingestad, T. (1987). Root:shoot ratio as a balance between nitrogen productivity and photosynthesis. *Plant, Cell Environ.* 10: 579–586.

Barta, A. L. (1976). Transport and distribution of $^{14}CO_2$ assimilate in *Lolium perenne* in response to varying nitrogen supply to halves of a divided root system. *Physiol. Plant.* 38: 48–52.

Bassman, J. H., and Dickmann, D. I. (1982). Effects of defoliation in the developing leaf zone on young *Populus* x *euramericana* plants. I. Photosynthetic physiology, growth, and dry weight partitioning. *For. Sci.* 28: 599–612.

Bassman, J. H., and Dickmann, D. I. (1985). Effects of defoliation in the developing leaf zone on young *Populus* x *euramericana* plants. II. Distribution of ^{14}C photosynthate after defoliation. *For. Sci.* 31: 358–366.

Bassman, J., Myers, W., Dickmann, D., and Wilson, L. (1982). Effects of simulated insect damage on early growth of nursery-grown hybrid poplars in northern Wisconsin. *Can. J. For. Res.* 12: 1–9.

Borchert, R. (1978). Feedback control and age-related changes of shoot growth in seasonal and nonseasonal climates. *In* "Tropical Trees as Living Systems" (P. B. Tomlinson and M. H. Zimmermann, eds.), pp. 497–515. Proc. 4th Cabot Symp., Harvard Forest, Petersham, Massachusetts, 26–30 April 1976.

Brouwer, R. (1983). Functional equilibrium: sense or nonsense? *Neth. J. Agric. Sci.* 31: 335–348.

Bunce, J. A. (1978). Interrelationships of diurnal expansion rates and carbohydrate accumulation and movement in soya beans. *Ann. Bot.* 42: 1463–1466.

Carmi, A. (1986). Effects of cytokinins and root pruning on photosynthesis and growth. *Photosynthetica* 20: 1–8.

Ceulemans, R., Impens, I., and Steenackers, V. (1988). Genetic variation in aspects of leaf growth of *Populus* clones, using the leaf plastochron index. *Can. J. For. Res.* 18:1069–1077.

Chapman, J. M., and Perry, R. (1987). A diffusion model of phyllotaxis. *Ann. Bot.* 60: 377–389.

Coley, P. D., Bryant, J. P., and Chapin, F. S., III. (1985). Resource availability and plant antiherbivore defense. *Science* 230: 895–899.

Constable, G. A., and Rawson, H. M. (1982). Distribution of ^{14}C label from cotton leaves: consequences of changed water and nitrogen status. *Aust. J. Plant Physiol.* 9: 735–747.

Crist, J. W., and Stout, G. J. (1929). Relation between top and root size in herbaceous plants. *Plant Physiol.* 4: 63–85.

Daie, J. (1988). Mechanism of drought-induced alterations in assimilate partitioning and transport in crops. *CRC Crit. Rev. Plant Sci.* 7: 117–137.

Dale, J. E., and Milthorpe, F. L., eds. (1983). "The Growth and Functioning of Leaves." Cambridge Univ. Press, London.

Darrall, N. M. (1989). The effect of air pollutants on physiological processes in plants. *Plant, Cell Environ.* 12: 1–30.

Denne, M. P., and Atkinson, C. J. (1987). Reactivation of vessel expansion in relation to budbreak in sycamore *(Acer pseudoplatanus)* trees. *Can. J. For. Res.* 17: 1166–1174.

de Wit, C. T., and Penning de Vries, F. W. T. (1983). Crop growth models without hormones. *Neth. J. Agric. Sci.* 31: 313–323.

Dickson, R. E. (1986). Carbon fixation and distribution in young *Populus* trees. *In* "Proc. Crown and Canopy Structure in Relation to Productivity" (T. Fujimori and D. Whitehead, eds.), pp. 409–426. Forestry and Forest Products Research Institute, Ibaraki, Japan.

Dickson, R. E., and Nelson, E. A. (1982). Fixation and distribution of ^{14}C in *Populus deltoides* during dormancy induction. *Physiol. Plant.* 54: 393–401.

Dickson, R. E., and Shive, J. B., Jr. (1982). $^{14}CO_2$ fixation, translocation, and carbon metabolism in rapidly expanding leaves of *Populus deltoides*. *Ann. Bot.* 50: 37–47.

Drew, A. P., and Ledig, F. T. (1980). Episodic growth and relative shoot:root balance in loblolly pine seedlings. *Ann. Bot.* 45: 143–148.

Eis, S. (1986). Differential growth of individual components of trees and their interrelationships. *Can. J. For. Res.* 16: 352–359.

Eissenstat, D. M., and Caldwell, M. M. (1988). Seasonal timing of root growth in favorable microsites. *Ecology* 69: 870–873.

Eschrich, W., and Eschrich, B. (1987). Control of phloem unloading by source activities and light. *Plant Physiol. Biochem.* 25:625–634.

Fayle, D. C. F. (1968). Radial growth in tree roots. Tech. Rep. No. 9. Faculty of Forestry, Univ. of Toronto.

Fisher, D. G. (1986). Ultrastructure, plasmodesmata frequency, and solute concentration in green areas of variegated *Coleus blumei* Benth. leaves. *Planta* 169: 141–152.

Fisher, D. G., and Eschrich, W. (1985a). Import and unloading of ^{14}C assimilate into mature leaves of *Coleus blumei. Can. J. Bot.* 63: 1700–1707.

Fisher, D. G., and Eschrich, W. (1985b). Import and unloading of ^{14}C assimilate into nonphotosynthetic portions of variegated *Coleus blumei* leaves. *Can. J. Bot.* 63: 1708–1712.

Fisher, D. G., Larson, P. R., and Dickson, R. E. (1983). Phloem translocation from a leaf to its nodal region and axillary branch in *Populus deltoides. Bot. Gaz.* 144:481–490.

Flanagan, L. B., and Moser, W. (1985). Pattern of ^{14}C assimilate distribution in a clonal herb, *Aralia nudicaulis. Can. J. Bot.* 63: 2111–2114.

Geiger, D. R. (1987). Understanding interactions of source and sink regions of plants. *Plant Physiol. Biochem.* 25: 659–666.

Goffinet, M. C., and Larson, P. R. (1981). Structural changes in *Populus deltoides* terminal buds and in the vascular transition zone of the stems during dormancy induction. *Am. J. Bot.* 68: 118–129.

Good, J. E. G., and Williams, T. G. (1986). Growth responses of selected clones of birch (*Betula pendula* Roth., *B. pubescens* Ehrh.) and willow (*Salix caprea* L., *S. cinerea* L.) to nitrogen in solution culture. *Plant Soil* 92: 209–222.

Gregory, R. A., and Wargo, P. M. (1986). Timing of defoliation and its effect on bud development, starch reserves, and sap sugar concentration in sugar maple. *Can. J. For. Res.* 16: 10–17.

Gregory, R. A., Williams, M. W., Jr., Wong, B. L., and Hawley, G. J. (1986). Proposed scenario for dieback and decline of *Acer saccharum* in northeastern U.S.A. and southeastern Canada. *IAWA Bull. N.S.* 7: 357–369.

Groff, P. A., and Kaplan, D. R. (1988). The relation of root systems to shoot systems in vascular plants. *Bot. Rev.* 54: 387–422.

Hansen, E. A., and Dickson, R. E. (1979). Water and mineral nutrient transfer between root systems of juvenile *Populus. For. Sci.* 25: 247–252.

Hanson, P. J., Dickson, R. E., Isebrands, J. G., Crow, T. R., and Dixon, R. K. (1986). A morphological index of *Quercus* seedling ontogeny for use in studies of physiology and growth. *Tree Physiol.* 2: 273–281.

Hartnett, D. C., and Bazzaz, F. A. (1983). Physiological integration among interclonal ramets in *Solidago canadensis. Ecology* 64: 779–788.

Heichel, G. H., and Turner, N. C. (1983). CO_2 assimilation of primary and regrowth foliage of red maple (*Acer rubrum* L.) and red oak (*Quercus rubra* L.): response to defoliation. *Oecologia* 57: 14–19.

Heichel, G. H., and Turner, N. C. (1984). Branch growth and leaf numbers of red maple (*Acer rubrum* L.) and red oak (*Quercus rubra* L.): response to defoliation. *Oecologia* 62: 1–6.

Hicks, G. S. (1980). Control of primordia formation at the shoot apex. *In* "Control of Shoot Growth in Trees" (C. H. A. Little, ed.), pp. 143–156. Proc. IUFRO Workshop, Fredericton, New Brunswick.

Hoffman, Von G., and Lyr, H. (1973). Charakterisierung des Wachstumsverhaltens von Pflanzen durch Wachstumsschemata. *Flora* 162: 81–98.

Huber, S. C. (1983). Relation between photosynthetic starch formation and dry-weight partitioning between the shoot and root. *Can. J. Bot.* 61: 2709–2716.

Hunt, E. R., Jr., Weber, J. A., and Gates, D. M. (1985). Effects of nitrate application on *Amaranthus powellii* Wats. III. Optimal allocation of leaf nitrogen for photosynthesis and stomatal conductance. *Plant Physiol.* 79: 619–624.

Hunt, R., and Nicholls, A. O. (1986). Stress and the course control of growth and root-shoot partitioning in herbaceous plants. *Oikos* 47: 149–158.

Isebrands, J. G. (1982). Toward a physiological basis of intensive culture of poplar. *In* "Proc. TAPPI R&D Conference," pp. 81–90. Asheville, North Carolina, 29 Aug–1 Sept. 1982.

Isebrands, J. G., and Larson, P. R. (1977). Organization and ontogeny of the vascular system in the petiole of eastern cottonwood. *Am. J. Bot.* 64: 65–77.

Isebrands, J. G., and Larson, P. R. (1980). Ontogeny of the major veins in the lamina of *Populus deltoides* Bartr. *Am. J. Bot.* 67: 23–33.

Isebrands, J. G., and Michael, D. A. (1986). Effects of leaf morphology and orientation on solar radiation interception and photosynthesis in *Populus*. *In* "Proc. Crown and Canopy Structure in Relation to Productivity" (T. Fujimori and D. Whitehead, eds.), pp. 359–381. Forestry and Forest Products Research Institute, Ibaraki, Japan.

Isebrands, J. G., Nelson, N. D., Dickmann, D. I., and Michael, D. A. (1983). Yield physiology of short-rotation, intensively cultured poplars. *U.S.D.A. For. Serv. Gen. Tech. Rep. NC* 91, pp. 77–93.

Isebrands, J. G., Rauscher, H. M., Crow, T. R., and Dickmann, D. I. (1990). Whole-tree growth process models based on structural–functional relationships. *In* "Process Modelling of Forest Growth Responses to Environmental Stress" (R. K. Dixon, R. S. Meldahl, G. A. Ruark, and W. G. Warren, eds.), pp. 96–111. Timber Press, Portland, Oregon.

Johnson-Flanagan, A. M., and Owens, J. N. (1985). Development of white spruce (*Picea glauca*) seedling roots. *Can. J. Bot.* 63: 456–462.

Kerr, P. S., Rufty, T. W., Jr., and Huber, S. C. (1985). Changes in nonstructural carbohydrates in different parts of soybean (*Glycine max* L. Merr.) plants during a light/dark cycle and in extended darkness. *Plant Physiol.* 78: 576–581.

King, R. W. (1983). The shoot apex in transition: flowers and other organs. *In* "The Growth and Functioning of Leaves" (J. E. Dale and F. L. Milthorpe, eds.), pp. 109–144. Cambridge Univ. Press, London.

Kozlowski, T. T., and Winget, C. H. (1963). Patterns of water movement in forest trees. *Bot. Gaz.* 124: 301–311.

Kremer, A., and Larson, P. R. (1982). The relation between first-season bud morphology and second-season shoot morphology of jack pine seedlings. *Can. J. For. Res.* 12: 893–904.

Kremer, A., and Larson, P. R. (1983). Genetic control of height growth components in jack pine seedlings. *For. Sci.* 29: 451–464.

Kuehnert, C. C., and Larson, P. R. (1983). Development and organization of the primary vascular system in the phase II leaf and bud of *Osmunda cinnamomea* L. *Bot. Gaz.* 144: 310–317.

Kuhns, M. R., Garrett, H. E., Teskey, R. O., and Hinckley, T. M. (1985). Root growth of black walnut trees related to soil temperature, soil water potential, and leaf water potential. *For. Sci.* 31: 617–629.

Kummerow, J., Kummerow, M., and Da Silva, W. S. (1982). Fine root growth dynamics in cacao *(Theobroma cacao)*. *Plant Soil* 65: 193–201.

Lambers, H. (1983). "The functional equilibrium," nibbling on the edges of a paradigm. *Neth. J. Agric. Sci.* 31: 305–311.

Lambers, H., Simpson, R. J., Beilharz, V. C., and Dalling, M. J. (1982). Growth and transloca- tion of C and N in wheat *(Triticum aestivum)* grown with a split root system. *Physiol. Plant.* 56: 421–429.

Lamoreaux, R. J., Chaney, W. R., and Brown, K. M. (1978). The plastochron index: A review after two decades of use. *Am. J. Bot.* 65: 586–593.

Landsberg, J. J. (1986). "Physiological Ecology of Forest Production." Academic Press, London.

Larigauderie, A., Hilbert, D. W., and Oechel, W. C. (1988). Effects of CO_2 enrichment and nitrogen availability on resource acquisition and resource allocation in a grass, *Bromus mollis*. *Oecologia* 77: 544–549.

Larson, P. R. (1977). Phyllotactic transitions in the vascular system of *Populus deltoides* Bartr. as determined by ^{14}C labeling. *Planta* 134: 241–249.

Larson, P. R. (1980). Control of vascularization by developing leaves. In "Control of Shoot Growth in Trees" (C. H. A. Little, ed.), pp. 157–172. Proc. IUFRO Workshop, Frederic- ton, New Brunswick.

Larson, P. R. (1982). The concept of cambium. In "New Perspectives in Wood Anatomy" (P. Baas, ed.), pp. 85–121. Martinus Nijhoff, The Hague.

Larson, P. R. (1983). Primary vascularization and the siting of primordia. In "The Growth and Functioning of Leaves" (J. E. Dale and F. L. Milthorpe, eds.), pp. 25–51, Cambridge Univ. Press, Cambridge.

Larson, P. R. (1984). The role of subsidiary trace bundles in stem and leaf development of the Dicotyledoneae. In "Contemporary Problems in Plant Anatomy" (R. A. White and W. C. Dickison, eds.), pp. 109–143. Academic Press, New York.

Larson, P. R. (1986b). Vascularization of a multilacunar species: *Polyscias quilfoylei* (Aralia- ceae). I. The stem. *Am. J. Bot.* 73: 1620–1631.

Larson, P. R. (1986b). Vascularization of a multilacunar species: *Polyscias quilfoylei* (Aralia- ceae). II. The leaf base and rachis. *Am. J. Bot.* 73: 1632–1641.

Larson, P. R., and Dickson, R. E. (1986). ^{14}C translocation pathways in honeylocust and green ash: Woody plants with complex leaf forms. *Physiol. Plant.* 66: 21–30.

Larson, P. R., and Fisher, D. G. (1983). Xylary union between elongating lateral branches and the main stem in *Populus deltoides*. *Can. J. Bot.* 61: 1040–1051.

Larson, P. R., and Isebrands, J. G. (1971). The plastochron index as applied to developmental studies of cottonwood. *Can. J. For. Res.* 1: 1–11.

Larson, P. R., and Isebrands, J. G. (1974). Anatomy of the primary-secondary transition zone in stems of *Populus deltoides*. *Wood Sci. and Technol.* 8: 11–26.

Larson, P. R., and Pizzolato, T. D. (1977). Axillary bud development in *Populus deltoides* I. Origin and early ontogeny. *Am. J. Bot.* 64: 835–848.

Leary, R. A. (1988). Some factors that will affect the next generation of growth models. *U.S.D.A. For. Serv. Gen. Tech. Rep. NC* 120, pp. 22–32.

Lyr, H., and Hoffmann, G. (1967). Growth rates and growth periodicity of roots. *Int. Rev. For. Res.* 2: 181–236.

MacAdam, J. W., Volenec, J. J., and Nelson, C. J. (1989). Effects of nitrogen on mesophyll cell division and epidermal cell elongation in tall fescue leaf blades. *Plant Physiol.* 89: 549–556.

McDonald, A. J. S., Lohammar, T., and Ericsson, A. (1986). Growth response to step-decrease in nutrient availability in small birch *(Betula pendula* Roth). *Plant, Cell Environ.* 9: 427–432.

Magda, D., Warembourg, F. R., and Labeyrie, V. (1988). Physiological integration among ramets of *Lathyrus sylvestris* L. *Oecologia* 77: 255–260.

MaGuire, D. A., and Hann, D. W. (1987). A stem dissection technique for dating branch mortality and reconstructing past crown recession. *For. Sci.* 33: 858–871.

Milligan, S. P., and Dale, J. E. (1988). The effects of root treatments on growth of the primary leaves of *Phaseolus vulgaris* L.: Biophysical analysis. *New Phytol.* 109: 35–40.

Minchin, P. E. H., and McNaughton, G. S. (1987). Xylem transport of recently fixed carbon within lupin. *Aust. J. Plant Physiol.* 14: 325–329.

Mladenoff, D. J. (1987). Dynamics of nitrogen mineralization and nitrification in hemlock and hardwood treefall gaps. *Ecology* 68: 1171–1180.

Mooney, H. A. (1972). The carbon balance of plants. *Annu. Rev. Ecol. Syst.* 3: 315–346.

Nambiar, E. K. S. (1987). Do nutrients retranslocate from fine roots? *Can. J. For. Res.* 17: 912–918.

Norby, R. J., Pastor, J., and Melillo, J. M. (1986). Carbon-nitrogen interactions in CO_2-enriched white oak: Physiological and long-term perspectives. *Tree Physiol.* 2: 233–241.

Pastor, J., and Post, W. M. (1988). Response of northern forests to CO_2-induced climate change. *Nature* 334: 55–58.

Peoples, M. B., Pate, J. S., and Atkins, C. A. (1985). The effect of nitrogen source on transport and metabolism of nitrogen in fruiting plants of cowpea (*Vigna unguiculata* L. Walp.). *J. Exp. Bot.* 36: 567–582.

Philipson, J. J. (1988). Root growth in sitka spruce and douglas-fir transplants: Dependence on the shoot and stored carbohydrate. *Tree Physiol.* 4: 101–108.

Pieters, G. A. (1983). Growth of *Populus euramericana*. *Physiol. Plant.* 57: 455–462.

Pieters, G. A. (1986). Dimensions of the growing shoot and the absolute growth rate of a poplar shoot. *Tree Physiol.* 2: 283–288.

Pieters, G. A., and van den Noort, M. E. (1985). Leaf area coefficient of some *Populus euramericana* cultivars grown at various irradiances and NO_3 supply. *Photosynthetica* 19: 188–193.

Pieters, G. A., and van den Noort, M. E. (1988). Effect of irradiance and plant age on the dimensions of the growing shoot of popular. *Physiol. Plant.* 74: 467–472.

Pizzolato, T. D., and Larson, P. R. (1977). Axillary bud development in *Populus deltoides*. II. Late ontogeny and vascularization. *Am. J. Bot.* 64: 849–860.

Rauscher, H. M., Isebrands, J. G., Crow, T. R., Dickson, R. E., Dickmann, D. I., and Michael, D. A. (1988). Simulating the influence of temperature and light on growth of juvenile poplars in their establishment year. *U.S.D.A. For. Serv. Gen. Tech. Rep. NC* 120, pp. 331–339.

Reich, P. B. (1987). Quantifying plant response to ozone: A unifying theory. *Tree Physiol.* 3: 63–91.

Reich, P. B., and Schoettle, A. W. (1988). The role of phosphorus and nitrogen in photosynthetic and whole-plant carbon gain and nutrient use efficiency in eastern white pine. *Oecologia* 77: 25–33.

Reich, P. B., Teskey, R. O., Johnson, P. S., and Hinckley, T. M. (1980). Periodic root and shoot growth in oak. *For. Sci.* 26: 590–598.

Richards, J. H., and Larson, P. R. (1982). The initiation and development of secondary xylem in axillary branches of *Populus deltoides*. *Ann. Bot.* 49: 149–163.

Russell, C. R., and Morris, D. A. (1983). Patterns of assimilate distribution and source-sink relationships in the young reproductive tomato plant (*Lycopersicon esculentum* Mill.). *Ann. Bot.* 52: 357–363.

Schulze, E.-D. (1983). Root–shoot interactions and plant life forms. *Neth. J. Agric. Sci.* 31: 291–303.

Singleton, P. W., and Van Kessel, C. (1987). Effect of localized nitrogen availability to soybean half-root systems on photosynthate partitioning to roots and nodules. *Plant Physiol.* 83: 552–556.

Sleigh, P. A., Collin, H. A., and Hardwick, K. (1984). Distribution of assimilate during the flush cycle of growth in *Theobroma cacao* L. *Plant Growth Regul.* 2: 381–391.

Stieber, J., and Beringer, H. (1984). Dynamic and structural relationships among leaves, roots, and storage tissue in the sugar beet. *Bot. Gaz.* 145: 465–473.

Syvertsen, J. P. (1987). Nitrogen content and CO_2 assimilation characteristics of *Citrus* leaves. *HortScience* 22: 289–291.

Thomas, L. P., and Watson, M. A. (1988). Leaf removal and the apparent effects of architectural constraints on development in *Capsicum annuum. Am. J. Bot.* 75: 840–843.

Thrower, S. L., and Thrower, L. B. (1980). Translocation into mature leaves — the pathway of assimilate movement. *New Phytol.* 86: 145–154.

Tietema, T., Hoekstra, S. M. R., and Van Die, J. (1972). Translocation of assimilates in *Fritillaria imperialis* L. II. Downward movement of ^{14}C-labeled photosynthates into the developing bulb and their subsequent distribution among the scale parts. *Acta Bot. Neerl.* 21: 395–399.

Tolley-Henry, L., and Raper, C. D., Jr. (1986). Expansion and photosynthetic rate of leaves of soybean plants during onset of and recovery from nitrogen stress. *Bot. Gaz.* 147: 400–406.

Tschaplinski, T. J., and Blake, T. J. (1989a). Photosynthetic reinvigoration of leaves following shoot decapitation and accelerated growth of coppice shoots. *Physiol. Plant.* 75: 157–165.

Tschaplinski, T. J., and Blake, T. J. (1989b). The role of sink demand in carbon partitioning and photosynthetic reinvigoration following shoot decapitation. *Physiol. Plant.* 75: 166–173.

Turgeon, R. (1987). Phloem unloading in tobacco sink leaves: Insensitivity to anoxia indicates a symplastic pathway. *Planta* 171: 73–81.

Van Den Driessche, R. (1987). Importance of current photosynthate to new root growth in planted conifer seedlings. *Can. J. For. Res.* 17: 776–782.

Vogelmann, T. C., Larson, P. R., and Dickson, R. E. (1982). Translocation pathways in the petioles and stem between source and sink leaves of *Populus deltoides* Bartr. ex Marsh. *Planta* 156: 345–358.

Vogelmann, T. C., Dickson, R. E., and Larson, P. R. (1985). Comparative distribution and metabolism of xylem-borne amino compounds and sucrose in shoots of *Populus deltoides. Plant Physiol.* 77: 418–428.

Wargo, P. M., and Montgomery, M. E. (1983). Colonization by *Armillaria mellea* and *Agrilus bilineatus* of oaks injected with ethanol. *For. Sci.* 29: 848–857.

Watson, M. A., and Casper, B. B. (1984). Morphogenetic constraints on patterns of carbon distribution in plants. *Annu. Rev. Ecol. Syst.* 15: 233–258.

Wu, A., and Thrower, L. B. (1973). Translocation into mature leaves. *Plant Cell Physiol.* 14: 1225–1228.

Zhang, J., and Davies, W. J. (1989). Abscisic acid produced in dehydrating roots may enable the plant to measure the water status of the soil. *Plant, Cell Environ.* 12: 73–81.

2

Ecological Scaling of Carbon Gain to Stress and Resource Availability

Christopher B. Field

I. Scaling and Stress

The challenge of understanding global change must be met with a program for integrating research at many scales, from the biochemical and molecular to the global (Committee on Global Change, 1988). For many of the critical questions concerning the responses of ecosystems to climate change and, especially, feedbacks of vegetation to climate, understanding the responses of plant carbon balance and growth to stress is a critical prerequisite. A bottom-up physiological approach, synthesizing results of detailed

35

studies on the mechanistic basis of stress responses, is one important strategy for developing the necessary understanding. Here, I develop the concept and discuss some of the evidence relevant to an ecological–evolutionary strategy, which has the potential both to simplify the research agenda for the physiological, bottom-up approach and to extend its usefulness. The underlying idea is simple: if selection consistently favors a limited suite of mechanisms for coping with a given stress, and if many or all stress responses share general characteristics, then the mapping of stress or resource availability to response may be much simpler than one would predict from the entire range of physiological possibilities. To the extent it exists, ecological simplification or convergence of stress responses may be a useful concept for several reasons: it potentially simplifies the task of predicting vegetation responses to global change, it constrains the challenge of remote sensing of ecosystem processes from limited information about the status of the vegetation, and it can guide future research. Conversely, if species respond uniquely to every stress, and if every stress must be evaluated from a different perspective, the scaling problem is greatly complicated, perhaps beyond all hope of practical solution.

Physiological studies at the leaf, plant, and canopy levels, and remote-sensing studies at the landscape level, all point to the generality of stress and resource effects on carbon gain. This chapter explores the hypothesis that a single suite of mechanisms underlies the consistencies at all levels. I first discuss the concept of ecological scaling and develop two hypotheses based on it. Then I review three empirical relationships at different levels of organization in relation to these hypotheses: (1) the relationship between leaf nitrogen and photosynthetic capacity, (2) the relationship between leaf nitrogen and light availability within a canopy, and (3) the relationship between ecosystem primary production and light interception. Finally, I consider connections among these relationships and some implications of these connections for the remote sensing of photosynthesis.

Part of the motivation for this approach is applied. Understanding the productive status of Earth's plant communities must be a key component of research into global change, both to quantify the consequences of the changes and because plant processes play important roles in the biogeochemical cycles of several compounds implicated in global change, including carbon dioxide, methane, other hydrocarbons, and oxides of nitrogen. Quantifying the amount and distribution of plant biomass, as well as production and physiological activity, is a central challenge in plant research related to global change. From an applied perspective, the fundamental unanswered question about stress effects on photosynthesis is as follows: If we want accurate remote-sensing indices of plant photosynthetic production, do we need to measure the levels of stress, independent from the structure and organization of the photosynthetic machinery, or does the photosynthetic apparatus respond so that its structure and organization intrinsically yield information about the levels of chronic stresses?

Another part of the motivation concerns basic questions about how natural selection operates. With quantitative analyses of the return in photosynthate for a given investment in resources, one can identify constraints on the operation of selection and begin to test the hypothesis that natural selection shapes plant processes so that the return in fitness on investments in resources tends toward the maximum consistent with existing constraints. If building and maintaining photosynthetic machinery is expensive, then plants should never have unusable excess photosynthetic capacity. When a plant's ability to use photosynthate is limited by one or more stresses — even though some or all the stresses may not directly affect the reactions of photosynthesis — selection should favor reduced investment in photosynthetic machinery.

The basic and applied inquiries are obviously parallel. At the core of each is the question, Does the physiological status of a leaf, plant, or landscape scale upward to yield useful predictions of carbon gain and growth, and scale downward to allow inferences of resource availability and stress?

The current status of stress studies in photosynthesis research has been reviewed by Osmond *et al.* (1987), Woodrow and Berry (1988), and Björkman (1989). Recent reviews of the responses of photosynthesis to single stresses are given by Schulze (1986: water stress), Weiss and Berry (1988: high-temperature stress), Chapin *et al.* (1986: nutrient stress), Powles (1984: high-light stress), and Winner *et al.* (1985: SO_2 stress). My goal in this chapter is to consider a conceptual framework for interpreting stress responses.

II. Ecological Consistencies

A. The Functional Convergence Hypothesis

If natural selection has molded plant function the way a manager organizes a firm, then investment in any function should cease when the return on that investment falls below the return on alternative investments (Bloom *et al.*, 1985). Mooney and Gulmon (1979, 1982) used economic concepts to predict how biochemical capacity for carbon dioxide fixation should vary with resource availability. Their general conclusion can be stated as the functional convergence hypothesis: biochemical capacity for CO_2 fixation should be curtailed whenever a limitation in the availability of any resource prevents the efficient exploitation of additional capacity. Constraints on efficient use may affect photosynthesis directly (e.g., low light) or indirectly, at sites close to or distant from the reactions of photosynthesis. Examples of effects close to the reactions of photosynthesis include phosphorus deficiency, mediated through limitations on the use of photosynthate (Sharkey, 1985), and water stress, which affects photosynthesis both directly (Sharkey, 1984) and indirectly through decreased stomatal conductance, resulting in decreased CO_2 availability (Farquhar and Sharkey, 1982). An evolutionary limitation on maximum potential growth rate (Chapin, 1980) that results from selection for long-term survival in habitats deficient in resources required for growth but not for photosynthesis (e.g., calcium) is an example of a potentially important constraint mechanistically remote from the reactions of photosynthesis.

The functional convergence hypothesis implies that the biochemical capacity for CO_2 fixation should be highly correlated with gross primary production (GPP), that is, plant- or ecosystem-level photosynthetic carbon gain. This is not a simple outcome forced by the role of photosynthetic capacity in GPP. The biochemical capacity for CO_2 fixation places an upper but not a lower limit on GPP. If plants and canopies with high biochemical capacity but low GPP are rare, then the explanation must be outside the realm of short-term physiological and biochemical regulation and within the realm of evolutionary effects on the structure and organization of the photosynthetic apparatus.

If evolution has shaped plants so that the biochemical capacity for CO_2 fixation reflects the availability of all the resources required for plant growth, then biochemical capacity may be a good predictor of stress, resource availability, GPP, or growth potential. Biochemical capacity for CO_2 fixation may be a master integrator of the environment. Functional convergence requires the accurate regulation of biochemical capacity in response to resource availability. Inaccuracy in sensing the resource environment, or in mounting the response, potentially leads to errors. So do temporal and spatial heterogeneity in resource availability. Biochemical capacity for CO_2 fixation responds to changes in the environment with a time constant of one to several days (Gauhl, 1976; Osmond *et al.*, 1988), largely restricting the sphere of functional convergence to chronic and predictable components of the resource environment.

Across a broad range of species, the regulation of biochemical capacity may involve components that are plastic responses to perceived or predicted resource availability as well as components that are genetically fixed. The relative importance of the two kinds of components may reflect a compromise between environmental variability and the cost of physiological plasticity (Field, 1983). For genetically fixed responses, the regulation of biochemical capacity can be effected only through differential growth, survival, and reproduction. Weak or moderate selection for functional convergence, as well as anthropogenic factors like the creation of novel habitats, could also allow mismatches between resource availability and biochemical capacity for CO_2 fixation.

B. The Superleaf Hypothesis

Functional convergence predicts that investment in biochemical capacity for CO_2 fixation scales with average light availability. Light availability decreases not only with depth in a canopy, but also with depth in individual leaves (Vogelmann *et al.*, 1989). The spatial scale and the fine structure of the light gradients in leaves and canopies differ, but the phenomena are fundamentally parallel. To the extent that biochemical capacity for CO_2 fixation is tailored to local resource availability, then the photosynthetic response of a canopy should depend on the quantity of light absorbed and on the availability of the other resources required for growth. The biochemical capacity of the canopy should be largely independent of leaf area index. With respect to light use, canopies should behave like superleaves. If both canopies and leaves match the capacity for CO_2 fixation to the light profile, then the details of the profile should be irrelevant for many questions. For some analyses, canopies might be effectively parameterized as big leaves or as totally unstructured diffuse absorbers—nicknamed green slime in the global modeling community.

The big-leaf analogy has an extensive history in hydrometeorology, where canopies have traditionally been characterized by a single average stomatal conductance and saturation deficit (Jarvis and McNaughton, 1986). The big-leaf model represented by the Penman-Monteith equation (Monteith, 1965), although not specifically relevant to the gradient of intercepted radiation, builds from the concept that similar physical and physiological factors constrain leaf and canopy function.

Parallels between the light gradients in canopies and in leaves suggest that canopies should behave like superleaves, but two other kinds of considerations suggest that the convergence should be less than perfect. First, canopy structure directly affects some aspects of stress related to leaf temperature and gas exchange, especially the turbulent transport of mass and energy into and out of canopies. These differences could easily drive differences in GPP that could feed back to modify the investment in biochemical capacity. Second, cells within a leaf may differ in exposure to light, but they share access to water and nutrients and are exposed to largely common stresses. Within a canopy comprising multiple individuals, species, or growth forms, individuals may differ in exposure to stress and in access to resources other than light.

The functional convergence and superleaf hypotheses are derived from purely ecological considerations. Yet they make specific, and therefore testable, predictions about the structure and function of individual plants and plant communities. If supported, these hypotheses provide general and potentially powerful simplifications for integrating effects of multiple stresses and resource availability. Such simplification can provide critical tools for scaling ecophysiological concepts to the level of the canopy and the landscape, where they have the potential for significant contributions to research into global change.

III. Evidence from Three Levels of Organization

Both the functional convergence and the superleaf hypotheses apply directly to the whole-plant level. The basic idea underlying functional convergence is that the resources necessary for biochemical capacity for CO_2 fixation should be invested elsewhere whenever the alternatives are expected to yield higher returns. The concept of returns on alternative investments makes sense only in the context of the whole plant. In practical terms, expected fitness returns on alternative resource investments are very difficult to quantify, and a broad-based assessment of the hypotheses must rely on simplified criteria.

The relationship between leaf nitrogen (N) and photosynthetic capacity (A_{max}, determined as the maximum rate of CO_2 uptake under conditions of normal CO_2 and otherwise favorable conditions) provides a test on some aspects of the functional convergence hypothesis. Functional convergence predicts that investments in biochemical capacity for CO_2 fixation should cease when stress or resource availability limits the return on further investments. Biochemical capacity for CO_2 fixation involves several components (Farquhar *et al.*, 1980; Woodrow and Berry, 1988), all of which contain nitrogen. Uniformity—across taxa, habitats, resources, and stress —in the A_{max}–N relationship is sufficient, but not necessary, to support the functional convergence hypothesis. Uniformity in the A_{max}–N relationship is not a necessary criterion for two reasons. First, much of the leaf nitrogen is invested in compounds unrelated to photosynthesis (Chapin *et al.*, 1986; Evans and Seemann, 1989). Second, changes in the relative availability of resources may change the amount of nitrogen required to generate a given biochemical capacity for CO_2 fixation (Seemann *et al.*, 1987; Evans and Seemann, 1989).

The superleaf hypothesis predicts that the scaling of biochemical capacity to light availability should be similar in leaves and canopies and that both should be structured to maintain a high return on invested resources. Gradients of light and biochemical capacity within individual leaves have been studied recently (Vogelmann *et al.*, 1989; Terashima and Inouye, 1985) but are still poorly known relative to canopy-level gradients. Since within-leaf gradients are poorly known, one powerful approach to testing the superleaf hypothesis is by comparing canopy-level gradients with those predicted to maximize the returns on nitrogen investments. The limited data on intraleaf gradients provide another level of comparison.

The A_{max}–N relationship can serve as a probe for the efficiency of nitrogen investments in terms of instantaneous returns, measured under favorable conditions, but it places few constraints on the relationship between investments in biochemical capacity and long-term GPP or carbon gain. To examine the longer-term facets of the functional convergence and superleaf hypotheses, I consider one example of a relationship between a resource and plant growth: the relationship between intercepted radiation and primary production. Another example, parallel in many ways to the relationship between intercepted radiation and primary production, is the relationship between a plant's nitrogen content and its relative growth rate (Ågren and Ingestad, 1987). These relationships should not be considered complete substitutes for the much less available relationships based on GPP, but they can be used for initial tests of the hypotheses. Tight relationships between nitrogen and growth and light and growth are sufficient to support the hypothesis predicting the scaling of investment in infrastructure to resource availability, but they are not necessary. The fate of fixed carbon, which varies widely between ecosystems and with ecosystem development, controls the relationship between integrated photosynthesis and growth.

A. Mechanistic/Leaf Level

For many years, discovering whether and how plants are adapted to the habitats they occupy and identifying the adaptive consequences of functional or structural differences among species or ecotypes have been major objectives in physiological ecology (Field and Davis, 1989). In one sense, these traditional objectives contrast fundamentally with tests of the hypothesis that consistent trends in stress responses transcend taxonomic boundaries. In a discipline that has traditionally emphasized differences among plants, analysis directed toward identifying and understanding common mechanisms must shift the focus.

At a deeper level, the search for consistent stress responses is a direct extension of the dominant paradigms in physiological ecology. If plants have been selected to maximize the return on available resources (Mooney and Gulmon, 1979, 1982; Bloom *et al.*, 1985), then the spatial distribution of capacity should reflect the spatial distribution of limiting resources. Plants of a single species in a range of sites, as well as plants of different species at a single site, access different pools or resources. Some of the differences are driven by spatial variation in resource availability or through site-to-site differences and differences in canopy height and rooting depth; others are driven by plant characteristics, for example, root or canopy architecture or preferences for different forms of nitrogen; and still others are driven through biotic interactions, for example, effects of mycorrhizae and nitrogen-fixing bacteria on nutrient availability and effects of herbivores and pathogens on the effective duration of roots and leaves. Independent of the nature of the controls on resource availability, however, plants should still maximize the efficiency of function with an allocation program that makes all resources equally limiting and prevents the development of unusable physiological capacity (Bloom *et al.*, 1985).

1. The A_{max}-N Relationship   Many studies have reported strong correlations between a leaf's photosynthetic capacity and its nitrogen content (Field and Mooney, 1986). This A_{max}-N relationship is typically linear for naturally growing plants but may be saturating under conditions of artificially high nutrient availability. The correlation is sometimes, but not always, stronger when both photosynthesis and nitrogen are expressed on the basis of leaf mass rather than leaf area (Evans, 1989). It persists across species and growth forms and is relatively robust to the source of variation in leaf nitrogen, whether it is leaf aging (Field and Mooney, 1983; Field *et al.*, 1983), nutrient availability (Gulmon and Chu, 1981; Evans, 1983; Sage and Pearcy, 1987), light availability (DeJong and Doyle, 1985; Seemann *et al.*, 1987; Evans and Seemann, 1989; Hirose *et al.*, 1989), or intrinsic differences among species (Field and Mooney, 1986). Important deviations from a single universal A_{max}-N relationship include the generally greater sensitivity of A_{max} to nitrogen in C_4 than in C_3 species (Pearcy *et al.*, 1982; Sage and Pearcy, 1987), a frequently greater A_{max} at any nitrogen level in sun species than in shade species (Seemann *et al.*, 1987), and a lower A_{max} at any nitrogen level in species with sclerophytic rather than mesophytic leaves (Field and Mooney, 1986; Evans, 1989).

a. Ecological Interpretation   The A_{max}-N relationship does not have a simple mechanistic basis (Field and Mooney, 1986). Under the conditions at which A_{max} is determined, the primary carboxylating enzyme of C_3 photosynthesis, ribulose 1,5-bisphosphate carboxylase/oxygenase, or rubisco, typically imposes the major biochemical limitation to photosynthesis (Woodrow and Berry, 1988). Rubisco contains a significant proportion of the leaf nitrogen (5 to 30%; Seemann *et al.*, 1987; Evans, 1989), but this proportion is too small to explain the A_{max}-N relationship. Overall, the most likely explanation for the generality of the A_{max}-N relationship is that plants regulate investments in the biochemical capacity for photosynthesis so that the limitation by factors other than biochemical capacity is held below a level that depends on the costs of increasing photosynthetic capacity.

The evidence for this interpretation comes from two sources. One is the unsuitability of alternative interpretations based on a simple physiological or biochemical control of A_{max}. The other is broad ecological patterns in the A_{max}-N relationship, especially the sources of variation. Three aspects of the general pattern merit examination: (1) the robustness of the A_{max}-N relationship across genotypes, growth forms, and habitats, (2) the sensitivity of the A_{max}-N relationship to variation in the identity of the limiting resources, and (3) the robustness of the A_{max}-N relationship in the face of environmental stress.

Field and Mooney (1986) demonstrated consistencies in the A_{max}–N relationship, over variation in taxa, habitat of origin, and growth form. For the data presented by Field and Mooney (1986) and Evans (1989), the consistencies appear much stronger for data expressed on the basis of leaf mass rather than leaf area. Evans (1989) argues that the analysis based on leaf area is more revealing from the perspective of understanding the role of nitrogen in photosynthesis. From the perspective of understanding tissue costs, stand-level carbon balance, and biogeochemical cycling, the mass-based analysis is more useful. The primary difference between the two expressions is that evergreen sclerophylls are outliers in an area-based but not in a mass-based analysis (Field and Mooney, 1986). For an area-based relationship without evergreen sclerophylls, A_{max} varies two- to three-fold

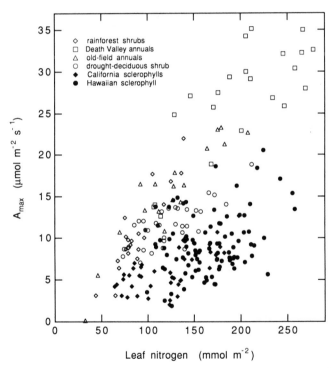

Figure 1   The A_{max}–N relationship for naturally growing C_3 species representing a range of growth forms, habitats, and stress levels. Sclerophyll species are shown with open symbols, nonsclerophylls with closed symbols. A_{max} and nitrogen are both expressed per unit of leaf area. For all species, $A_{max} = 0.189 + 0.076N$ ($r = 0.539$, $p < 0.001$). For the nonsclerophylls only, $A_{max} = -1.39 + 0.126N$ ($r = 0.887$, $p < 0.001$). For the sclerophylls only, $A_{max} = 1.746 + 0.043N$ ($r = 0.496$, $p < 0.001$). [From data of Mooney *et al.* (1981: Death Valley and old-field annuals), Field and Mooney (1983: a drought-deciduous shrub), Field *et al.* (1983: California sclerophylls), Chazdon and Field (1987: rainforest shrubs), and Vitousek *et al.* (1990: an Hawaiian sclerophyll).]

at any leaf nitrogen level (Fig. 1). If sclerophylls are included, the variation at one nitrogen level may be five-fold or higher.

One of the most interesting aspects of the multispecies, multihabitat A_{max}–N relationship in Figure 1 is the lack of obvious saturation. Saturation in the relationship has been reported in studies of plants exposed to high nutrient availability under greenhouse or growth chamber conditions (Evans, 1983), but has not been observed in others (Gulmon and Chu, 1981; Sage and Pearcy, 1987). I know of no studies showing clear saturation in the A_{max}–N relationship for naturally growing plants. This lack of saturation is one of the strongest pieces of evidence indicating that nitrogen investment is regulated to match the availability of other resources. If it were not, we would expect to see saturation in the A_{max}–N relationship in plants stressed with respect to other resources. Schulze and Chapin (1987) argue that the A_{max}–N relationship presented by Field and Mooney (1986) does show the initial stages of saturation, with each growth form dropping from the pattern just as it begins to saturate. This interpretation suggests that it may be possible to use the kinetics of the initial stages of saturation to assess the magnitude and operating range of the growth-form and habitat-specific factors that limit returns on further investments in leaf nitrogen. In most of the existing data, however, the scatter is too great to allow confident identification of saturation, independent of quantifying kinetics.

b. Interacting Environmental Factors   Variation in the A_{max}–N relationship is best understood for plants grown in contrasting light environments. In general, shade-grown plants invest a lower proportion of leaf nitrogen in rubisco and photosynthetic electron transport and a higher proportion in light-harvesting complexes than do sun-grown plants (Evans and Seemann, 1989). Species vary substantially in the flexibility of different compartments. Seemann *et al.* (1987), comparing two species inhabiting a range of light environments, found that bean has a very flexible investment in rubisco but a relatively conservative investment in chlorophyll and chlorophyll proteins (Fig. 2). Alocasia follows the opposite pattern. Most species increase nitrogen allocation to the thylakoids when grown under low light, but pea and spinach do not (Evans and Seemann, 1989).

The variation in the A_{max}–N relationship with growth irradiance is also consistent with the hypothesis that nitrogen investments are regulated to yield high returns. In low light, increased investment in light harvesting should receive higher priority than CO_2 processing, at least until CO_2 processing becomes limiting. Interspecific differences in the flexibility of nitrogen-containing compartments may reflect typical resource availabilities over the range of sites where the species are successful.

In addition to variation in the A_{max}–N relationship caused by variation in the allocation of nitrogen to the components of photosynthesis, some

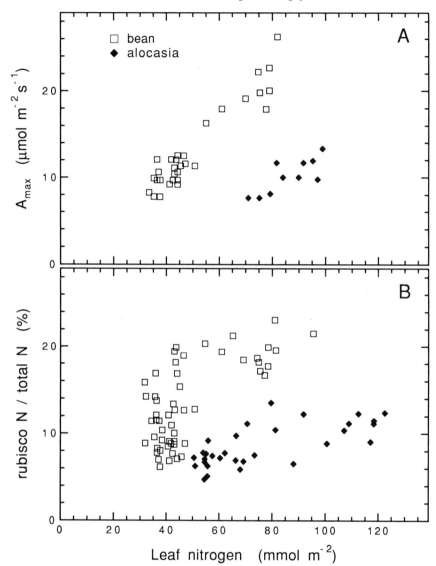

Figure 2 (A) The relationship between CO_2-saturated photosynthetic capacity and leaf nitrogen for bean (a predominantly sun species) and alocasia (a predominantly shade species) grown under two levels of light and two levels of nitrogen availability; (B) the fraction of the leaf nitrogen in rubisco in bean and alocasia grown under two levels of light and two levels of nitrogen. [Redrawn from Seemann *et al.* (1987).]

variation results from genetically based differences in the specific activity of rubisco (Evans, 1989). These differences appear in the $A_{max}-$N relationship in some cases (Seemann and Berry, 1982) but not in others, where the differences in rubisco specific activity are compensated for by differences in stomatal conductance (Evans, 1989) or by differences in resistance to internal diffusion (Evans and Seemann, 1984).

Another important source of variation in the $A_{max}-$N relationship concerns the relatively low photosynthesis per unit of leaf nitrogen in plants with sclerophyllous leaves. Part of the difference may be because investment of nitrogen in processes other than photosynthesis must increase with leaf mass, whereas investments in photosynthetic capacity are potentially independent of mass (Field and Mooney, 1986). In addition, sclerophylly may impose internal diffusion resistances that effectively decrease the concentration of CO_2 at the sites of carboxylation. Large resistances to CO_2 diffusion inside sclerophyllous leaves were predicted by Parkhurst (1986) and measured by Parkhurst *et al.* (1988). Vitousek *et al.* (1990) report a pattern of carbon isotope discrimination consistent with a positive relationship between leaf mass per unit area and internal diffusion resistance.

c. Stress Effects   Many studies address the response of the $A_{max}-$N relationship to controlled variation in light and nitrogen availability, but few address the response to water, temperature, pollution, or non-nitrogen nutrient stress. Field studies have examined the $A_{max}-$N relationship in species and sites characterized by a range of stresses. Controlled-environment studies are useful for characterizing plant responses over a range of conditions. Measurements under natural conditions protect the integrity of the full range of ecological mechanisms potentially involved in scaling photosynthetic capacity to resource availability. Although both kinds of studies are clearly necessary, three potentially important considerations constrain the utility of controlled-environment studies for addressing ecological questions.

The first concerns phenotypic plasticity. Plants differ dramatically in the extent to which they adjust nutrient levels, photosynthesis, and growth in response to resource availability (Grime, 1979), but no plants succeed across the entire resource spectrum. Results of a controlled-environment experiment inconsistent with an ecological hypothesis may indicate either ecologically important deviations or experimental conditions outside the range of accurate plastic adjustment. The second consideration concerns dynamics. Adjusting investments in the biochemical machinery for photosynthesis has a time constant of a day or more (Gauhl, 1976; Osmond *et al.*, 1988). Clearly, adjustments in investment cannot provide an effective response to rapid or even diurnal variation in stress and resource availability. The evolutionary response of adjusting investments in response to resource availability must be limited to chronic or predictable stress and long-term resource availability. The precise limits to what is chronic and what is acute are not known. Maintaining chronic conditions of stress or resource limitation in controlled environments can be very difficult, and it is possible that unrealistic dynamics in controlled environments introduce unusual features in the A_{max}–N relationship. The final difficulty involves environmental signals. The challenge of adjusting investments to match the availability of limiting resources is analogous to trying to hit a moving target with a slow, expensive bullet. Information that makes it possible to predict the target's future position should be critical. Evidence documenting adjustments of photosynthesis to signals, in contrast to resource availability directly, is scarce. To the extent, however, that adjustments are based on predictors rather than on resource availability, adjustments might be different in natural and controlled environments.

In general, the integrity of the A_{max}–N relationship is at least partially maintained across stress exposure in controlled-environment studies. As a typical minimum, leaf nitrogen and A_{max} both decrease as stress increases, whether the stress is drought (Ehleringer, 1983; Walters and Reich, 1989), nitrogen limitation (Gulmon and Chu, 1981; Evans, 1983; Seemann et al., 1987; Walters and Reich, 1989; Sage and Pearcy, 1987), pollutants (Reich and Schoettle, 1988), or low light (Gulmon and Chu, 1981; Seemann et al., 1987; Evans and Seemann, 1989). Even this pattern, however, has some exceptions. For example, Mulligan (1989) found that in *Eucalyptus* seedlings native to low-phosphorus sites, decreasing phosphorus availability led to increased leaf nitrogen. Reich and Schoettle (1988) found no relationship between A_{max} and leaf nitrogen in pines grown on soils of varying phosphorus availability. Among the studies reporting some scaling of nitrogen investment to stress, some report a single A_{max}–N relationship over a range of stress levels (e.g., nitrogen: Sage and Pearcy, 1987), stress factors (e.g., light and nitrogen: Gulmon and Chu, 1981) and species (e.g., Field and Mooney, 1986). Others, however, report that the slope of the A_{max}–N relationship changes with stress (e.g., drought: Ehleringer, 1983) and that it is different with interacting than with single stresses (e.g., water and nitrogen: Reich et al., 1989).

The pattern in naturally growing plants appears to be substantially simpler, with plants adapted to a range of light, water, and nutrient availabilities falling on a more-or-less single A_{max}–N relationship (see Fig. 1). Specific comparisons of plants adapted to sites of varying light availability also support the concept of a consistent relationship (Walters and Field, 1987; Chazdon and Field, 1987). The nutrient story is complicated by the strong increase in leaf mass per area with decreasing nutrient availability (Loveless, 1961). Across a wide range of naturally growing plants, sclerophylls consistently realize lower A_{max} at any leaf nitrogen content than other plants, especially when the data are expressed on the basis of leaf area (Field and Mooney, 1986; Evans, 1989). Among sclerophylls, those adapted to sites of very low phosphorus availability appear to fall on the same A_{max}–N relationship as plants adapted to relatively phosphorus-rich sites (Fig. 3). Water stress has a more complex effect on the A_{max}–N relationship, as a consequence of the potential for a trade-off in the returns on investment of water and nitrogen for the acquisition of photosynthate. By opening the stomata, plants increase photosynthesis for any leaf nitrogen level but at the cost of increased water loss. Stomatal closure increases photosynthesis per unit of water lost but decreases it per unit of nitrogen invested.

In a business firm, maximizing profit requires more than distributing resources to make a given resource equally limiting to each stage of the production process. When a process requires several resource inputs, maximizing profit also requires an investment pattern that makes each resource equally limiting (Bloom *et al.*, 1985). A trade-off between returns on nitrogen investments and water use is the pattern observed by Field *et al.* (1983) in plants that naturally occur in sites of varying moisture availability and by Reich *et al.* (1989) in elm seedlings grown under a range of water

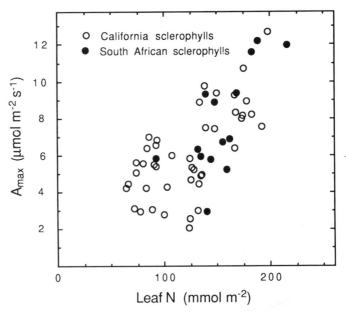

Figure 3   The area-based A_{max}–N relationship for two groups of naturally growing evergreen sclerophylls. The South African sclerophylls occur in sites of extreme chronic phosphorus deficiency. The California sclerophylls occur in sites of no more than modest phosphorus deficiency. [Redrawn from data of Field *et al.* (1983) and Mooney *et al.* (1983).]

and nitrogen availabilities. Plants exposed to dry conditions tend to compromise photosynthesis per unit of nitrogen, resulting in increased photosynthesis per unit of water transpired.

B. Whole-Plant/Canopy Level

1. The Distribution of Nitrogen   Most plants respond to light availability during growth by scaling photosynthetic capacity to light (Björkman and Holmgren, 1963; Björkman, 1981). The range of flexibility within a species or ecotype is usually greater for taxa found in sunny habitats than for those in shady habitats (Björkman and Holmgren, 1963), although the flexibility may be independent of habitat of origin (Clough *et al.*, 1979; Ferrar and Osmond, 1986). Some studies indicate that decreased investment in photosynthetic capacity leads to increased daily photosynthesis at low light (Björkman and Holmgren, 1963; Field, 1981). In all cases, increasing photosynthetic capacity above moderate levels in low-light sites leads to strongly decreasing marginal returns.

The superleaf hypothesis predicts that intraorganism gradients of photosynthetic capacity, at two levels of organization, should parallel the patterns characteristic of different individuals. Both within individual leaves and within canopies, biochemical capacity for photosynthesis should scale with some measure of integrated light availability, perhaps with adjustments to the temporal dynamics as well as the total amount of radiation (Takenaka, 1989; Pearcy, 1990).

Farquhar (1989) developed a formal version of the superleaf hypothesis and solved it for temporally constant incoming photon flux density of photosynthetically action radiation (PFD). His solution applies equally to light gradients within leaves and within the canopies of individual plants. If I_n and J_{maxn} are the absorbed PFD and the capacity for photosynthetic electron transport in layer n, then electron transport is maximized when

$$I_1 : I_2 : \ldots : I_n = J_{max1} : J_{max2} : \ldots : J_{maxn} \tag{1}$$

or when electron transport capacity varies in the same proportion as absorbed PFD. The returns on investments of nitrogen in photosynthesis are maximized when, on average, electron transport and carboxylation are equally limiting (Farquhar *et al.*, 1980). Thus, gradients of biochemical capacity for CO_2 fixation should match the gradients of electron transport capacity, meaning that the biochemical capacity for CO_2 fixation should also scale with absorbed PFD.

The distribution of capacity that maximizes photosynthesis under constant PFD may be suboptimal under naturally varying PFD. Takenaka (1989) developed a simple model to demonstrate that the relationship between daily photosynthesis and photosynthetic capacity depends on both the average PFD and its frequency distribution. Photosynthetic capacity is optimally distributed, through a leaf or a canopy, when any redistribution decreases daily photosynthesis. As long as

$$\delta^2 A_{day}/\delta A_{max}^2 < 0 \qquad (2)$$

where A_{day} is daily photosynthesis and A_{max} is the biochemical capacity for photosynthesis, this occurs when

$$\delta A_{day}/\delta A_{max} = \lambda \qquad (3)$$

for every leaf or every leaf layer, where λ is an unspecified Lagrange multiplier (Field, 1983; Farquhar, 1989). Equation (2) requires diminishing returns on investments in biochemical capacity for CO_2 fixation. Since leaf nitrogen is often a good index of total investments in biochemical capacity for CO_2 fixation (see Fig. 1), we can substitute nitrogen for A_{max} in Equations (2) and (3).

The $A_{max}-N$ relationship in Figure 1 shows little evidence of diminishing returns on nitrogen investments. Still, a version of Equation (2) with nitrogen substituted for A_{max} is supported by the evidence. This equation concerns the relationship between daily photosynthesis and leaf nitrogen. The returns in photosynthesis from nitrogen investments are linear, or nearly linear, under high light but diminish rapidly at low light (Gulmon and Chu, 1981). The diurnal cycle of PFD ensures diminishing returns under a wide range of nitrogen levels.

Only a few studies describe the intraleaf radiation regime, and even fewer describe intraleaf gradients in biochemical capacity for CO_2 fixation. Terashima and Inouye (1985) assayed several components of photosynthesis in transverse sections of spinach leaves and found a gradient from sun-type features at the top of the leaf to shade-type features near the bottom (Fig. 4). A partial return to sun-type characteristics at the bottom, observed by Terashima and Inouye (1985), is consistent with greater illumination resulting from scattering from soil and canopy elements. Several studies have used gas-exchange analysis to test for intraleaf gradients in photosynthetic capacity. The theoretical prediction that intraleaf gradients should make photosynthesis at intermediate light intensities higher in leaves illuminated from above than in those lit from below is generally supported (Terashima, 1989), but with some interesting variations. Photosynthesis is essentially independent of direction of illumination in conifer needles (Leverenz and Jarvis, 1979) and in isobilateral leaves adapted to illumination from both sides (Wong *et al.*, 1985).

Several studies have tested the hypothesis that individual plant canopies should be constructed so that the distribution of photosynthetic capacity parallels the gradient of PFD. Field (1983), DeJong and Doyle (1985), Hirose and Werger (1987), and Hollinger (1989) documented the scaling of leaf nitrogen to either daily PFD or position in the canopy (Fig. 5).

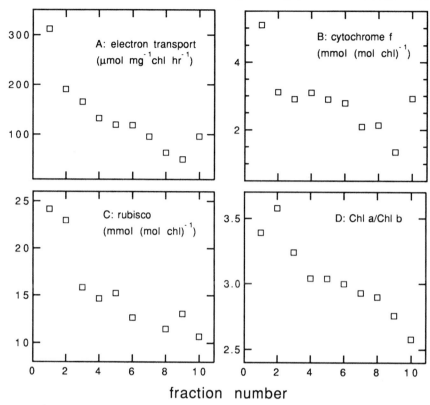

Figure 4 Intraleaf gradients in the biochemical determinants of photosynthesis in spinach. Fraction number refers to position in the leaf, with fraction 1 representing the top layer and fraction 10 the bottom. For each of the parameters, high values are characteristic of a sun-type physiology, and low values are characteristic of a shade-type physiology. [Redrawn from Terashima and Inoue (1985).]

Gutschick and Wiegel (1988) reached the same result using leaf mass per area as the index for the biochemical capacity for CO_2 fixation. The gradient in biochemical capacity with depth in the canopy often reflects a combination of acclimation to the light environment and leaf aging (Field, 1983). In experiments using a sedge canopy, with the youngest leaves at the bottom, Hirose *et al.* (1989) elegantly demonstrated that it is possible to have a close match between PFD and leaf nitrogen through a canopy, even when the age gradient is opposite the PFD gradient.

The quantitative significance of an optimal distribution of photosynthetic capacity with depth in a canopy depends on leaf area index (LAI). At low leaf area index, an optimal canopy yields less than 5% more daily photosynthesis than a canopy with a uniform distribution of capacity (Field,

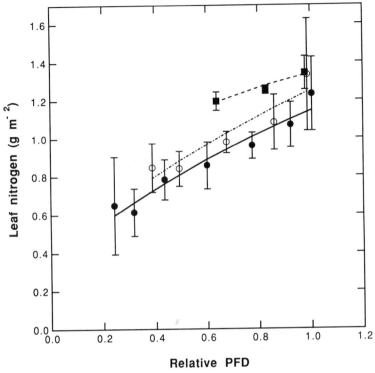

Figure 5 Relationship between leaf nitrogen per unit of leaf area and relative photon flux density (PFD) in a *Carex acutiformis* canopy at three dates: ○ 6 May 1987, ■ 11 June 1987, ● 23 July 1987. In this species, the youngest leaves are at the bottom of the canopy, in the darkest microsites. [Redrawn from Hirose *et al.* (1989).]

1983). At high nitrogen levels and high leaf area indices, the advantage may be more than 100% (Fig. 6). Real canopies tend to have photosynthetic capacities that approximate the optimal but whose gradients of capacity are slightly too shallow. For example, Hirose and Werger (1987) calculated that a canopy of *Solidago altissima* realized a daily carbon gain 4.7% less than that predicted for the optimal distribution of capacity. This was, however, more than 20% greater than that expected for a canopy with no gradient of photosynthetic capacity. Possible explanations for the deviation from optimality include the difficulty of predicting future PFD and the costs of nitrogen translocation (Field, 1983), considerations related to competition between individuals for light (Gutschick and Wiegel, 1988; Hollinger, 1989), and differences in the constraints on light acclimation between leaves and between plants (Leverenz and Jarvis, 1980).

Overall, biochemical capacity for CO_2 fixation generally scales with PFD,

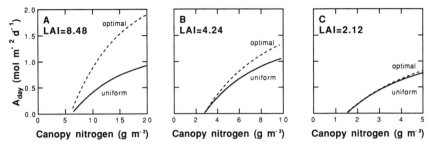

Figure 6   The relationship between simulated daily canopy photosynthesis and total leaf nitrogen in the canopy with optimal and uniform distributions of the nitrogen among leaves, for canopies with three different leaf area indices (LAI). The optimum was determined from the criterion in equation 3 in the text. [Redrawn from Hirose and Werger (1987).]

within leaves, single plants, and single-species canopies. The scaling is consistent with optimizing the distribution of biochemical capacity for the maximization of photosynthesis, subject to the constraint that the total biochemical capacity is fixed (Farquhar, 1989). For a single leaf or a single plant, it is reasonable to argue that, everything else being equal, selection should favor the individuals that realize the greatest photosynthesis for a given investment in biochemical capacity. In many ecosystems, the canopy is diverse in the horizontal plane, but single individuals tend to display leaves in the entire range of vertical microsites (e.g., Hirose and Werger, 1987; Hollinger, 1989; Hirose *et al.*, 1989). In these ecosystems, optimization at the individual level extends directly to the canopy. When the vertical structure of the canopy involves more than one individual, however, the implications of the scaling are less clear. Since it is unlikely that whole-canopy photosynthesis is subject to selection, interindividual adjustment of biochemical capacity to PFD probably reflects the PFD dependence of the marginal returns on investments in photosynthetic capacity. If marginal returns fall consistently with PFD, and if the marginal costs of building capacity are similar among individuals and species, then the adjustment of biochemical capacity to local PFD should be as strong in complex as in simple canopies.

C. Landscape/Global Level

1. The APAR–Yield Relationship   Several studies in the 1960s indicated that dry matter accumulation in crops is roughly proportional to integrated radiation interception, at least for crops growing under unstressful conditions (Monteith, 1977). Since then, many other studies have addressed the relationship between growth and absorbed photosynthetically active radiation (APAR) (Russell *et al.*, 1989). Published ratios of dry matter production to the integrated energy absorbed (ε in g MJ^{-1}) vary more than twofold, but much of the variation is due to differences in technique (e.g., whether and how carefully roots were harvested), production dynamics (e.g., fine root turnover not detected in a harvest), and the distribution of light absorption between photosynthetic and nonphotosynthetic structures. The average value of ε for unstressed C_3 crops is 1.4 g MJ^{-1} (Monteith, 1977), with some evidence for variation due to photosynthetic pathway (Bonhomme *et al.*, 1982), canopy architecture (Heath and Hebblethwaite, 1985), developmental stage (Garcia *et al.*, 1988), planting density and stress by drought (Munchow, 1985), high temperature (Mohamed *et al.*, 1988), or ozone exposure (Unsworth *et al.*, 1984). However, ε often varies much less than APAR and may be nearly constant across a range of chronic stresses (Munchow, 1985; Garcia *et al.*, 1988).

Kumar and Monteith (1981) developed a framework for extending the APAR–yield relationship to remote sensing with a theory indicating that APAR should be simply related to a vegetation index combining canopy reflectance in the red and near infrared. This prediction has been tested for several grassland and savannah sites with data from hand-held (Tucker *et al.*, 1981) as well as satellite radiometers, especially the AVHRR (advanced

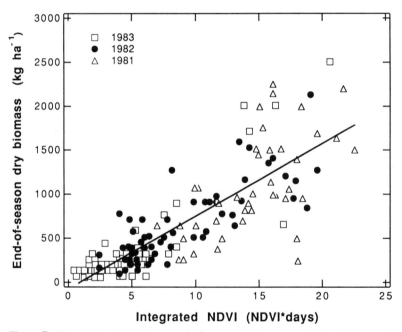

Figure 7   The relationship between end-of-season dry biomass and integrated normalized difference vegetation index (NDVI) from the AVHRR sensor for three years in the Sengalese Sahel. The equation for the regression line (for all three years) is Biomass = −220 + 89*NDVI. [Redrawn from Tucker *et al.* (1985).]

very high resolution radiometer) on several NOAA satellites (Fig. 7). Goward *et al.* (1985) extended the relationship between a vegetation index and production beyond herbaceous systems to ecosystems ranging from desert to forest by comparing the AVHRR vegetation index with net primary production from Whittaker and Likens (1975) (Fig. 8). The general conclusion from the remote-sensing studies is that APAR and production are highly correlated. The limited scatter places strong constraints on the possible variation in ε.

a. Biophysical Interpretation   Monteith (1972) developed a model of crop growth in which production is estimated from the time integral of the product of the amount of radiation absorbed and ε, the efficiency with which absorbed radiation is converted into biomass. This model is a canopy version of simple plant growth models that calculate growth from the product of photosynthesis per unit of leaf area and total leaf area. In a canopy, however, mutual shading at all but the lowest leaf area indices breaks the relationship between leaf area and growth. APAR is potentially a

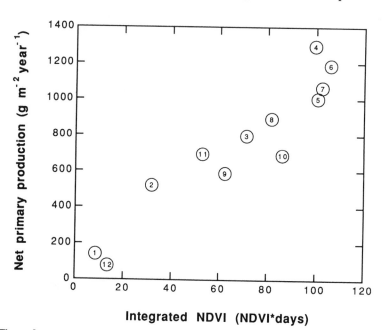

Figure 8   Relationship between mean net primary productivity for several biomes and the seasonally integrated normalized difference vegetation index (NDVI) from the AVHRR sensor. The primary productivity values are published estimates considered representative for (1) tundra, (2) tundra-coniferous ecotone, (3) boreal coniferous, (4) temperate moist coniferous, (5) coniferous-deciduous ecotone, (6) deciduous, (7) oak-pine subclimax, (8) pine subclimax, (9) grassland, (10) cultivated, (11) woodland and scrub, and (12) desert. [Redrawn from Goward *et al.* (1985).]

measure of activity-weighted leaf area. If ε is constant or nearly constant, then it should be possible to predict production from APAR only.

The data from experiments on crops, as well as from remote sensing, indicate that ε is not a constant, but it is surprisingly invariant. The explanation for the robustness of ε could involve biophysical constraints or ecological scaling. One potentially important biophysical constraint is energy availability. Monteith (1977) developed a simple model of crop photosynthesis based on the assumption of a Blackman-type light response of photosynthesis, with a constant photon requirement up to a sudden saturation. Using the model of exponential light extinction proposed by Monsi and Saeki (1953), Monteith observed that when light extinction is complete (APAR = 1), canopy photosynthesis is relatively insensitive to photosynthetic capacity, increasing only 30% for a doubling of capacity. This general conclusion is robust as long as the fraction of leaves above light saturation remains low, as a result of either high photosynthetic capacities or steep leaf angles at the top of the canopy. Although these conditions for maintaining the robustness of the relationship are not uncharacteristic of natural or managed vegetation, they are but a small segment of the possible parameter space. If real plant canopies typically fall within this segment, it must be a result of ecological and evolutionary factors, discussed below.

Sellers (1985, 1987), using more realistic formulations for radiation penetration and photosynthesis, extended Monteith's (1977) theory to address the APAR–yield relationship for APAR assessed from satellite vegetation indices. Assuming a photosynthetic capacity that is invariant with depth in the canopy, Sellers (1985) concludes that canopy photosynthesis is an essentially linear function of APAR because the saturation kinetics in the relationships between APAR and leaf area index and between photosynthesis and leaf area index essentially cancel (Fig. 9). Sellers (1985, 1987) further concludes that APAR is often a linear or nearly linear function of SR, the simple ratio of radiance in the infrared and red AVHRR bands. Nonlinearities in the relationship between SR and photosynthesis can be introduced by vegetation clumping, soil reflectance, and photosynthetic pathway (Sellers, 1985; Choudhury, 1987). Seller's (1985) analysis, like Monteith's (1972) model, predicts a nearly constant ε over a range of reasonable canopy structures and photosynthetic responses. But ε remains constant under environmental stress only if decreases in canopy-level photosynthetic capacity scale with changes in light absorption. That scaling is in the realm of ecological and evolutionary responses.

A second biophysical constraint on the efficiency of radiation use involves the aerodynamic resistance to the transfer of mass between plant canopies and the mixed atmosphere. As photosynthetic capacity increases, the potential for aerodynamic constraints on CO_2 supply also increases, especially in short crops, which tend to be poorly coupled to the atmo-

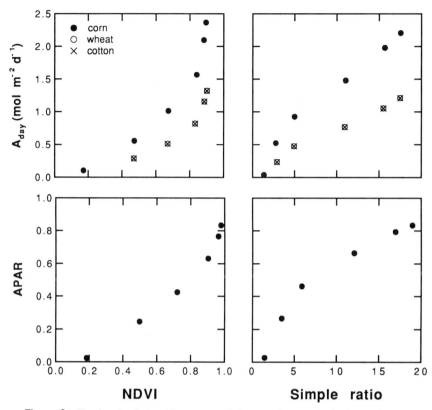

Figure 9 Simulated relationships among daily net photosynthesis (A_{day}), fraction of absorbed photosynthetically active radiation (APAR), the normalized difference vegetation index (NDVI), and the simple ratio (SR) for wheat, corn, and cotton. [Redrawn from Choudhury *et al.* (1987).]

sphere (Jarvis and McNaughton, 1986). Little evidence directly indicates significant aerodynamic constraints on CO_2 supply. The lack of evidence may, however, be more a reflection of ecological than biophysical processes. CO_2 concentrations in the canopies of C_4 crops are frequently in the range of 150 to 200 μmol mol^{-1}, levels saturating for C_4 plants but low enough to impose large limitations on C_3 photosynthesis (von Caemmerer and Farquhar, 1981). Since C_3 and C_4 crops of similar stature have similar aerodynamic resistances, the failure of C_3 crops to attain photosynthetic capacities sufficient to generate large intracanopy CO_2 depletions is likely to reflect the scaling of investment in biochemical capacity for photosynthesis to the availability of a limiting resource—in this case CO_2.

A third biophysical constraint on the efficiency of light use involves a circularity inherent in measuring integrated APAR. At least for canopies growing on current photosynthate, APAR measures not only the potential for future photosynthesis but also the consequences of past photosynthesis. Since seasonal APAR integrals are dominated by the highest observations, this retrospective component of the APAR–yield relationship may be quantitatively important. One way to separate the retrospective and predictive components is to relate growth increments for part of the growing season to preceding APAR measurements. This analysis, available for only a limited number of communities, indicates the high predictive value of an APAR measurement and supports the concept of a relatively invariant ϵ. In ecosystems with evergreen canopies or those that produce new canopies on stored reserves, APAR is a poor measure of past photosynthesis.

b. Ecological Interpretation   From an ecological perspective, APAR measures investment in light capture. Observed regularities in ε, in chronically stressed as well as largely unstressed habitats, require more than biophysical constraints; they must reflect ecological scaling of the potential for light capture to the availability of the resources limiting growth. This interpretation is the remote-sensing analog of the functional convergence hypothesis. It is not a simple restatement because light capture involves more than the biochemical capacity for CO_2 fixation; it also involves leaf-area development, canopy architecture, and the physiological responses of individual leaves.

The mechanistic role of APAR in providing the energy for photosynthesis places an upper but not a lower limit on GPP. Ecological rather than biophysical factors must be responsible for the scaling of photosynthetic capacity to light with depth in a canopy (Section III,B,1) as well as for reductions in canopy development and alterations to canopy architecture in response to stress. If the APAR–yield relationship is as general as the available data indicate, then the accuracy of the scaling must be largely independent of the nature of the stress or limiting resource.

Although the available data indicate strong convergence on a single APAR–yield relationship, or a nearly constant ε, the relationship is likely to be supported in some ecosystems more strongly than in others. It should be weakest in seasonal habitats dominated by evergreen species, which may spend several months of every year doing little or no photosynthesis while APAR remains high (Mooney *et al.*, 1975; Running and Nemani, 1988). Only if investments in light harvesting reflect the annual integral of the availability of other resources should the annual integral of APAR still predict growth accurately. The evidence that the APAR–yield relationship for evergreens falls on basically the same line as the relationship for seasonal deciduous and nonseasonal ecosystems (see Fig. 8) suggests that resources are integrated over long time constants. Year-to-year variability in resource availability may limit plants' ability to scale the capacity for resource uptake to availability in any given year. Another limitation on the functional convergence and superleaf hypotheses is that APAR is unlikely to track photosynthesis over an interval shorter than a day or to respond to changes in moisture availability that last only a few days.

IV. Integration across Levels

A. The Status of the Hypotheses

Relationships at three levels of organization — A_{max}–N at the leaf level, N–PFD at the whole-plant level, and APAR–yield at the ecosystem level —all indicate that, under normal circumstances, plants tend to restrict investments in the capacity for photosynthesis when some resource becomes limiting for growth. The investment may be restricted at the level of canopy development, biochemical capacity, or both.

Across a broad range of species, growth forms, ecosystems, habitats, and stress factors, the pattern of investments scaled to resources is clearly evident. It is also clear, however, that the A_{max}–N, N–PFD, and APAR–yield relationships all admit substantial variation. For particular groups of species or particular habitats, the relationships may disappear altogether (Running and Nemani, 1988) or become too species specific to allow useful generalizations (Evans, 1989).

Is this variation sufficient to reject the functional convergence and superleaf hypotheses? Two considerations are relevant. First, the relationships discussed here may be too simplistic to allow a fair evaluation of the hypotheses. Second, the importance of the variation may depend on the scale of the questions being asked. For example, variation that is critical for documenting that interspecific differences in the specific activity of rubisco affect photosynthesis may be relatively unimportant when estimating global carbon balance.

The simplicity of the relationships clearly contributes to the variation at some levels. For the A_{max}–N relationship, complications that potentially affect A_{max} without compromising the scaling of nitrogen investments to resource availability include changes in the distribution of nitrogen among compounds (Evans and Seemann, 1989) and trade-offs between the returns on investments in nitrogen and water (Field *et al.*, 1983), or nitrogen and other nutrients (Reich and Schoettle, 1988), with changes in the relative availability of resources. At the whole-plant level, costs of nutrient retranslocation potentially constrain scaling nitrogen investments to the PFD gradient (Field, 1983) without implying that nitrogen profiles are not optimal. At the ecosystem level, the hypotheses predict a relationship between APAR and GPP, but differences in allocation and turnover of plant parts immediately challenge the relevance of yield and other growth descriptors available from many past studies.

Factors important at small scales may be unimportant at larger scales for two reasons. First, the amount of difference needed to answer the critical questions at small scales may represent minor noise at larger scales. Especially at the level of species interactions and population dynamics, small differences tend to have large consequences. A one percent difference in selection coefficient, a measure of the intensity of natural selection, is gigantic in relation to the values used in traditional population genetics models (Lewontin, 1974). Yet, a difference of this magnitude is essentially impossible to identify with statistical confidence, either at the level of selection or at the level of physiology. Second, functional differences at one scale may be compensated for by other differences at other scales. For example, the specific activity of rubisco is greater in wheat than in rice, but photosynthetic capacity per unit of nitrogen is nearly the same in the two species because of differences in nitrogen allocation and stomatal conductance (Evans, 1989).

Other sources of variation in the $A_{max}-N$, $N-PFD$, and $APAR-GPP$ relationships cannot be subsumed under either of these caveats. At the leaf level, intrinsic differences in the efficiency of nitrogen use for photosynthesis between C_3 and C_4 plants (Sage and Pearcy, 1987), sun and shade plants (Seemann *et al.*, 1987), and sclerophylls and nonsclerophylls (Field and Mooney, 1986; Evans, 1989) clearly fall outside the realm of scaling investments to resources. At the whole plant level, selection for shading competitors (Gutschick and Wiegel, 1988; Hollinger, 1989) and interspecific differences in access to nitrogen could disrupt the scaling of nitrogen to PFD. At the ecosystem level, seasonally inactive evergreen vegetations clearly depart from the GPP predicted from APAR (Running and Nemani, 1988).

B. Implications for Remote Sensing

To the extent that natural selection has molded plant communities so that the biochemical capacity for CO_2 fixation is scaled to the availability of the resource most limiting for growth, the challenge of assessing some terrestrial processes with remote sensing is greatly simplified. It should be possible to use a canopy-level character like APAR not only as a surrogate for time-integrated carbon gain, but also as an index of the combined effects of multiple stresses.

Biophysical analyses, beginning with Monteith (1977) and continuing through Sellers (1985, 1987) and Choudhury (1987), as well as simulation studies like those of Running and Nemani (1988), indicate why the time integral of APAR should be functionally related to ecosystem GPP. From an ecological perspective, factors that potentially depress GPP, growth, or both are not necessarily reflected in APAR. The functional convergence and superleaf hypotheses provide an evolutionary mechanism with the potential to scale APAR to these other factors.

From the level of the leaf to the level of the ecosystem, much of the available evidence supports the functional convergence and superleaf hypotheses. At each level, however, some evidence is contradictory or confusing. Currently, neither the theoretical nor the practical implications of the contradictory evidence is understood. It is clear, however, that an ecological–evolutionary perspective has the potential to contribute to the challenge of global ecosystem analysis.

Acknowledgments

Research leading to the ideas discussed here was supported by NSF grant BSR 8717422 and a grant from the Mellon Foundation to the Carnegie Institution of Washington. Thanks to N. Chiariello and D. Schimel and J. Gamon for stimulating discussions. This is CIWDPB Publication number 1070.

References

Ågren, G. I., and Ingestad, T. (1987). Root:shoot ratio as a balance between nitrogen productivity and photosynthesis. *Plant, Cell Environ.* 10: 579–586.

Björkman, O. (1981). Responses to different quantum flux densities. *In* "Encyclopedia of Plant Physiology, New Series" (O. L. Lange, P. S. Nobel, C. B. Osmond, and H. Ziegler, eds.), Vol. 12A, pp. 57–107. Springer-Verlag, Berlin.

Björkman, O. (1989). Some viewpoints on photosynthetic response and adaptation to environmental stress. *In* "Photosynthesis" (W. R. Briggs, ed.), pp. 45–58. Allan R. Liss, New York.

Björkman, O., and Holmgren, P. (1963). Adaptability of the photosynthetic apparatus to light intensity in ecotypes from exposed and shaded habitats. *Physiol. Plant.* 16: 889–914.

Bloom, A. J., Chapin, F. S., III, and Mooney, H. A. (1985). Resource limitation in plants—an economic analogy. *Annu. Rev. Ecol. Syst.* 16: 363–392.

Bonhomme, R., Ruget, F., Derieux, M., and Vincourt, P. (1982). Relations entre production de matière sèche aérienne et énergie interceptée chez différents génotypes de maïs. *Comptes Rendus Acad. Sci. Paris* 294: 393–398.

Chapin, F. S., III. (1980). The mineral nutrition of wild plants. *Annu. Rev. Ecol. Syst.* 11: 233–260.

Chapin, F. S., III, Vitousek, P. M., and Van Cleve, K. (1986). The nature of nutrient limitation in plant communities. *Am. Nat.* 127: 48–58.

Chazdon, R. L., and Field, C. B. (1987). Determinants of photosynthetic capacity in six rainforest *Piper* species. *Oecologia* 73: 222–230.

Choudhury, B. J. (1987). Relationships between vegetation indices, radiation absorption, and net photosynthesis evaluated by a sensitivity analysis. *Remote Sens. Environ.* 22: 209–233.

Clough, J. M., Teeri, J. A., and Alberte, R. S. (1979). Photosynthetic adaptation of *Solanum dulcamara* L. to sun and shade environments. 1. A comparison of sun and shade populations. *Oecologia* 38: 13–21.

Committee on Global Change. (1988). "Toward an Understanding of Global Change." National Academy Press, Washington, D.C.

DeJong, T. M., and Doyle, J. F. (1985). Seasonal relationships between leaf nitrogen content (photosynthetic capacity) and leaf canopy light exposure in peach *(Prunus perspicata)*. *Plant, Cell Environ.* 8: 701–706.

Ehleringer, J. (1983). Ecophysiology of *Amaranthus palmeri*, a Sonoran desert summer annual. *Oecologia* 57: 107–112.

Evans, J. R. (1983). Nitrogen and photosynthesis in the flag leaf of wheat (*Triticum aestivum* L.). *Plant Physiol.* 72, 297–302.

Evans, J. R. (1989). Photosynthesis and nitrogen relationships in leaves of C₃ plants. *Oecologia* 78: 9–19.

Evans, J. R., and Seemann, J. R. (1984). Differences between wheat genotypes in the specific activity of ribulose 1,5-bisphosphate carboxylase and the relationship to photosynthesis. *Plant Physiol* 74, 759–765.

Evans, J. R., and Seemann, J. R. (1989). The allocation of protein nitrogen in the photosynthetic apparatus: costs, consequences, and control. *In* "Photosynthesis" (W. R. Briggs, ed.), pp. 183–205. Allan R. Liss, New York.

Farquhar, G. D. (1989). Models of integrated photosynthesis of cells and leaves. *Philos. Trans. R. Soc. Lond. B Biol. Sci.* 323: 357–367.

Farquhar, G. D., and Sharkey, T. D. (1982). Stomatal conductance and photosynthesis. *Annu. Rev. Plant Physiol.* 33: 317–345.

Farquhar, G. D., von Caemmerer, S., and Berry, J. A. (1980). A biochemical model of photosynthetic CO_2 assimilation in leaves of C₃ species. *Planta* 149: 78–90.

Ferrar, P. J., and Osmond, C. B. (1986). Nitrogen supply as a factor influencing photoinhibition and photosynthetic acclimation after transfer of shade-grown *Solanum dulcamara* to bright light. *Planta* 168: 563–570.

Field, C. (1981). Leaf-age effects on the carbon balance of individual leaves in relation to microsite. *In* "Components of Productivity of Mediterranean Regions — Basic and Applied Aspects" (N. A. Margaris and H. A. Mooney, Eds.), pp. 41–50. Dr. W. Junk, The Hague.

Field, C., (1983). Allocating leaf nitrogen for the maximization of carbon gain: Leaf age as a control on the allocation program. *Oecologia* 56: 341–347.

Field, C. B., and Davis, S. D. (1989). Physiological ecology. *Nat. Hist. Mus. Los Angel. Cty. Sci. Ser.* 34: 154–164.

Field, C., and Mooney, H. A. (1983). Leaf age and seasonal effects on light, water, and nitrogen use efficiency in a California shrub. *Oecologia* 56: 348–355.

Field, C., and Mooney, H. A. (1986). The photosynthesis–nitrogen relationship in wild plants. *In* "On the Economy of Plant Form and Function" (T. J. Givnish, ed.), pp. 25–55. Cambridge Univ. Press, Cambridge.

Field, C., Merino, J., and Mooney, H. A. (1983). Compromises between water-use efficiency and nitrogen-use efficiency in five species of California evergreens. *Oecologia* 60: 384–389.

Garcia, R., Kanemasu, E. T., Blad, B. L., Bauer, A., Hatfield, J. L., Major, D. J., Reginato, R. J., and Hubbard, K. G. (1988). Interception and use efficiency of light in winter wheat under different nitrogen regimes. *Agric. For. Meteorol.* 44: 175–186.

Gauhl, E. (1976). Photosynthetic responses to varying light intensity in ecotypes of *Solanum dulcamara* L. from shaded and exposed habitats. *Oecologia* 22: 275–286.

Goward, S. N., Tucker, C. J., and Dye, D. G. (1985). North American vegetation patterns observed with the NOAA-7 advanced very high resolution radiometer. *Vegetatio* 64: 3–14.

Grime, J. P. (1979). "Plant Strategies and Vegetation Processes." Wiley, Chichester.

Gulmon, S. L., and Chu, C. C. (1981). The effects of light and nitrogen on photosynthesis, leaf characteristics, and dry matter allocation in the chaparral shrub, *Diplacus aurantiacus*. *Oecologia* 49: 207–212.

Gutschick, V. P., and Wiegel, F. W. (1988). Optimizing the canopy photosynthetic rate by patterns of investment in specific leaf mass. *Am. Nat.* 132: 67–86.

Heath, M. C., and Hebblethwaite, P. D. (1985). Solar radiation intercepted by leafless, semi-leafless and leafed peas *(Pisum sativum)* under contrasting field conditions. *Ann. Appl. Biol.* 107: 309–318.

Hirose, T., and Werger, M. J. A. (1987). Maximizing daily canopy photosynthesis with respect to the leaf nitrogen allocation pattern in the canopy. *Oecologia* 72: 520–526.

Hirose, T., Werger, M. J. A., and van Rheenen, J. W. A. (1989). Canopy development and leaf nitrogen distribution in a stand of *Carex acutiformis. Ecology* 70: 1610–1618.

Hollinger, D. Y. (1989). Canopy organization and foliage photosynthetic capacity in a broad-leaved evergreen montane forest. *Funct. Ecol.* 3: 53–62.

Jarvis, P. G., and McNaughton, K. G. (1986). Stomatal control of transpiration: Scaling up from leaf to region. *Adv. Ecol. Res.* 15: 1–49.

Kumar, M., and Monteith, J. L. (1981). Remote sensing of crop growth. *In* "Plants and the Daylight Spectrum" (H. Smith, ed.), pp. 133–144. Academic Press, London.

Leverenz, J. W., and Jarvis, P. G. (1979). Photosynthesis in Sitka spruce. VIII. The effects of light flux density and direction on the rate of net photosynthesis and the stomatal conductance of needles. *J. Appl. Ecol.* 16: 919–932.

Leverenz, J. W., and Jarvis, P. G. (1980). Acclimation to quantum flux density within and between trees. *J. Appl. Ecol.* 17: 697–708.

Lewontin, R. C. (1974). "The Genetic Basis of Evolutionary Change." Columbia Univ. Press, New York.

Loveless, A. R. (1961). A nutritional interpretation of sclerophylly based on differences in the chemical composition of sclerophyllous and mesophytic leaves. *Ann. Bot.* 25: 168–183.

Mohamed, H. A., Clarke, J. A., and Ong, C. K. (1988). Genotypic differences in the temperature responses of tropical crops. III. Light interception and dry matter production of pearl millet *(Pennisetum typhoides* S. & H.). *J. Exp. Bot.* 39: 1137–1143.

Monsi, M., and Saeki, T. (1953). Über den Lichtfaktor in den Pflanzengesellschaften und seine Bedeutung für die Stoffproduktion. *Jpn. J. Bot.* 14: 22–52.

Monteith, J. L. (1965). Evaporation and environment. *Symp. Soc. Exp. Biol.* 19, 205–234.

Monteith, J. L. (1972). Solar radiation and productivity in tropical ecosystems. *J. Appl. Ecol.* 9: 747–766.

Monteith, J. L. (1977). Climate and the efficiency of crop production in Britain. *Philos. Trans. R. Soc. Lond. B Biol. Sci.* 281: 277–294.

Mooney, H. A., and Gulmon, S. L. (1979). Environmental and evolutionary constraints on photosynthetic characteristics of higher plants. *In* "Topics in Plant Population Biology" (O. T. Solbrig, S. Jain, G. B. Johnson, and P. H. Raven, eds.), pp. 316–337. Columbia Univ. Press, New York.

Mooney, H. A., and Gulmon, S. L. (1982). Constraints on leaf structure and function in relation to herbivory. *BioScience* 32: 198–206.

Mooney, H. A., Harrison, A. T., and Morrow, P. A. (1975). Environmental limitations of photosynthesis on a California evergreen shrub. *Oecologia* 19: 293–301.

Mooney, H. A., Field, C., Gulmon, S. L., and Bazzaz, F. A. (1981). Photosynthetic capacity in relation to leaf position in desert versus old-field annuals. *Oecologia* 50: 109–112.

Mooney, H. A., Field, C., Gulmon, S. L., Rundel, P., and Kruger, F. J. (1983). Photosynthetic characteristics of South African sclerophylls. *Oecologia* 58: 398–401.

Mulligan, D. R. (1989). Leaf phosphorus and nitrogen concentrations and net photosynthesis in Eucalyptus seedlings. *Tree Physiol.* 5: 149–157.

Munchow, R. C. (1985). An analysis of the effects of water deficits on grain legumes grown in a semi-arid tropical environment in terms of radiation interception and its efficiency of use. *Field Crop Res.* 11: 309–323.

Osmond, C. B., Austin, M. P., Berry, J. A., Billings, W. D., Boyer, J. S., Dacey, J. W. H., Nobel, P. S., Smith, S. D., and Winner, W. E. (1987). Stress physiology and the distribution of plants. *BioScience* 37: 38–48.

Osmond, C. B., Oja, V., and Laisk, A. (1988). Regulation of carboxylation and photosynthetic oscillations during sun-shade acclimation in *Helianthus annuus* measured with a rapid-response gas-exchange system. *In* "Ecology of Photosynthesis in Sun and Shade" (J. R. Evans, S. von Caemmerer, and W. W. Adams III, eds.), pp. 239–251. CSIRO, Melbourne.

Parkhurst, D. F. (1986). Internal leaf structure: a three-dimensional perspective. *In* "On the Economy of Plant Form and Function" (T. J. Givnish, ed.), pp. 215–249. Cambridge Univ. Press, Cambridge.

Parkhurst, D. F., Wong, S. C., Farquhar, G. D., and Cowan, I. R. (1988). Gradients of intercellular CO_2 levels across the leaf mesophyll. *Plant Physiol.* 86: 1032–1037.

Pearcy, R. W. (1990). Sunflecks and photosynthesis in plant canopies. *Annu. Rev. Plant Physiol.* 41: 421–453.

Pearcy, R. W., Osteryoung, K., and Randall, D. (1982). Carbon dioxide exchange characteristics of C_4 Hawaiian *Euphorbia* species native to diverse habitats, *Oecologia* 55: 333–341.

Powles, S. B. (1984). Photoinhibition of photosynthesis induced by visible light. *Annu. Rev. Plant Physiol.* 35: 15–44.

Reich, P. B., and Schoettle, A. W. (1988). Role of phosphorus and nitrogen in photosynthetic and whole-plant carbon gain and nutrient use efficiency in eastern white pine. *Oecologia* 77: 25–33.

Reich, P. B., Walters, M. B., and Tabone, T. J. (1989). Response of *Ulmus americana* seedlings to varying nitrogen and water status. 2. Water and nitrogen use efficiency. *Tree Physiol.* 5: 173–184.

Running, S. W., and Nemani, R. R. (1988). Relating seasonal patterns of the AVHRR vegetation index to simulated photosynthesis and transpiration of forests in different climates. *Remote Sens. Environ.* 24: 347–367.

Russell, G., Jarvis, P. G., and Monteith, J. L. (1989). Absorption of radiation by canopies and stand growth. *In* "Plant Canopies: Their Growth, Form, and Function" (G. Russell, B. Marshall, and P. G. Jarvis, eds.), pp. 21–39. Cambridge Univ. Press, Cambridge.

Sage, R. F., and Pearcy, R. W. (1987). The nitrogen use efficiency of C_3 and C_4 plants. II. Leaf nitrogen effects on the gas exchange characteristics of *Chenopodium album* (L.) and *Amaranthus retroflexus* (L.). *Plant Physiol.* 84: 959–963.

Schulze, E.-D. (1986). Carbon dioxide and water vapor exchange in response to drought in the atmosphere and in the soil. *Annu. Rev. Plant. Physiol.* 37: 247–274.

Schulze, E.-D., and Chapin, F. S., III. (1987). Plant specialization to environments of different resource availability. *In* "Potentials and Limitations in Ecosystem Analysis" (E.-D. Schulze and H. Zwölfer, eds.), pp. 120–148. Springer-Verlag, Berlin.

Seemann, J. R., and Berry, J. A.. (1982). Interspecific differences in the kinetic properties of RuBP carboxylase protein. *Carnegie Inst. Washington Year Book* 81: 78–83.

Seemann, J. R., Sharkey, T. D., Wang, J. L., and Osmond, C. B. (1987). Environmental effects on photosynthesis, nitrogen-use efficiency, and metabolite pools in leaves of sun and shade plants. *Plant Physiol.* 84: 796–802.

Sellers, P. J. (1985). Canopy reflectance, photosynthesis, and transpiration. *Int. J. Remote Sens.* 6: 1335–1372.

Sellers, P. J. (1987). Canopy reflectance, photosynthesis, and transpiration. II. The role of biophysics in the linearity of their interdependence. *Remote Sens. Environ.* 21: 143–183.

Sharkey, T. D. (1984). Transpiration-induced changes in the photosynthetic capacity of leaves. *Planta* 160: 143–150.

Sharkey, T. D. (1985). Photosynthesis in intact leaves of C_3 plants: Physics, physiology and rate limitations. *Bot. Rev.* 51: 53–105.

Takenaka, A. (1989). Optimal leaf photosynthetic capacity in terms of utilizing a natural light environment. *J. Theor. Biol.,* 139: 517–529.

Terashima, I. (1989). Productive structure of a leaf. *In* "Photosynthesis" (W. R. Briggs, ed.), pp. 207–226. Alan R. Liss, New York.

Terashima, I., and Inouye, T. (1985). Vertical gradient in photosynthetic properties of spinach chloroplasts dependent on intra-leaf light environment. *Plant Cell Physiol.* 26: 781–785.

Tucker, C. J., Holben, B. N., Elgin, J. H., and McMurtrey, J. E. (1981). Remote sensing of total dry-matter accumulation in winter wheat. *Remote Sens. Environ.* 11: 171–189.

Unsworth, M. H., Lesser, V. M., and Heagle, A. S. (1984). Radiation interception and the growth of soybeans exposed to ozone in open-top field chambers. *J. Appl. Ecol.* 21: 1059–1079.

Vitousek, P. M., Field, C. B., and Matson, P. A. (1990). Variation in $\delta^{13}C$ in Hawaiian *Metrosideros polymorpha:* A case of internal resistance? *Oecologia* 84: 362–370.

Vogelmann, T. C., Bornman, J. F., and Josserand, S. (1989). Photosynthetic light gradients and spectral regime within leaves of *Medicago sativa. Philos. Trans. R. Soc. Lond. B Biol. Sci.* 323: 411–421.

von Caemmerer, S., and Farquhar, G. D. (1981). Some relationships between the biochemistry of photosynthesis and the gas exchange of leaves. *Planta* 153: 376–387.

Walters, M. B., and Field, C. B. (1987). Photosynthetic light acclimation in two rainforest *Piper* species with different ecological amplitudes. *Oecologia* 72: 449–456.

Walters, M. B., and Reich, P. B. (1989). Response of *Ulmus americana* seedlings to varying nitrogen and water status. 1. Photosynthesis and growth. *Tree Physiol.* 5: 159–172.

Weis, E., and Berry, J. A. (1988). Plants and high temperature stress. *Symp. Soc. Exp. Biol.* 42: 329–346.

Whittaker, R. H., and Likens, G. E. (1975). Primary production: The biosphere and man. *In* "Primary Productivity of the Biosphere" (H. Lieth and R. H. Whittaker, eds.), pp. 305–328. Springer-Verlag, Berlin.

Winner, W. E., Mooney, H. A., Williams, K., and von Caemmerer, S. (1985). Measuring and assessing SO_2 effects on photosynthesis and plant growth. *In* "Sulfur Dioxide and Vegetation" (W. E. Winner and H. A. Mooney, eds.), pp. 118–132. Stanford Univ. Press, Stanford, California.

Wong, S. C., Cowan, I. R., and Farquhar, G. D. (1985). Leaf conductance in relation to rate of CO_2 assimilation. II. Effects of short-term exposure to different photon flux densities. *Plant Physiol.* 78: 826–829.

Woodrow, I. E., and Berry, J. A. (1988). Enzymatic regulation of CO_2 fixation in higher plants. *Annu. Rev. Plant Physiol.* 39: 533–594.

3

Effects of Multiple Environmental Stresses on Nutrient Availability and Use

F. Stuart Chapin III

I. Introduction

Nutrient availability is one of the major environmental determinants of productivity in natural and agricultural ecosystems. Any thorough consideration of the impact of stress on vegetation must, therefore, consider the interactions of various stresses with plant nutrition. How a stress affects plants depends on the nature of the stress. Grime (1977) defined stress as any factor that restricts growth. It is useful, however, to distinguish between (1) stress caused by insufficient resources (e.g., water, light, nutrients) for supporting maximal growth and (2) stress caused by direct physiological damage to the plant due to disruption of metabolism. The second stress differs from Grime's (1977) concept of disturbance, which results in destruction or removal of plant parts.

Plant response to insufficient resources is quite predictable. It involves compensatory changes in allocation to maximize acquisition of those resources that most directly limit growth (Bloom *et al.*, 1985; see also Chapters 1, 2, and 6). For example, light-limited plants increase allocation to leaf and stem production at the expense of root production, whereas nutrient- or water-limited plants increase allocation to roots at the expense of shoots. All plants can counteract stresses from low resource supply, although their success in doing so depends on the range of resource availabilities to which they are adapted (Chapin, 1988). The evolutionary response to insufficient resources is to fix genetically the patterns of allocation that maximize acquisition of the limiting resource and to reduce overall resource requirement by reducing growth rate.

On the other hand, changes in patterns of resource acquisition cannot readily ameliorate stresses resulting from direct damage to the plant. The physiological and evolutionary responses to stresses such as mineral toxicity, flooding, ozone, and ultraviolet damage are highly specific and cannot be predicted from general principles of compensatory allocation. The most likely effect of such direct-damage stresses is to increase rates of tissue loss (and therefore increase the requirement for resources to replace these tissues) or to reduce growth rate and, therefore, the rate of acquisition of all resources. Because these two predictions are contradictory, we need to examine carefully the time course and dose response to stress before we can make reliable predictions.

Stress must be defined with respect to the physiological requirements of a species. No habitat can be defined as stressful in an absolute sense. For example, Australian heath plants occur on soils of extremely low nitrogen and phosphorus availability and accumulate toxic levels of these elements by luxury consumption if grown on normal topsoil (Heddle and Specht, 1975; Groves and Keraitis, 1976). Such topsoil might support maximal growth of wild plants from more fertile soils, but it might be deficient in nitrogen and phosphorus for growth of certain nutrient-demanding crops (Chapin, 1988). Thus soils of intermediate nutrient availability are toxic to some species but represent a low-resource stress to others. Similarly, wet soils may depress growth of some species from dry environments, yet provide too little water for hydrophytes. Light intensities that limit photo-synthesis and growth of desert annuals may cause photoinhibition and reduce growth in rainforest understory species.

Most plants exhibit near-optimal growth over a broad range of mesic conditions. Outside this range, we can recognize certain habitats as stressful because they supply either insufficient or toxic levels of resources to most plants. Nonetheless, those plants that occupy these extreme habitats generally exhibit specific physiological adaptations that minimize the degree to which the extreme factors affect growth, so we cannot assume that extreme environments are stressful to the species occupying those environments.

This chapter deals with the effects of multiple environmental stresses on nutrient availability and use by plants. The study of plant responses to stress is complicated for the reasons outlined above and also because stresses have multiple effects on plants. For example, water stress reduces the rate at which nutrients become available in soils, the flux of nutrients to the root surface, the rate of nutrient uptake by plant roots, and the requirement of plants for nutrients. Often these indirect effects on plant nutrition are as important as the direct effects of water stress. Only by recognizing the multiple effects of environmental stresses can we begin to predict plant responses to these stresses. This chapter emphasizes the multiple ways that stress affects plant mineral nutrition and the influence of nutrient status on plant susceptibility to other environmental stresses.

II. Nutrient Availability

A. Direct Controls

1. Parent Material   Parent material, i.e., the bedrock that gives rise to soil, determines the proportions of minerals that are available in the soil and strongly influences weathering rate. For example, granite is resistant to weathering and generally has lower concentrations of phosphorus and many cations required by plants than does limestone. At the opposite extreme, weathering of certain parent materials such as serpentine can cause toxic accumulation of heavy metals that are either not required by plants or are required in very small amounts. Parent material thus directly determines the rate of nutrient input from rock to soil, thereby exerting a major influence over vegetation composition and productivity (e.g., Beadle, 1954). However, other factors (climate, topography, vegetation, and time) substantially modify the relationship between parent material and nutrient availability.

2. Water Availability   When little water is available, all ions become less mobile in the soil because air spaces replace water in the pores between soil particles, making the pathway from the soil to the root surface less direct (Nye and Tinker, 1977). These effects on nutrient mobility are important even over ranges of soil water content that would have little effect on plant water relations (Fig. 1). Ion mobility can decrease by two orders of magnitude between -0.01 MPa and -1.0 MPa, a range in soil water potential that does not strongly restrict water uptake by most plants (Nye and Tinker, 1977). Because the rate of ion diffusion to the root surface is usually the rate-limiting step in nutrient uptake by plants, reduction in water availability may affect plant growth substantially. In fact, two lines of evidence suggest that the effects of low soil water content on nutrient availability may be nearly as important as the direct effects of water stress on plant growth: (1) Tissue concentrations of growth-limiting nutrients (nitrogen and phosphorus) often decline during water stress (Fig. 2; Day, 1981), whereas one would expect these elements to increase in concentration if water directly restricted growth more strongly than it affected nutrient uptake (Chapin, 1988). (2) Experimental manipulations indicate that adding nutrients enhances growth of some desert annuals more than adding water (Romney *et al.*, 1978; Gutierrez and Whitford, 1987).

3. Soil Acidity   Acid rain might have three general effects on nutrient availability: (1) increasing the atmospheric input of potentially limiting nutrients, particularly nitrogen; (2) increasing the weathering rate of parent material; and (3) altering nutrient availability indirectly through various effects of soil acidity. The productivity of most ecosystems is limited in part by nitrogen availability, and one of the major consequences of acid rain has been to reduce the degree of nitrogen limitation because of the large input of nitric acid (Nihlgard, 1985; Oren *et al.*, 1988a). Unfortunately, the

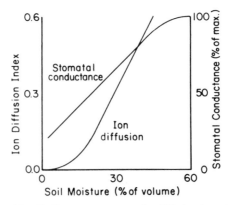

Figure 1   Relationship of soil water content to the diffusion impedence factor (an index of ion diffusion rate) for chloride and leaf (stomatal) conductance to water vapor for plants grown in a sandy loam soil. [Redrawn from Nye and Tinker (1977); Schulze (1986).]

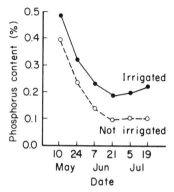

Figure 2 Phosphorus concentration (% dry mass) in barley grown under irrigated and unirrigated conditions. [Redrawn from Day (1981).]

deleterious effects of acid rain generally outweigh any benefits from reduced nitrogen limitation, often by creating a nutrient imbalance.

Acid rain has surprisingly little effect on the weathering rate of parent material (Johnson *et al.*, 1972). Thus the rate at which soil cations are replenished from parent material is often insufficient to meet increased losses due to leaching (see Section II,B,1).

In ecosystems receiving substantial acid rain, base cations required for plant growth (e.g., potassium, calcium, and magnesium) are displaced by hydrogen from ion exchange sites on clay particles and soil organic matter. These base cations are then liable to be lost by leaching. Moreover, because the exchange complex of acidified soils is dominated by hydrogen, the soil has less capacity to store cations that enter via normal weathering and cycling. In situations where sulfate input exceeds biological demand and the capacity of the soil to adsorb it, sulfate can be leached below the rooting horizon, carrying with it (to maintain charge balance) mobile mineral cations (e.g., potassium, calcium, and magnesium) and leaving behind a predominance of hydrogen and aluminum ions (Reuss and Johnson, 1986). This process is a major mechanism of soil acidification and can reduce cation availability to the point that plant growth declines (Meyer *et al.*, 1988; Oren *et al.*, 1988a).

In nitrogen-limited ecosystems, nitrate and its associated cations are less susceptible than sulfate to leaching loss, despite the high mobility of nitrate in the soil, because plants and microorganisms readily absorb the nitrate (Reuss and Johnson, 1986). Only when the biological demands for nitrogen are exceeded does nitrate leaching become a major mechanism of soil acidification and cation loss. Most soils that are naturally acid have low rates of nitrification (Reuss and Johnson, 1986). Contrary to expectations, however, soil acidification caused by acid rain may not reduce rates of nitrification (van Breeman and Jordans, 1983; Reuss and Johnson, 1986).

Soil acidity also has profound effects on the solubility of other ions affecting plant growth. Below pH 6.0, phosphate availability in the soil declines substantially (Black, 1968). In areas heavily influenced by acid rain, soil pH has gone well below this value, presumably making phosphate less available to plants. Aluminum solubility increases in acid soils, so that aluminum concentration in the soil solution can reach potentially toxic concentrations (Meyer *et al.*, 1988). This effect of acidity on aluminum solubility is least pronounced in organic soils because aluminum ions are adsorbed to carboxyl groups on soil organic matter (Ulrich, 1987). Other effects of acidity on soils processes are discussed in detail elsewhere (e.g., Ulrich and Pankrath, 1983; Reuss and Johnson, 1986).

B. Indirect Controls

Environmental stresses also influence nutrient availability indirectly through their impacts on vegetation and microbial processes. Nutrients absorbed by plants come from three major sources: rocks (via weathering), the atmosphere (via precipitation, fallout, and nitrogen fixation), and recirculation of nutrients through plants and back to soil. Of these sources, recirculation within the ecosystem is by far the most important (Table I). Thus, any factor that influences the rate of nutrient loss from plants or the resistance of plant litter to decomposition will strongly influence nutrient availability. These are probably the major mechanisms by which multiple environmental stresses influence nutrient availability.

1. Leaching Leaching accounts for about 15% of the annual nitrogen and phosphorus return from plants to soil and 25–50% of the cycling of potassium, calcium, and magnesium (Henderson *et al.*, 1977; Morton, 1977; Table II). Some of this nutrient input comes from dry deposition on

Table I Major Sources of Nutrients Entering the Soil That Are Available for Plant Uptake[a]

Nutrient	Source of plant nutrient (% of total)		
	Atmosphere	Weathering	Recycling
Tundra (Barrow)			
N	4	0	96
P	4	<1	96
Temperate forest (Hubbard Brook)			
N	7	0	93
P	1	<10?	>89
K	2	10	88
Ca	4	31	65

[a] Calculated from Chapin *et al.* (1980) and Whittaker *et al.* (1979).

the surface of the leaves; some is leached from inside the leaf. Acid rain substantially increases leaching loss from plants for reasons discussed in Section IV. The effect of increased leaching of nutrients from plants is to speed nutrient return to the available soil pool (Reuss and Johnson, 1986). Johnson *et al.* (1985) speculate that the rates of potassium, calcium, and magnesium cycling in the deciduous forests of eastern Tennessee have doubled as a result of increased leaching caused by acid rain. High leaching losses add nutrients to the available soil pool but also increase plant demand for these nutrients.

2. Litterfall   Litterfall is the major avenue of nutrient return from plants to the soil (Table II). Ecosystems characterized by low availability of soil resources (water or nutrients) or by a toxic excess of salts or heavy metals have low productivity, low rates of litter production, and therefore a low flux of nutrients, particularly nitrogen and phosphorus, back to the available pool. (The physiological basis for this is discussed in Section IV). Some environmental stresses, such as sulfur dioxide or ozone, may cause direct tissue damage and increase tissue turnover rates (Lechowicz, 1987) and thereby enhance nutrient availability in the short term. In the long term, however, pollutants reduce production and therefore slow the cycling of organically bound nutrients.

3. Litter Quality   The nutrient concentration of litter is determined by the nutrient concentration in live leaf tissue and the degree to which these nutrients are resorbed before leaf abscission. Nutrient concentration in live leaves is largely a function of nutrient availability (Vitousek, 1982; Chapin, 1988). Thus factors that enhance nutrient availability enhance nutrient cycling and the return of nutrients to the available pool by

Table II Nutrients Leached from the Canopy (Throughfall) as a Percentage of the Total Aboveground Nutrient Return from Plants to the Soil

Nutrient	Throughfall (% of annual return)	
	Evergreen	Deciduous
N	14 ± 3	15 ± 3
P	15 ± 3	15 ± 3
K	59 ± 6	48 ± 4
Ca	27 ± 6	24 ± 5
Mg	33 ± 6	38 ± 5

[a] Calculated from Cole and Rapp (1981) for 12 deciduous and 12 evergreen forests.

increasing (1) the quantity of litter produced (see above) and (2) the concentrations of nutrients in that litter. (The physiological determinants of resorption efficiency and therefore of nutrient concentration are discussed in Section IV.)

Once litter reaches the soil surface, some nutrients are lost immediately by leaching, and the remainder are released by decomposition. Potassium is released from litter entirely by leaching, so variations in litter quality have minimal effect on its rate of return to the soil. About 80–90% of the nitrogen and phosphorus in litter must be released through microbial mineralization, which is largely determined by the litter's lignin : nitrogen ratio (Fig. 3; e.g., Aber and Melillo, 1982; Flanagan and Van Cleve, 1983). Other plant secondary metabolites, such as tannins and phenolics, also retard decomposition (Middleton, 1984; White, 1988). Thus, any factor that influences the nitrogen or secondary metabolite content of plant material will also influence nutrient availability.

Environmental stress has predictable effects on concentrations of many plant secondary metabolites (Chapter 12) and therefore on the decomposition rate of litter produced by the plant. When conditions are favorable for growth, the concentrations of many carbon-based secondary metabolites (i.e., those containing no nitrogen) decrease because carbon is allocated preferentially to growth processes (Fig. 4; Bryant *et al.*, 1983; Lorio, 1986). This is particularly true for immobile secondary metabolites like lignin and

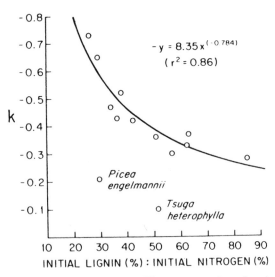

Figure 3 The decomposition constant *(k)* of litter, expressed as a function of the ratio of initial lignin concentration to initial nitrogen concentration. [Reprinted from Melillo *et al.* (1982) with permission from the Ecological Society of America.]

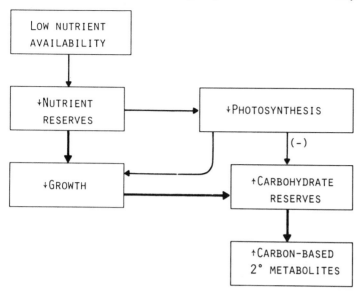

Figure 4   Effect of nutrient availability on concentrations of carbon-based secondary metabolites.

many tannins that are the final end-products of metabolism (Coley *et al.,* 1985) and are most strongly implicated as inhibitors of decomposition. Environmental stresses that inhibit growth potential more than photosynthetic potential lead to increased carbohydrate concentrations. This carbohydrate is often funneled into secondary metabolite production (Bryant *et al.* 1983; Lorio, 1986). Thus nutrient stress, water stress, heavy metal toxicity, and pollutants that have their major direct impact on growth (rather than on photosynthesis) can be expected to reduce litter quality by increasing concentrations of lignin and other secondary metabolites (Fig. 5). In contrast, shade stress and factors that enhance nitrogen availability (e.g., acid rain) would reduce concentrations of carbon-based secondary metabolites, thereby improving litter quality.

It is fairly clear that nitrogen fertilization enhances litter quality, both by increasing nitrogen content and by reducing tannin and lignin content (Table III; Flanagan and Van Cleve, 1983; Waring *et al.,* 1985; Bryant *et al.,* 1987). Similarly, along natural fertility gradients, plants growing in infertile soils produce litter with low nitrogen concentrations (Vitousek, 1982) and high lignin and tannin content (Flanagan and Van Cleve, 1983). This pattern occurs because of the phenotypic response of individuals to nitrogen availability and because infertile soils are occupied by slow-growing species, particularly evergreen shrubs and trees, which produce tissues that are characteristically high in lignin and tannin, in part to deter herbivores

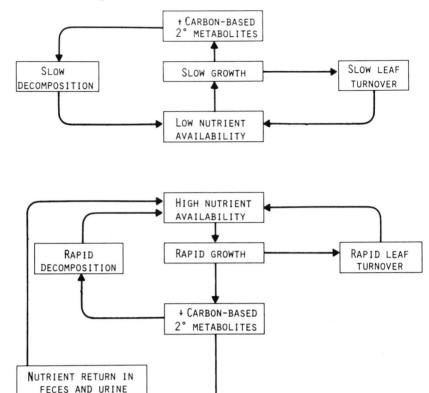

Figure 5 Feedbacks by which nutrient-cycling processes exaggerate natural differences in soil fertility.

Table III Effect of Nutrient Availability on Tissue
Concentration of Nitrogen, Starch, and Two Carbon-based
Secondary Metabolites in Willow[a]

Tissue concentration	High nutrient availability	Moderate nutrient availability
Nitrogen (mg dm^{-2})	21.5	14.0
Starch (%)	5.1	20.7[b]
Tannin (relative units)	65	100[b]
Lignin (%)	20.8	24.5[b]

[a] Data from Waring *et al.* (1985).
[b] Significant ($p < 0.05$) treatment effect.

(Coley *et al.*, 1985). Thus, the indirect effects of nutrient availability on litter quality (and therefore the rate of mineralization) provide a positive feedback that exaggerates any natural differences in soil fertility (Fig. 5).

Those environmental stresses that exert their influence primarily through soil fertility should have predictable effects on litter quality. Thus, to the extent that acid rain acts as a nitrogen fertilizer (Nihlgard, 1985), it should increase soil nitrogen availability directly and enhance decomposition rates by increasing tissue nitrogen and decreasing tissue lignin.

Grazing enhances litter quality through three mechanisms. First, grazers remove leaves before nitrogen and other minerals have been resorbed. Second, they feed preferentially on leaves that are high in nitrogen and low in tannin (e.g., Owen-Smith and Cooper, 1987), causing plants to produce new leaves with high nitrogen and low tannin. And third, in the case of homeotherm herbivores, much of the carbon contained in ingested vegetation is burned off in respiration so that the "litter" returned as feces and urine contains high concentrations of available nitrogen. For these three reasons, the immediate effect of grazing is generally to enhance litter quality relative to that which is shed by plants. On the other hand, over evolutionary time, particularly in infertile soils, herbivory selects for increased levels of plant secondary metabolites, which deter feeding but also retard decomposition and nutrient recycling. Thus herbivores tend to exaggerate any inherent differences in nutrient availability: they concentrate their activity in areas of high nutrient availability and so enhance litter quality in these sites, whereas in areas of infertile soils they constitute a selective pressure to increase levels of secondary metabolites that retard decomposition and reduce nutrient availability (Fig. 5; Chapin and McNaughton, 1989).

Other environmental stresses can also strongly influence litter quality through their effects on plant carbon–nutrient balance. Decreased light availability decreases the carbon available for carbon-based secondary metabolites (Bryant *et al.*, 1983), thereby increasing litter quality (Table IV)

Table IV   Effect of Light Intensity on Tissue Concentrations of Nitrogen, Starch, and Two Carbon-based Secondary Metabolites in Willow[a]

Tissue concentration	High light	Low light
Nitrogen (mg dm^{-2})	21.5	13.5[b]
Starch (%)	5.1	5.3
Tannin (relative units)	65	64
Lignin (%)	20.8	13.4[b]

[a] Data from Waring *et al.* (1985).
[b] Statistics as in Table III.

and decomposition rate. Water stress tends to reduce tissue nitrogen content and increase tissue lignin, in part because of a relative carbon surplus (Lorio, 1986) and in part because of the lignification of xylem vessels (Orians and Solbrig, 1977). Ozone and sulfur dioxide stress reduce leaf nitrogen concentration (Lechowicz, 1987) and therefore litter quality. The effects of these pollutants on tissue carbohydrate and secondary metabolite concentrations are inconsistent (Lechowicz, 1987).

III. Uptake and Assimilation

A. Nutrient Availability

The rate of nutrient uptake by plants depends on both the rate of nutrient supply (largely determined by mineralization of organic matter in the case of nitrogen and phosphorus) and the plant's uptake potential. In many ecosystems, the uptake of nitrogen is tightly correlated with its rate of mineralization. It is difficult to know, however, to what extent this indicates that supply is the major determinant of uptake or that plant nutrient status is an important determinant of mineralization (see Section II,B,3). When light and water availability are optimal for plant growth, nutrient uptake is a linear function of nutrient supply over a broad range (Ingestad, 1981) because supply is the major limitation on growth. In laboratory experiments where growth conditions other than nutrient supply are varied, the plant's demand for nutrients is generally a more important determinant of uptake rate than is nutrient supply (Clarkson and Hanson, 1980). This suggests that uptake rate is a function of a plant's integrated response to its environment, not an isolated response to nutrient supply. The general principle governing resource acquisition in response to stress is that the plant allocates resources preferentially to those functions that most strongly limit growth (Bloom *et al.,* 1985; Chapin *et al.,* 1987).

The potential of a plant to absorb nutrients depends on the quantity of absorbing root surface and the activity of those roots. Under conditions where a single nutrient limits growth, plants increase their root : shoot ratio (Brouwer, 1966; Chapter 1; Chapter 6) as well as their potential to absorb the specific growth-limiting nutrient. For example, plants exhibit a high potential to absorb phosphate under conditions of phosphate stress (Harrison and Helliwell, 1979), a high potential to absorb nitrogen under conditions of nitrogen stress, a high potential to absorb potassium under conditions of potassium stress, and so on (Lee, 1982; Table V). This compensatory response is specific to the nutrient that limits growth: nitrogen stress reduces the potential of roots to absorb phosphate (Smith and Jackson, 1987), and phosphate stress reduces the potential of roots to absorb nitrogen.

Table V Effect of Environmental Stresses on
Nutrient Uptake Rate[a]

Stress	Ion absorbed	Uptake rate by stressed plant[b] (% of control)
Nitrogen	Ammonium	209
	Nitrate	206
	Phosphate	56
	Sulfate	56
Phosphorus	Phosphate	400
	Sulfate	70
	Nitrate	35
Sulfur	Sulfate	895
	Phosphate	32
	Nitrate	69
Water	Phosphate	13[c]
Light	Nitrate	73

[a] Per gram root. Calculated from Lee (1982) and Lee
and Rudge (1987).

[b] Values for barley, except for water stress.

[c] Value for tomato.

B. Water

The effects of other environmental factors on nutrient uptake follow
logically from the tendency of plants to minimize overwhelming limitation
by any one environmental factor. Water stress causes an increase in
root:shoot ratio (Chapter 6) which coincidentally increases the root surface area for nutrient uptake. However, the uptake potential of roots
produced during water stress is generally less than that of unstressed roots
(Fig. 6; Table V). As discussed above, low water availability can also directly

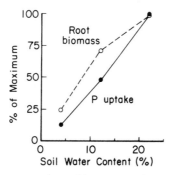

Figure 6 Effect of water stress on root biomass and phosphate uptake per unit of root
weight in tomato. Calculated from Table 7.7 in Nye and Tinker (1977).

reduce nutrient availability, so that water stress is often accompanied by nutrient stress. Water-stressed plants often have lower concentrations of tissue nitrogen and phosphorus than do unstressed plants (see Fig. 2; Day, 1981), and nutrients can limit the growth of some desert plants as much as water does (Romney *et al.*, 1978; Gutierrez and Whitford, 1987).

C. Light

Low light availability generally reduces nutrient uptake both by reducing root:shoot ratio (Mooney and Winner, Chapter 6) and by reducing the potential of roots to absorb nutrients (Fig. 7). The effect of light intensity on nutrient uptake is strongest when light limits growth. Nutrient uptake by nutrient-limited plants is relatively insensitive to light intensity, presumably because carbohydrate supply seldom limits root activity in these plants. In the case of nitrogen uptake, the effect of light is more complex because nitrogen can be acquired by (1) symbiotic nitrogen fixation; (2) nitrate uptake, reduction, and assimilation; or (3) ammonium uptake and assimilation. Symbiotic nitrogen fixation has the greatest energy cost, ammonium uptake the lowest (Bloom, 1986). Nitrogen fixation is therefore largely restricted to situations where light availability is high and nitrogen availability is low. As nitrogen availability increases, nitrogen-fixing plants switch to ammonium or nitrate uptake for their major nitrogen source. Most plants, given a choice of ammonium or nitrate, show preferential absorption of ammonium, the energetically least expensive source of nitrogen (van den Driessche, 1971; Haynes and Goh, 1978). In species that reduce nitrate in leaves, however, assimilation can be partially driven by photosynthetic electron transport. When light is not limiting to photosynthesis, nitrate assimilation in these species diverts relatively little energy away from photosynthetic carbon assimilation, and thus the carbon cost of nitrate assimilation in high light is minor (Bloom, 1986).

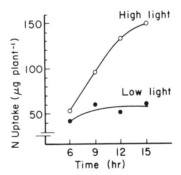

Figure 7   Uptake of ammonium by young rice plants at two light intensities. [Redrawn from Mengel and Viro (1978).]

The effects of grazing on nutrient uptake depend on plant carbon–nutrient balance. When nutrients are strongly limiting, grazing may deplete nutrient reserves more than carbohydrate reserves, thus increasing nutrient demand and, therefore, the potential of roots to absorb nutrients (Chapin and Slack, 1979). Where the environment is more favorable for growth, or where light limits growth (as in many growth chamber experiments), grazing or clipping reduces the potential of roots to absorb nutrients (Jameson, 1963; Clement *et al.*, 1978). Little is now known about which of these grazing effects is more common in nature.

D. Anthropogenic Stresses

The effects of anthropogenic stresses on nutrient uptake are generally consistent with the effects expected from altered plant carbon–nutrient balance. Heavy metals generally restrict growth and therefore nutrient demand, and plants grown under these conditions generally have a low potential for absorbing nutrients, even though heavy metals do not directly inhibit the uptake process (except when there is direct competitive inhibition between chemically similar elements, e.g., strontium and calcium). To the extent that pollutants exert their major effect through a reduction in growth or in carbon assimilation, they appear to reduce the potential of roots to absorb nutrients, as if they had reduced plant demand for nutrients. Acid rain can inhibit root growth as a result of increased soil aluminum concentration (Meyer *et al.*, 1988). The reduced root growth in turn limits the potential of plants to absorb other potentially limiting nutrients and makes them more susceptible to drought. Similarly, root : leaf ratio declines consistently in plants under ozone and sulfur dioxide stress, perhaps explaining the low leaf nitrogen concentrations of plants exposed to these pollutants (Lechowicz, 1987).

E. Evolutionary Response to Stress

One of the major evolutionary responses of plants to low availability of major resources (light, water, and nutrients) has been a low relative growth rate among perennials (Grime and Hunt, 1975; Grime, 1977; Chapin, 1988). This can be seen among morphologically similar species of grasses, forbs, shrubs, and trees (Chapin, 1988). An important consequence of a low relative growth rate is a low nutrient demand and therefore a low potential to absorb major macronutrients such as phosphate (Chapin *et al.*, 1982, 1986). Thus, plants adapted to drought and low light are likely to have a relatively low potential to absorb phosphate, whereas plants whose nutrient demand is usually high because of regular intensive grazing may have a greater potential to absorb nutrients (McNaughton and Chapin, 1985).

IV. Nutrient Loss

A. Leaching

Nutrient loss is just as important as nutrient uptake in determining the nutrient budget of perennial plants. Nutrient loss occurs primarily through leaching and litterfall. As mentioned in Section II, leaching accounts for about 15% of the annual nitrogen and phosphorus returned from plants to the soil. It is not known whether low availability of light, water, or nutrients has any substantial effect on the susceptibility of leaves to leaching. Crops grown at extremely high nutrient availability are prone to loss of nutrients by foliar leaching (Tukey, 1970). The likelihood that macronutrients will be lost by leaching decreases in the order: potassium > calcium > nitrogen = phosphorus (Tukey, 1970; Baker *et al.*, 1985). Loveless (1961) suggested that the thick cuticles produced by plants growing on phosphorus-deficient soils were adaptive because they minimized leaching loss. There is no clear evidence, however, that thickness of cuticle or degree of sclerophylly correlates with the susceptibility of leaves to leaching loss.

Acid rain substantially increases leaching loss (Fig. 8) because hydrogen ions in the rain exchange with cations held on the cuticular exchange surface and because acidity alters the chemical nature of the cuticle so that it is more susceptible to diffusion and mass flow of nutrients to the leaf surface (Shriner and Johnston, 1985; Reuss and Johnson, 1986).

B. Litterfall

Litterfall is the major avenue of nutrient loss from plants. Sites with low availability of water or nutrients have low rates of litter production because

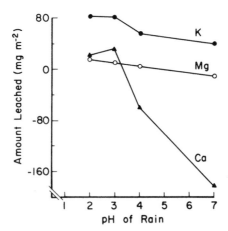

Figure 8 Effects of pH of simulated rain on leaching of calcium, potassium, and magnesium from spruce crowns. [Redrawn from Abrahamson *et al.* (1976).]

they produce new biomass slowly and because each unit of biomass turns over slowly as well (Turner, 1977; Shaver, 1981; Chapin, 1988). Consequently, nutrients incorporated into plant biomass are retained by the plant for a long time before being lost.

Some environmental stresses, such as sulfur dioxide or ozone, damage tissues directly and increase tissue turnover rates, thereby increasing nutrient loss in litter (Lechowicz, 1987). In the long term, pollutants reduce production and therefore the rate of nutrient loss.

Nutrient loss in litter is also strongly influenced by the litter's nutrient concentration, which is determined by the tissue concentration in live leaves and the degree to which nutrients are resorbed before leaf abscission. The nutrient concentration in live leaves is largely a function of nutrient availability (Vitousek, 1982; Chapin, 1988). Of the major plant nutrients, calcium is minimally resorbed, potassium is resorbed some, and nitrogen and phosphorus are substantially resorbed (Epstein, 1972; Baker *et al.*, 1985). The degree of nitrogen and phosphorus resorption varies widely among species, and the factors controlling the resorption of these two elements are not well understood. Under laboratory conditions optimal for growth, nutrient deficiency increases the proportion of nutrients withdrawn from senescing foliage (Shaver and Melillo, 1984). In natural communities, however, the proportion of nitrogen and phosphorus resorbed from leaves before senescence shows no consistent relationship to plant nutrient status (e.g., Flanagan and Van Cleve, 1983; Chapin and Kedrowski, 1983; Boerner, 1984; Birk and Vitousek, 1986; Lajtha, 1987). Various other environmental stresses may also influence the degree of nutrient resorption. Stresses that damage tissues directly may prevent resorption of nutrients altogether and thus increase nutrient loss from plants.

Many evergreen species shed leaves at the time when new leaves are actively growing. In these species, any factor that promotes growth and the requirement of the plant for nutrients will maximize the resorption of nutrients from senescing leaves (Nambiar and Fife, 1987; Oren *et al.*, 1988b). Thus nutrient stress, water stress, or low temperatures could reduce the proportion of nutrients resorbed from leaves of evergreen species through their effect on growth.

V. Nutrient Stress and the Susceptibility of Plants to Other Stresses

Under most circumstances, moderate nutrient stress reduces the susceptibility of plants to other stresses. Plants that are growing rapidly because of high nutrient availability are generally more susceptible to frost and water stress than are plants whose growth is nutrient limited. Similarly, plants growing under optimal nutrition produce fewer secondary metabolites and are therefore more susceptible to herbivore or pathogen attack than are nutrient-limited plants (Bryant *et al.*, 1983; Jones and Coleman, Chapter 12).

VI. General Conclusions

Many of the patterns of nutrient availability and uptake observed with respect to environmental stresses are predictable from the integrated responses of plants to stress.

Nutrient availability is a function of the nutrient content and weathering rate of parent material, but this relationship between rock and soil fertility is strongly altered by environmental stress. Water stress reduces ion mobility and therefore the rate of diffusion to the root surface. Acid rain may increase the input (and therefore availability) of nitrogen, but it strongly reduces availability of other essential nutrients, particularly cations like magnesium and calcium, which are required for growth. Perhaps the most important effects of environmental stresses on nutrient availability act through nutrient recycling. Because acid rain increases leaching losses from leaves, it has doubled the annual cycling of potassium, calcium, and magnesium in some forests. Stresses that reduce growth more than photosynthesis (e.g., nutrient stress and mild drought) result in higher concentrations of secondary metabolites in plant tissues. These secondary metabolites retard decomposition, thereby slowing the release of plant-available nutrients.

Any stress that reduces plant growth and nutrient demand also reduces nutrient uptake. Stresses that reduce availability of a particular nutrient increase the potential of roots to absorb that nutrient; for example, nitrogen stress increases the potential of roots to absorb nitrogen. However, low availability of water and nutrients reduce the potential of roots to absorb nonlimiting nutrients. Low light availability generally reduces nutrient uptake both by reducing root : shoot ratio and by reducing the potential of roots to absorb nutrients. Most anthropogenic stresses reduce plant growth rate and therefore nutrient demand. Plants growing under these conditions generally have a low potential to absorb nutrients.

The evolutionary response to environmental stress has generally been a low relative growth rate and associated low nutrient requirement. Plants with these traits may prove to be the most successful in resisting anthropogenic stresses. The obvious cost of such stress resistance is reduced overall productivity. Unless the sources of anthropogenic pollutants and other stresses are curtailed, a decline in productivity of agricultural crops, forests, and other natural ecosystems seems likely.

Acknowledgments

Research leading to these generalizations was supported by National Science Foundation grants BSR-8300397 and DEB-8205344.

References

Aber, J. D., and Melillo, J. M. (1982). Nitrogen immobilization in decaying hardwood leaf litter as a function of initial nitrogen and lignin content. *Can. J. Bot.* 60: 2263–2269.

Abrahamson, G., Bjor, K., Horntvedt, R., and Tveite, B. (1976). Effects of acid precipitation on coniferous forest. *In* "Impact of Acid Precipitation on Forest and Freshwater Ecosystems in Norway" (F. H. Braekke, ed.), pp. 37–63. Research Report SNSF, Project FR6/76. Norwegian Forest Research Institute, Aas.

Baker, J. G., Hodgkiss, P. D., and Oliver, G. R. (1985). Accession and cycling of elements in a coastal stand of *Pinus radiata* D. Don in New Zealand. *Plant Soil* 86: 303–307.

Beadle, N. C. W. (1954). Soil phosphate and the delimitation of plant communities in eastern Australia. *Ecology* 35: 370–375.

Birk, E. M., and Vitousek, P. M. (1986). Nitrogen availability and nitrogen use efficiency in loblolly pine stands. *Ecology* 67: 69–79.

Black, C. A. (1968). "Soil–Plant Relationships," 2nd ed. Wiley, New York.

Bloom, A. (1986). Plant economics. *Trends Ecol. & Evol.* 1: 93–100.

Bloom, A. J., Chapin, F. S., III, and Mooney, H. A. (1985). Resource limitation in plants—an economic analogy. *Annu. Rev. Ecol. Syst.* 16: 363–392.

Boerner, R. E. J. (1984). Foliar nutrient dynamics and nutrient use efficiency of four deciduous tree species in relation to site fertility. *J. Appl. Ecol.* 21: 1029–1040.

Brouwer, R. (1966). Root growth of grasses and cereals. *In* "The Growth of Cereals and Grasses" (F. L. Milthorpe and J. D. Ivins, eds.), pp. 153–166. Butterworth, London.

Bryant, J. P., Chapin, F. S., III, and Klein, D. R. (1983). Carbon–nutrient balance of boreal plants in relation to vertebrate herbivory. *Oikos* 40: 357–368.

Bryant, J. P., Chapin, F. S., III, Reichardt, P. B., and Clausen, T. P. (1987). Response of winter chemical defense in Alaska paper birch and green alder to manipulation of plant carbon–nutrient balance. *Oecologia* 72: 510–514.

Chapin, F. S., III. (1988). Ecological aspects of plant mineral nutrition. *Adv. Miner. Nutr.* 3: 161–191.

Chapin, F. S., III, and Kedrowski, R. A. (1983). Seasonal changes in nitrogen and phosphorus fractions and autumn retranslocation in evergreen and deciduous taiga trees. *Ecology* 64: 376–391.

Chapin, F. S., III, and McNaughton, S. J. (1989). Lack of compensatory growth under phosphorus stress in grazing-adapted grasses from the Serengeti Plains. *Oecologia* 79: 551–557.

Chapin, F. S., III, and Slack, M. (1979). Effect of defoliation upon root growth, phosphate absorption, and respiration in nutrient-limited tundra graminoids. *Oecologia* 42: 67–79.

Chapin, F. S., III, Miller, P. C., Billings, W. D., and Coyne, P. I. (1980). Carbon and nutrient budgets and their control in coastal tundra. *In* "An Arctic Ecosystem: The Coastal Tundra at Barrow, Alaska" (J. Brown, P. C. Miller, L. L. Tieszen, and F. L. Bunnell, eds.), pp. 458–482. Dowden, Hutchinson & Ross, Stroudsburg, Pennsylvania.

Chapin, F. S., III, Follett, J. M., and O'Connor, K. F. (1982). Growth, phosphate absorption, and phosphorus chemical fractions in two *Chionochloa* species. *J. Ecol.* 70: 305–321.

Chapin, F. S., III, Van Cleve, K., and Tryon, P. R. (1986). Relationship of ion absorption to growth rate in taiga trees. *Oecologia* 69: 238–242.

Chapin, F. S., III, Bloom, A. J., Field, C. B., and Waring, R. H. (1987). Plant responses to multiple environmental factors. *BioScience* 37: 49–57.

Clarkson, D. T., and Hanson, J. B. (1980). The mineral nutrition of higher plants. *Annu. Rev. Plant Physiol.* 31: 239–298.

Clement, C. R., Hopper, M. J., Jones, L. H. P., and Leafe, E. L. (1978). The uptake of nitrate by *Lolium perenne* from flowing nutrient solution. II. Effect of light, defoliation, and relationship to CO_2 flux. *J. Exp. Bot.* 29: 1173–1183.

Cole, D. W., and Rapp, M. (1981). Elemental cycling in forest ecosystems. *In* "Dynamic Properties of Forest Ecosystems" (D. E. Reichle, ed.), pp. 341–409. Cambridge Univ. Press, Cambridge.

Coley, P. D., Bryant, J. P., and Chapin, F. S., III. (1985). Resource availability and plant anti-herbivore defense. *Science* 230: 895–899.

Day, W. (1981). Water stress and crop growth. *In* "Physiological Processes Limiting Plant Growth" (C. B. Johnson, ed.), pp. 199–215. Butterworth, London.

Epstein, E. (1972). "Mineral Nutrition of Plants: Principles and Perspectives." Wiley, New York.

Flanagan, P. W., and Van Cleve, K. (1983). Nutrient cycling in relation to decomposition and organic matter quality in taiga ecosystems. *Can. J. For. Res.* 13: 795–817.

Grime, J. P. (1977). Evidence for the existence of three primary strategies in plants and its relevance to ecological and evolutionary theory. *Am. Nat.* 111: 1169–1194.

Grime, J. P., and Hunt, R. (1975). Relative growth rate: Its range and adaptive significance in a local flora. *J. Ecol.* 63: 393–422.

Groves, R. H., and Keraitis, K. (1976). Survival and growth of seedlings of three sclerophyll species at high levels of phosphorus and nitrogen. *Aust. J. Bot.* 24: 681–690.

Gutierrez, J. R., and Whitford, W. G. (1987). Chihuahuan desert annuals: Importance of water and nitrogen. *Ecology* 68: 2032–2045.

Harrison, A. F., and Helliwell, D. R. (1979). A bioassay for comparing phosphorus availability in soils. *J. Appl. Ecol.* 16: 497–505.

Haynes, R. J., and Goh, K. M. (1978). Ammonium and nitrate nutrition of plants. *Biol. Rev.* 53: 465–510.

Heddle, E. M., and Specht, R. L. (1975). Dark Island Heath (Ninety-Mile Plain, South Australia). VIII. The effect of fertilizers on composition and growth, 1950–1972. *Aust. J. Bot.* 23: 151–164.

Henderson, G. S., Harris, W. F., Todd, D. E., Jr., and Grizzard, T. (1977). Quantity and chemistry of throughfall as influenced by forest type and season. *J. Ecol.* 65: 365–371.

Ingestad, T. (1981). Nutrition and growth of birch and grey alder seedlings in low conductivity solutions and at varied relative rates of nutrient addition. *Physiol. Plant.* 52: 454–466.

Jameson, D. A. (1963). Responses of individual plants to harvesting. *Bot. Rev.* 29: 532–594.

Johnson, D. W., Richter, D. P., Lovett, G. M., and Lindberg, S. E. (1985). Effects of acid deposition on cation nutrient cycling in two deciduous forests. *Can. J. For. Res.* 15: 772–782.

Johnson, M. N., Reynolds, R. C., and Likens, G. E. (1972). Atmospheric sulfur: Its effect on the chemical weathering of New England. *Science* 177: 514–515.

Lajtha, K. (1987). Nutrient resorption efficiency and the response to phosphorus fertilization in the desert shrub *Larrea tridentata* (D.C.) Cov. *Biogeochemistry* 4: 265–276.

Lechowicz, M. J. (1987). Resource allocation by plants under air pollution stress: Implications for plant–pest–pathogen interactions. *Bot. Rev.* 53: 281–300.

Lee, R. B. (1982). Selectivity and kinetics of ion uptake by barley plants following nutrient deficiency. *Ann. Bot.* 50: 429–449.

Lee, R. B., and Rudge, K. A. (1987). Effects of nitrogen deficiency on the absorption of nitrate and ammonium by barley plants. *Ann. Bot.* 57: 471–486.

Lorio, P. L. (1986). Growth-differentiation balance: A basis for understanding southern pine beetle–tree interactions. *For. Ecol. Manage.* 14: 259–273.

Loveless, A. R. (1961). A nutritional interpretation of sclerophylly based on differences in chemical composition of sclerophyllous and mesophytic leaves. *Ann. Bot.* 25: 168–184.

McNaughton, S. J., and Chapin, F. S., III (1985). Effects of phosphorus nutrition and defoliation on C_4 graminoids from the Serengeti Plains. *Ecology* 66: 1617–1629.

Melillo, J. M., Aber, J. D., and Muratore, J. F. (1982). Nitrogen and lignin control of hardwood leaf litter decomposition dynamics. *Ecology* 63: 621–626.

Mengel, K., and Viro, M. (1978). The significance of plant energy status for the uptake and incorporation of NH_4-nitrogen by young rice plants. *Soil Sci. Plant Nutr.* 24: 407–416.

Meyer, J., Schneider, B. U., Werk, K. S., Oren, R., and Schulze, E.-D. (1988). Performance of two *Picea abies* (L.) Karst. stands at different stages of decine. V. Root tip and ectomycorrhiza development and their relationship to aboveground and soil nutrients. *Oecologia* 77: 7–13.

Middleton, J. (1984). Are plant toxins aimed at decomposers? *Experientia* 40: 299–301.

Morton, A. J. 1977. Mineral nutrient pathways in a *Molinietum* in autumn and winter. *J. Ecol.* 65: 993–999.

Nambiar, E. K. S., and Fife, D. N. (1987). Growth and nutrient retranslocation in needles of radiata pine in relation to nitrogen supply. *Ann. Bot.* 60: 147–156.

Nihlgard, B. (1985). The ammonium hypothesis—an additional explanation to forest dieback in Europe. *Ambio* 14: 2–8.

Nye, P. H., and Tinker, P. B. (1977). "Solute Movement in the Soil–Root System." Univ. of California Press, Berkeley.

Oren, R., Werk, K. S., Schulze, E.-D., Meyer, J., Schneider, B. U., and Schramel, P. (1988a). Performance of two *Picea abies* (L.) Karst. stands at different stages of decline. VI. Nutrient concentration. *Oecologia (Berlin)* 77: 151–162.

Oren, R., Schulze, E.-D., Werk, K. S., and Meyer, J. (1988b). Performance of two *Picea abies* (L.) Karst. stands at different stages of decline. VII. Nutrient relations and growth. *Oecologia* 77: 163–173.

Orians, G. H., and Solbrig, O. T. (1977). A cost-income model of leaves and roots with special reference to arid and semi-arid areas. *Am. Nat.* 111: 677–690.

Owen-Smith, N., and Cooper, S.N. (1987). Palatability of woody plants to browsing ruminants in a South African savanna. *Ecology* 68: 319–331.

Reuss, J. O., and Johnson, D. W. (1986). "Acid Deposition and the Acidification of Soils and Waters." Springer-Verlag, New York.

Romney, F. M., Wallace, A., and Hunter, R. B. (1978). Plant response to nitrogen fertilization in the northern Mojave Desert and its relationship to water manipulation. *In* "Nitrogen in Desert Ecosystems" (N. E. West and J. J. Skujens, eds.), pp. 232–243. Dowden, Hutchinson & Ross, Stroudsberg, Pennsylvania

Schulze, E.-D. (1986). Carbon dioxide and water vapor exchange in response to drought in the atmosphere and in the soil. *Annu. Rev. Plant Physiol.* 37: 247–274.

Shaver, G. R. (1981). Mineral nutrition and leaf longevity in an evergreen shrub, *Ledum palustre* ssp. *decumbens. Oecologia* 49: 362–365.

Shaver, G. R., and Melillo, J. M. (1984). Nutrient budgets of marsh plants: Efficiency concepts and relation to availability. *Ecology* 65: 1491–1510.

Shriner, D. S., and Johnston, J. W., Jr. (1985). Acid rain interactions with leaf surfaces: A review. *In* "Acid Deposition: Environmental, Economic, and Policy Issues" (D. D. Adams and W. P. Page, eds.), pp. 241–253. Plenum, New York.

Smith, F. W., and Jackson, W. A. (1987). Nitrogen enhancement of phosphate transport in roots of *Zea mays* L. I. Effects of ammonium and nitrate pretreatment. *Plant Physiol.* 84: 1314–1318.

Tukey, H. B., Jr. (1970). The leaching of substances from plants. *Annu. Rev. Plant Physiol.* 21: 305–324.

Turner, J. (1977). Effect of nitrogen availability on nitrogen cycling in a Douglas-fir stand. *For. Sci.* 23: 307–316.

Ulrich, B. (1987). Stability, elasticity, and resilience of terrestrial ecosystems with respect to matter balance. *In* "Potentials and Limitations of Ecosystem Analysis" (E.-D. Schulze and H. Zwölfer, eds.), pp. 11–49. Springer-Verlag, Berlin.

Ulrich, B., and Pankrath, J. (1983). "Effects of Accumulation of Air Pollutants in Forest Ecosystems." D. Reidel, Dordrecht.

van Breeman, N., and Jordans, E. R. (1983). Effects of atmospheric ammonium sulfate on calcareous and non-calcareous soils of woodlands in the Netherlands. *In* "Effects of Accumulation of Air Pollutants in Forest Ecosystems" (B. Ulrich and J. Pankrath, eds.), pp. 171–182. D. Reidel, Dordrecht.

van den Driessche, R. (1971). Response of conifer seedlings to nitrate and ammonium sources of nitrogen. *For. Sci.* 18: 126–132.

Vitousek, P. (1982). Nutrient cycling and nutrient use efficiency. *Am. Nat.* 119: 553–572.

Waring, R. H., McDonald, A. J. S., Laarson, S., Ericsson, T., Wiren, A., Arwidsson, E., Ericsson, A., and Lohammer, T. (1985). Differences in chemical composition of plants grown at constant relative growth rates with stable mineral nutrition. *Oecologia* 66: 157–160.

White, C. S. (1988). Nitrification inhibition by monoterpenoids: Theoretical mode of action based on molecular structures. *Ecology* 69: 1631–1633.

Whittaker, R. H., Likens, G. E., Bormann, F. H., Eaton, J. S., and Siccama, T. H. (1979). The Hubbard Brook ecosystem study: Forest nutrient cycling and element behavior. *Ecology* 60: 203–220.

4

Water and Nutrient Interactions with Plant Water Stress

E.-D. Schulze

I. Introduction
II. Coupling Leaf Water Potential to Transpiration
III. Changes in Resources within a Plant under Water Stress
IV. Acquisition of Water from the Soil
V. Conclusions
References

I. Introduction

Water is an essential resource for plant life and metabolism. At the cellular level, it is used in chemical reactions as well as to separate and arrange membranes and organelles. At the whole-plant level, it is the main carrier for substances traveling among plant organs and tissues. Moreover, because of the large difference in water potential between the hydrated plant cell and the dry atmosphere, no other substance in plants is replaced in the same quantities as water. Thus any limitation in the availability of water—generally known as water stress—affects almost all plant functions, including the ability of leaves to assimilate carbon dioxide and roots to take in nutrients (see the discussion by Kramer, 1988; Passioura, 1988a; and Schulze *et al.*, 1988c).

Response of Plants to Multiple Stresses. Copyright © 1991 by Academic Press, Inc. All rights of reproduction in any form reserved.

In this chapter, I will consider effects of water stress on the availability of other resources (for references see Lange *et al.,* 1982; Boyer, 1985; Schulze, 1986; and Passioura, 1988b). I wish to demonstrate that plant water use and plant water relations should not be viewed as independent parameters; they can only be understood from their interrelation with other resources at the whole-plant level. Thus only initially will I review basic hydraulic processes, concentrating in the following text on how other resources, such as nutrition, change within a plant under water stress. I conclude with an example of how plants may affect their own environment by water acquisition.

II. Coupling Leaf Water Potential to Transpiration

At a given soil water supply and hydraulic conductivity in the plant, leaf water potential is determined by transpiration (Fig. 1). The water potential reached at a certain transpiration rate depends primarily on the structure of the conducting tissue (Fig. 2). For example, at similar rates of water loss, conifers have a lower leaf water potential than herbaceous species, a fact that appears to have no obvious effect on metabolism as long as the supply

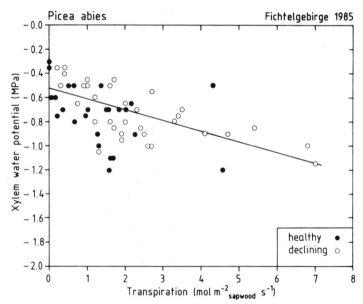

Figure 1   Relation between transpiration and leaf water potential in *Picea abies.* The slope of the line represents the hydraulic conductance for water flow from the soil to the needle. [Adapted from Schulze *et al.,* (1988a).]

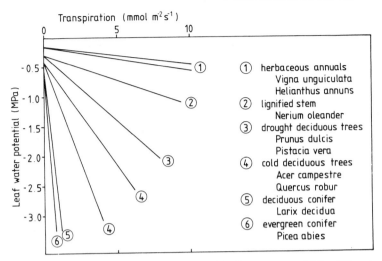

Figure 2 Relations between transpiration and leaf water potential in different plant life-forms. [Adapted from Schulze and Chapin (1987).]

of water from the soil can meet the flux to the atmosphere (Schulze and Chapin, 1987). When the availability of water in the soil decreases, the relation between water potential and transpiration becomes more complicated (Fig. 3). In contrast to the responses shown in Figure 1, which are generally observed during the course of a day and are reversible, a change in slope is not immediately reversible (Schulze and Hall, 1982), and this response is concurrent with numerous changes in plant functioning (Lange *et al.*, 1982).

As the soil dries during a season, however, predawn and daily minimum water potentials may change in parallel over quite a broad range of soil water potentials. In Figure 4, for example (Kappen *et al.*, 1975), a more-or-less constant daily range in leaf water potential was observed during increasing water stress. This may be typical for many situations, since a drop of −1.5 to −2 MPa in daily leaf water potential is commonly observed in plants, irrespective of life-form and habitat. It is not clear which plant functions regulate and maintain this drop in potential between predawn and midday. In Figure 4, predawn water potential closely follows the seasonal change of soil water potential, and minimum daily leaf water potential closely follows leaf osmotic potential, which decreases from −4.0 MPa in spring to almost −8.0 MPa in summer. As in herbaceous species (Turner *et al.*, 1978), osmotic adjustment in *Hammada scoparia* maintains plant functioning over an extended range of water stress (Schulze *et al.*, 1980). Nevertheless, the concept of osmotic adjustment does not identify the mechanism and "aim" of such regulation. One may hypothesize that if

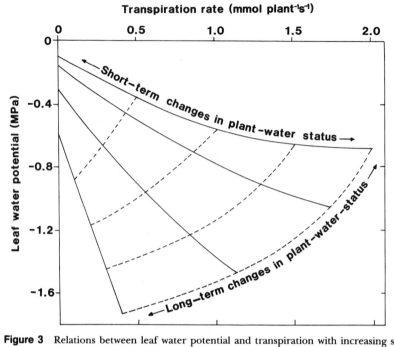

Figure 3 Relations between leaf water potential and transpiration with increasing soil drought. [After Schulze and Hall (1982).]

osmotic potential had already been lowered in the spring, when photosynthesis is high and water is abundant, the plant might have gained much more carbon, since potential transpiration is lower in spring than later in the dry season.

The integrated regulation of leaf water potential, osmotic adjustment, and transpiration is probably related not only to carbon and water relations in the leaf but to whole-plant functioning. In an integrated whole-plant system, transport of assimilates in the phloem and circulation of nutrients between xylem and phloem are central processes (see Nobel, 1974). Flow within the phloem must be maintained for plant integrity, and this requires not only positive pressure but also, at low water potentials, a pressure gradient between source and sink. According to Wheatherley (1982), the osmotic pressure of sieve-tube sap in physiologically active plants is about 1.5 to 2.5 MPa, which means that flow in the phloem will be affected if the daily variation of water potential exceeds this magnitude.

Table I shows the coupling of sieve-tube pressure and xylem water potential (Nobel, 1974). In a well-watered plant, the water potential of root phloem is probably close to zero. To maintain unloading of water from the phloem, the xylem water potential in roots should be somewhat lower and

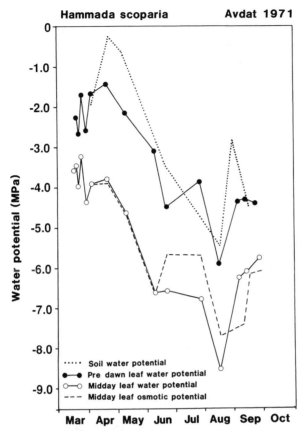

Figure 4   Seasonal changes in soil water potential, predawn water potential, midday water potential, and midday osmotic potential in the C₄ plant *Hammada scoparia*. [After Kappen *et al.* (1975).]

Table I   Water Relations Parameters (MPa) in the Xylem and Phloem of a Hypothetical Plant*

Plant part	Water potential Xylem	Water potential Phloem	Osmotic pressure phloem	Turgor pressure phloem
Root	−0.2	0	0.5	0.5
Leaf, morning	−0.3	−0.5	2.0	1.5
Leaf, noon	−1.5	−1.5	2.0	0.5

* Data from Nobel (1974).

decrease toward the leaf xylem and phloem. Because sucrose is loaded into the phloem in the leaves, osmotic pressure there could reach 2 MPa at that point. It decreases to about 0.5 MPa at the site where it is unloaded in the root. Thus, turgor pressure will be 1.5 MPa in the leaf phloem and 0.5 MPa in the root, and this difference maintains pressure-driven mass flow of water in the phloem. If leaf water potential decreases with transpiration under constant osmotic conditions in the phloem, leaf water potential could only decrease to −1.5 MPa before turgor in the leaf phloem would equal turgor in the root phloem. Leaf water potential could only decrease by 1.0 MPa before phloem mass flow would be affected, unless the osmotic potential at the unloading site decreased or the osmotic pressure in the leaf phloem increased. Thus xylem water potential cannot drop indefinitely as transpiration rate increases without perturbing the circulation of sap and water in the whole plant. Whole-plant functioning is seriously affected over a narrow range of water potentials set by osmotic pressure in the leaf phloem. Too few data are available on phloem solute concentrations to allow further generalization. It appears, however, that the daily variation in leaf water potential influences, and depends on, whole-plant circulation of water and other elements.

III. Changes in Resources within a Plant under Water Stress

Numerous resources besides water circulate within a plant. Yet, the absolute magnitude of this internal circulation is known for only few substances, e.g., nitrogen (Pate *et al.*, 1980). For most resources, not even an estimate of plant internal circulation exists, since phloem sap is quite inaccessible for routine measurements of concentrations and flow rates. This problem is apparent in recent field studies of 30-year-old spruce trees (Table II; Osonubi *et al.*, 1988; Schulze *et al.*, 1988b). Apical transport of potassium

Table II Nutrient Transport in the Xylem of 30-Year-Old Spruce Trees *(Picea abies)*[a]

	Calcium	Magnesium	Nitrogen	Potassium
Xylem sap[b]				
(mol m^{-3})	0.8–1.2	0.25–0.44	5.1–5.4	1.6–2.3
Apical transport[b]				
(mol m^{-2} yr^{-1})	0.24–0.36	0.074–0.131	1.5–1.6	0.47–0.68
Growth requirement[c]				
(mol m^{-2} yr^{-1})	0.25	0.042	1.3	0.14

[a] Seasonal transpiration was 0.298 m^{-3} m^{-2} (Schulze *et al.*, 1988b).
[b] Measurements of xylem concentrations from Osonubi *et al.* (1988).
[c] Measurements of growth from Oren *et al.* (1988).

and magnesium exceeds the shoot's growth requirements by factors of 4 and 2.4 respectively, and apical transport of nitrogen exceeds use for growth by a factor of 1.2. Calcium accumulates in the shoot.

Element concentrations in the xylem vary greatly in their response to water stress (Fig. 5). Magnesium concentration reaches a low and constant level by the time leaf water potential falls to -0.6 MPa, whereas potassium concentration continues to decrease with decreasing leaf water potential. Nitrogen did not respond initially but begins to decrease when magnesium concentrations have already reached their lowest level. A change in xylem sap concentration with drying soil indicates a decrease in plant internal circulation as well as a change in uptake or nutrient availability in the soil. The high concentration of nitrogen in the xylem sap as drought develops may indicate that the supply of nitrogen to the plant initially remains high and decreases when the mineralization rate is reduced because organic matter in the soil is also drying. In contrast to nitrogen, the concentrations of potassium and magnesium in xylem sap decrease with drought much earlier than those of nitrogen, which may reflect a change in plant internal circulation. It is known from experiments using ^{11}C labeling that phloem unloading may be very sensitive to water stress in the root (Goeschel et al., 1988). Thus nitrogen, magnesium, and potassium differ in their responses to plant water stress, depending on the contribution of uptake versus intercirculation.

As water stress develops, changes in the concentration of major nutrients in the phloem affect not only nutrition of the leaves, but also hormonal conditions in the xylem and the shoot. Increasing evidence indicates that stomata can respond to signals from the root (Schulze, 1986). For instance, the concentration of abscisic acid (ABA) in xylem sap increases with soil water stress (Gollan, 1987). Although stomata generally close with increasing xylem concentrations of ABA, the relationship is not direct (Fig. 6) because ABA's effect on stomata is modified by the ionic composition of the xylem sap. In particular, the ratio of cations to anions determines pH in the xylem sap and thus influences the protonation of ABA. In sunflower, xylem sap pH increases from about 6 to about 7 with progressive soil drying because of changes in the uptake of nitrate. At the same time, stomatal sensitivity to the xylem ABA supply increases. Gollan's findings (1987) indicate that such a root signal to the shoot is not just the result of a change in one compound, such as ABA. Rather, drying of the soil alters the ratio of nutrient uptake and circulation, in turn affecting pH and thus the transport form of ABA, which determines whether ABA in fact reaches the stomata. Thus the "root signal" that occurs with water stress in the soil has both a hormonal and a nutritional component. Plant nutrition modifies (regulates) the hormonal effect, in this case altering stomatal sensitivity to ABA in the xylem sap. A plant with high nitrate uptake will maintain

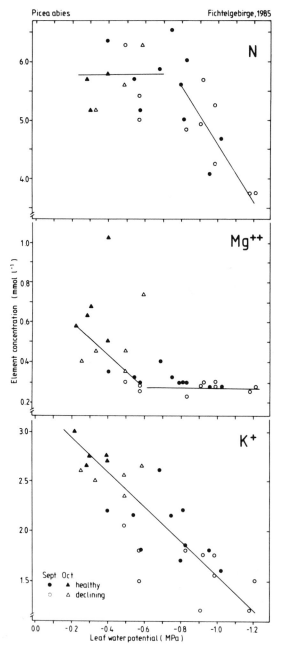

Figure 5   Changes in xylem sap concentrations of nitrogen (N), magnesium (Mg^{++}), and potassium (K^+) with decreasing leaf water potential. [After Osonubi *et al.* (1988).]

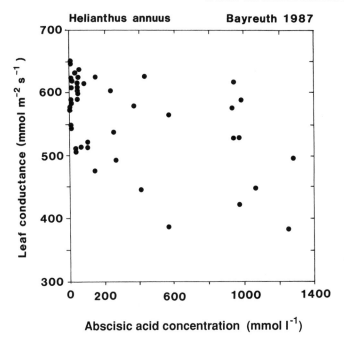

Figure 6   Relationship between stomatal conductance and the concentration of abscisic acid in the xylem of the sunflower *Helianthus annuus.* [After Gollan (1987).]

transpiration to lower soil water potentials than plants having a low supply of nitrate.

Hormonal and nutritional signals from the root appear to explain numerous observations that were previously considered contradictory (Schulze, 1986). Nevertheless root signals cannot explain all situations, since the maximum velocity of water in the xylem differs by several orders of magnitude among different plant forms (Ziegler, 1980; Table III). In herbaceous plants, water moves about 10 times faster than in conifers. Given the difference in their sizes, a root signal carried by xylem water in a 100-meter *Sequoia* tree may arrive after 10 days, whereas in an herbaceous plant, it will arrive within minutes. Thus, trees may not respond as readily to conditions in the root as herbaceous plants could, since a root signal transported in the xylem would be much too slow to be effective. Another mechanism must exist for communication between roots and shoots, especially in large perennial plants. Again we should consider the effect of plant internal circulation. If it is generally true that water stress influences the rate of phloem unloading in the root (Goeschel *et al.,* 1988), then a signal may also be mediated via pressure in the phloem. Such a signal would be

Table III   Maximum Velocity of Water in the
Xylem of Different Plant Forms[a]

Life-form	Velocity (m h^{-1})
Conifers	1.2
Sclerophylls	0.4–1.5
Deciduous trees	
Fagus-type	1.0–6.0
Quercus-type	4.0–44.0
Herbaceous plants	10.0–60
Lianas	150

[a] Data from Ziegler (1980).

very fast, even in a tall tree, because both water and sieve elements are quite incompressible.

In sum, the availability of water to the root affects plant processes in numerous ways. It alters the ratio of nutrient uptake and circulation and thus can markedly affect the xylem sap pH. This in turn alters hormone secretion levels and their chemical nature, which is eventually important for their function and their effect on growth. Changes in water availability can also change the pressure in the phloem, thus affecting unloading, and consequently altering carbon relations and sugar metabolism in the cytoplasm.

IV. Acquisition of Water from the Soil

The uptake of water from the soil has been reviewed by Passioura (1988b). To illustrate the interactions between plant water relations and the soil environment, I will therefore confine this discussion to the effect of water uptake on soil structure.

An experiment was conducted with *Prunus dulcis* in which plants were allowed to use the same amount of water each year; the treatments differed by the amount of water available to the plants each year (Schulze, unpublished data). After one year of growth in freshly sieved soil (powdery "loess"), carbon assimilation and leaf conductance declined to 50% of their maximum values when the soil had reached equilibrium with a predawn leaf water potential of -1.0 MPa (Fig. 7). With repeated drying cycles, however, predawn leaf water potentials were significantly higher. After three cycles in which the soil was dried and then rewetted to field capacity once per cycle, gas exchange fell to 50% when the soil had only dried to the equivalent of a -5.0 MPa predawn leaf water potential. The reason for this

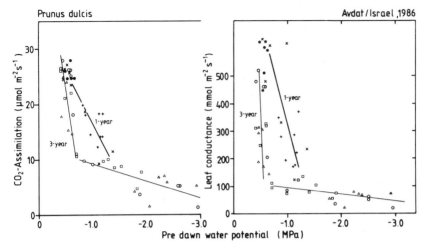

Figure 7 Relationship between stomatal conductance and predawn water potential in *Prunus dulcis* grown in lysimeters of different depth for one and three years. The plants were watered to field capacity once a year and then allowed to dry the soil by water uptake through the season. [After Wartinger (1991).]

change in response appears to be an effect of drought on soil structure (aggregation). The tensile strength of aggregates doubled within the time of observation (Horn *et al.*, 1987). Thus, a consequence of water withdrawal was to permanently increase soil aggregation, or compactness, which reduced the flow of water to the root. After three drying cycles, the roots did not receive enough water through the compacted soil at −0.5 MPa, and stomata closed at a much higher plant water status than when the soil was looser. The zone from which roots gather water had changed from a single, contiguous volume to a set of discrete, localized planes having less water capacity and greater resistance to water flow. In the long term, the root system responds to this change in water flow by lateral root branching (or increasing root density), which results in fragmentation of the soil crumbs, thus reducing the pathway for water flow. Soil strength limits this process, which indicates that eventually the cost of extracting water to a lower level may be quite high.

V. Conclusions

Plant water use and plant water relations can only be understood from their interaction with other processes at the whole-plant level, particularly with plant nutrition. Internal circulation of elements through phloem and xylem may not only determine the range over which water potentials and transpiration vary during the day, but also mediate hormonal signals from the root during soil water stress. Phloem transport may play a much larger role in mediating root signals than previously thought. Plant water uptake may by itself change the physical properties of the soil. Thus, by its very existence, the plant alters the availability of water in the soil, which may affect plant survival in the long run, especially if the availability of water is low.

References

Boyer, J. S. (1985). Water transport. *Annu. Rev. Plant Physiol.* 36: 473–516.

Goeschel, J. D., Fares, Y., Magnuson, C. E., Scheld, H. W., Strain, B. R., Jaeger, C. H., and Nelson, C. E. (1988). Short-lived isotope kinetics: A window to the inside. *In* "Research Instrumentation for the Twenty-first Century" (G. R. Beecher, ed.), pp. 21–52. Beltsville Symposia in Agricultural Research, Beltsville, Maryland.

Gollan, T. (1987). Wechselbeziehungen zwischen Abscisinsäure, Nährstoffhaushalt und pH im Xylemsaft und ihre Bedeutung für die stomatare Regulation bei Bodenaustrocknung. Ph.D. dissertation, Universität Bayreuth, Germany.

Horn, R., Stork, J., and Dexter, A. R. (1987). The influence of soil structure on penetration resistance. *Z. Pflanzenernahrung Bodenkunde* 150: 342–347.

Kappen, L., Oertli, J. J., Lange, O. L., Schulze, E.-D., Evenari, M., and Buschbom, U. (1975). Seasonal and diurnal courses of water relations of the arido-active plant *Hammada scoparia* in the Negev desert. *Oecologia* 21: 175–192.

Kramer, P. J. (1988). Changing concepts regarding plant water relations. *Plant, Cell Environ.* 11: 565–568.

Lange, O. L., Nobel, P. S., Osmond, C. B., and Ziegler, H., eds. (1982). "Encyclopedia of Plant Physiology, New Series," Vol. 12B, Plant Physiological Ecology II: Water Relations and Carbon Assimilation. Springer-Verlag, Berlin.

Nobel, P. S. (1974). "Biophysical Plant Physiology and Ecology." Freeman, San Francisco.

Oren, R., Schulze, E.-D., Werk, K. S., Meyer, J., Schneider, U., and Heilmeier, H. (1988). Performance of two *Picea abies* (L.) Karst. stands at different stages of decline. I. Carbon relations and stand growth. *Oecologia* 75: 25–37.

Osonubi, O., Oren, R., Werk, K. S., Schulze, E.-D., and Heilmeier, H. (1988). Performance of two *Picea abies* (L.) Karst. stands at different stages of decline. IV. Xylem sap concentrations of magnesium, calcium, potassium, and nitrogen. *Oecologia* 77: 1–6.

Passioura, J. B. (1988a). Response to P. J. Kramer's article, "Changing concepts regarding plant water relations," Volume 11, Number 7, pp. 565–568. *Plant, Cell Environ.* 11: 569–572.

Passioura, J. B. (1988b). Water transport in and to roots. *Annu. Rev. Plant Physiol.* 39: 245–265.

Pate, J. S., Layzell, D. B., and Atkins, C. A. (1980). Transport exchange of carbon, nitrogen, and water in the context of whole-plant growth and functioning: A case study of a nodulated legume. *Ber. Dtsch. Bot. Ges.* 93: 243–255.

Schulze, E.-D. (1986). Carbon dioxide and water vapor exchange in response to drought in the atmosphere and in the soil. *Annu. Rev. Plant Physiol.* 37: 247–274.

Schulze, E.-D., and Chapin, F. S., III. (1987). Plant specialization to environments of different resource availability. *In* "Potentials and Limitations of Ecosystem Analysis" (E.-D. Schulze and H. Zwölfer, eds.), pp. 120–148. Springer-Verlag, Berlin.

Schulze, E.-D., and Hall, A. E. (1982). Stomatal control of water loss. *In* "Encyclopedia of Plant Physiology, New Series" (O. L. Lange, P. S. Nobel, C. B. Osmond, and H. Ziegler, eds.), Vol. 12B, pp. 181–230. Springer-Verlag, Berlin.

Schulze, E.-D., Hall, A. E., Lange, O. L., Evenari, M., Kappen, L., Buschbom, U. (1980). Long-term effects of drought on wild and cultivated plants in the Negev desert. I. Maximal rates of net photosynthesis. *Oecologia* 45: 11–18.

Schulze, E.-D., Hantschel, R., Werk, K. S., and Horn, R. (1988a). Water relations of two *Picea abies* (L.) Karst. stands at different stages of decline. *In* "Air Pollution and Forest Decline: A Study of Spruce *(Picea abies)* on Acid Soils" (E.-D. Schulze, O. L. Lange, and R. Oren, eds.), pp. 341–351. Springer-Verlag, Berlin.

Schulze, E.-D., Oren, R., and Lange, O. L. (1988b). Nutrient relations in healthy and declining forest stands. *In* "Air Pollution and Forest Decline: A Study of Spruce *(Picea abies)* on Acid Soils" (E.-D. Schulze, O. L. Lange, and R. Oren, eds.), pp. 392–417. Springer-Verlag, Berlin.

Schulze, E.-D., Steudle, E., Gollan, T., and Schurr, U. (1988c). Response to P. J. Kramer's article, "Changing concepts regarding plant water relations," Volume 11, Number 7, pp. 565–568. *Plant, Cell Environ.* 11: 573–576.

Turner, N. C., Begg, J. E., and Tonnet, M. L. (1978). Osmotic adjustment of sorghum and sunflower crops in response to water deficits and its influence on the water potential at which stomata close. *Aust. J. Plant Physiol.* 5: 597–608.

Wartinger, A. (1991). Der Einfluß von Austrocknungszyklen auf Blattleitfähigkeit, CO_2— Assimilation, Wachstum und Wassernutzung von *Prunus dulcis* (Miller) D.D. Webb. Dissertation Bayreuth.

Weatherley, P. E. (1982). Water uptake and flow in roots. *In* "Encyclopedia of Plant Physiology, New Series" (O. L. Lange, P. S. Nobel, C. B. Osmond, and H. Ziegler, eds.), Vol. 12B, pp. 79–109. Springer-Verlag, Berlin.

Ziegler, H. (1980). Flüssigkeitsströme in Pflanzen. *In* "Biophysik" (W. Hoppe, W. Lohmann, H. Markl, and H. Ziegler, eds.), pp. 561–570. Springer-Verlag, Berlin.

5

Carbon Allocation and Response to Stress

Donald R. Geiger   **Jerome C. Servaites**

I. Introduction

A plant's ability to accumulate assimilated carbon is a function of its photosynthetic capacity and the pattern of carbon distribution among its parts, both of which appear to be genetically determined (Giffort *et al.,* 1984). In nature this capacity also depends on the plant's ability to maintain dry matter accumulation in the face of an ever-present variety of stresses (Boyer, 1982). Most plants fall far short of their full genetic potential for productivity because of environmental stress. For a variety of crops under field conditions, even agricultural yields are only 12 to 30% of record yields (Boyer, 1982).

The allocation and partitioning of assimilated carbon provide resources for acclimation to environmental stress. Allocation comprises the processes determining the biochemical fate of carbon that has become available for distribution, or partitioning, among plant parts. Without other specification, allocation refers to the initial determination of recently assimilated carbon. As a result of both biochemical conversions and compartmentation, newly assimilated carbon may be used immediately or set aside as a reserve to be mobilized and used later. Mobilization of reserves that were previously allocated to storage can also supply carbon needed for stress responses.

This chapter examines how allocation of assimilate allows plants to integrate responses to environmental stresses over both the short and the long term. Allocation of recently fixed carbon to export and to storage enables plants to maintain a steady supply of carbon both for development and for restoring and maintaining homeostasis under environmental stress. By understanding the ways in which allocation integrates stress responses in different plants we can recognize and, we hope, preserve the genetic diversity needed for acclimation to stress in native plants and also apply these genetic resources to improve crop yield.

II. Carbon Allocation in Plant Growth

A. Carbon Allocation Processes

Allocation of recently fixed carbon involves a number of biochemical pathways and compartmentation steps (Fig. 1). In a mature leaf, most of these processes support the export of assimilated carbon and its distribution among sinks (Geiger and Bestman, 1990). Triose-phosphate, the product of CO_2 fixation in the chloroplast (Fig. 1: process 1), can supply carbon for starch synthesis (Fig. 1: process 2), or it can exit the chloroplast and be used to make sucrose (Fig. 1: process 3) or other metabolites in the cytosol (Fig. 1: process 4). Regulation of carbon allocation (Fig. 1: A) between

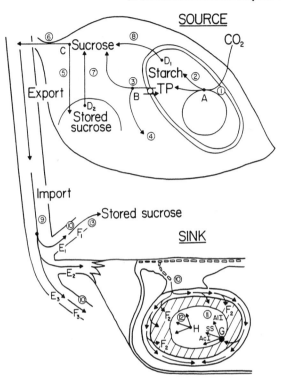

Figure 1 Some biochemical and compartmentation steps for allocation of recently fixed carbon. Numbers represent key processes, and letters mark decision points that help sustain translocation of assimilated carbon. Metabolic processes are: 1, CO_2 fixation in the chloroplast; 2, starch synthesis; 3, sucrose synthesis; 4, synthesis of metabolites in the cytosol; 5, vacuolar sucrose storage; 6, entrance of sucrose into sieve elements; 7, mobilization of stored sucrose; 8, starch degradation; 9, import of sucrose; 10, exit from sieve elements; 11, sucrose degradation; 12, use of sucrose in growth and metabolism; 13, membrane transport of sucrose. Points of regulation of carbon allocation or movement are: A, between starch synthesis and triose phosphate; B, between respiration and synthesis of compounds that remain in the leaf; C, between temporary storage or the sieve elements of minor veins; D_1, mobilization of stored starch; D_2 mobilization of stored sucrose; E_2, E_2, or E_3, partitioning among sinks such as storage roots, fruits, or developing leaves or roots; F_1, F_2, or F_3, sites of unloading from phloem; G, conversion of sucrose to pyruvate or to hexoses; H, use of products of sucrose breakdown for respiration or synthesis. AcI, acid invertase; AlI, alkaline invertase; SS, sucrose synthase; TP: triose phosphate.

starch synthesis and synthetic processes in the chloroplast and cytosol determines the amount of stored assimilate that is available for export or use in the leaf at times of low or no photosynthesis. Carbon that exits the chloroplast can be used (Fig. 1: B) for synthesis of sucrose (Fig. 1: process 3) or for respiration or for synthesis of compounds that remain in the leaf (Fig. 1: process 4). Regulation of two cytosolic enzymes, fructose bisphosphatase (EC 3.1.3.11) and sucrose phosphate synthase (EC 2.4.1.14), is particularly important in controlling the flux of carbon to sucrose (Stitt, 1986). Sucrose (Fig. 1: C) may either be stored temporarily in the cytosol and vacuole (Fig. 1: process 5) or move to the vicinity of the veins and enter the sieve elements (Fig. 1: process 6) to be translocated to sinks. At night and during daytime periods of low photosynthesis, stored sucrose is made available (Fig. 1: process 7, D_2; Servaites *et al.*, 1989b) or starch is degraded (Fig. 1: process 8, D_1; Fondy *et al.*, 1989; Servaites *et al.*, 1989a) to support sucrose export. The synthetic, transport, and regulatory processes involved in allocation of assimilate are the means by which source leaves maintain a relatively steady supply of carbon for use locally or in translocation sinks.

Sucrose imported from source regions (Fig. 1: process 9) is partitioned among various sinks such as storage roots, fruits, or developing leaves or roots (Fig. 1: E_1, E_2, or E_3). There, depending on the sink organ, solutes exit the sieve elements (Fig. 1: process 10) by one of several phloem-unloading processes (Fig. 1: F_1, F_2, or F_3; Geiger and Fondy, 1980). Flux of sucrose along the import and unloading path is maintained by sucrose degradation (Fig. 1: process 11) and used in sink organs (Fig. 1: process 12), including tissues along the translocation path (Geiger and Shieh, 1988) or by sucrose uptake by membrane transport (Fig. 1: process 13).

Sucrose may be converted to pyruvate or to hexoses in the sink tissue by one of at least three enzymatic pathways (Fig. 1: G). It may be hydrolyzed by acid or alkaline invertase or by sucrose synthase and enter metabolism through one of several parallel pathways (Sung *et al.*, 1988). These pathways offer alternative ways of using nucleotide triphosphates and pyrophosphate under strong internal regulation by fructose 2,6-bisphosphate. The products of sucrose breakdown can be used (Fig. 1: H) for respiration and for synthesis of other products. The uptake, conversion, and use of imported sucrose are the means by which sink organs sustain import and partitioning of carbon compounds.

B. Allocation Maintains a Steady Carbon Supply

Balanced allocation of carbon between immediate use and storage is essential for plant growth and for survival during stress. Both newly fixed and stored carbon are needed to maintain export from mature leaves as they alternate daily between periods of gradually increasing or decreasing photosynthesis and dark metabolism (Chapter 1). To provide for continued export, leaves must consume primarily stored carbohydrate during periods of low photosynthesis and allocate newly fixed carbon to export or storage during intervening periods of high photosynthesis (Servaites *et al.*, 1989a). It appears to be a high priority to allocate carbon so as to maintain a relatively steady supply of sucrose for export to plant sinks throughout a light–dark cycle (Fondy *et al.*, 1989; Servaites *et al.*, 1989a,b).

Mutant plants that lack an enzyme for synthesizing starch show impaired growth and metabolism. For example, when grown under continuous light, the phosphoglucomutase-deficient mutant of *Arabidopsis thaliana* was indistinguishable from the wild type in terms of growth and photosynthesis (Caspar *et al.*, 1985). When grown under 12 hours of light followed by 12 hours of dark, however, photosynthesis was 80% of that in the wild type, growth was reduced to 25% of normal, and the activities of a number of carbon metabolism enzymes changed. Clearly, a steady supply of carbon for night metabolism and translocation is important for normal plant growth. Growing plants under light–dark cycles similarly reduced growth in a low-starch mutant of *Clarkia xanthiana* (Jones *et al.*, 1986). Interestingly, reductions in growth were not seen in a starchless mutant of *Nicotiana sylvestris*, which was able to store sufficient sucrose in place of starch (Hanson and McHale, 1988). These examples demonstrate the importance of the regulated availability of reserve carbon in the day-to-day life of plants.

Further evidence of this priority for maintaining export at times of low or no photosynthesis is the constancy of the balanced allocation of newly assimilated carbon between sucrose and starch, even when the source-to-sink ratio is manipulated (Fondy and Geiger, 1980). Decreasing the amount of carbon available for export by shading all but one of the photosynthesizing leaves on a bean plant failed to increase export from the remaining undarkened leaf over a day. Increased export during the light would have come at the expense of starch and sucrose reserves that supply export at night. A change in allocation between export and starch storage in bean leaves was seen only when the sink tissue available for receiving translocated carbon was severely restricted, yet no decrease in photosynthesis was observed during the day of treatment (Fondy and Geiger, 1980).

III. Acclimation to Stress

Some responses to environmental stress may start quickly whereas others may be initiated only after intervals of hours or days (Fig. 2). Depending on the time at which a stress response occurs, the type of mechanism involved and its role in addressing the stress will differ. The progression in the type of response can be seen in the way a plant deals with water stress (see also Chapter 4).

Imposition of an environmental stress generally provokes changes in physical, biochemical, physiological, or developmental processes (Fig. 3).

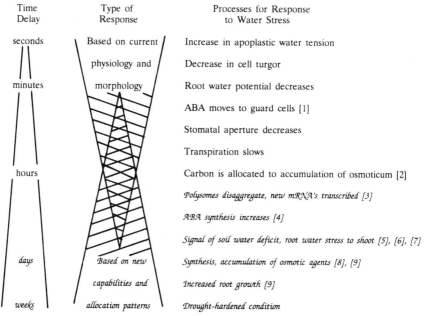

Time Delay	Type of Response	Processes for Response to Water Stress
seconds	Based on current	Increase in apoplastic water tension
	physiology and	Decrease in cell turgor
minutes	morphology	Root water potential decreases
		ABA moves to guard cells [1]
		Stomatal aperture decreases
		Transpiration slows
hours		Carbon is allocated to accumulation of osmoticum [2]
		Polysomes disaggregate, new mRNA's transcribed [3]
		ABA synthesis increases [4]
		Signal of soil water deficit, root water stress to shoot [5], [6], [7]
days	*Based on new*	*Synthesis, accumulation of osmotic agents [8], [9]*
	capabilities and	*Increased root growth [9]*
weeks	*allocation patterns*	*Drought-hardened condition*

Figure 2   How plants respond to stress over time. Progression from current to new capabilities illustrated by responses to water stress. Data from Cornish and Zeevaart (1985) [1]; Fox and Geiger (1984) [2]; Mason *et al.* (1988) [3]; Guerrero and Mullet (1988) [4]; Davies *et al.* (1986) [5]; Turner (1986) [6]; Gollan *et al.* (1986) [7]; Hanson and Hitz (1982) [8]; Meyer and Boyer (1981) [9].

With water stress, for instance, a sizeable drop in atmospheric humidity increases transpiration and decreases cell turgor. These changes alter the input into the existing configuration of the control system and result in a change in system output. Changed system inputs and outputs can bring about two types of responses that may allow the plant to acclimate.

A. Direct Responses

When direct regulation occurs, the stress-induced change in system input or output can signal the system to initiate an acclimation response. For instance, atmospheric humidity and transpiration rates are inputs for the system enabling the plant to achieve and maintain a balanced water status and avoid desiccation (Schulze, 1986; Fig. 3a). The system maintains plant turgor within a normative range. When the input, atmospheric humidity, changes (Fig. 3a: change 1), the original output moves outside the normative range of leaf cell turgor (Fig. 3a: change 2). This results in a system with a new input and a new output. Atmospheric humidity (Fig. 3a: regulatory signal 3a), or perhaps leaf turgor (Fig. 3a: regulatory signal 3b), triggers a

A   DIRECT

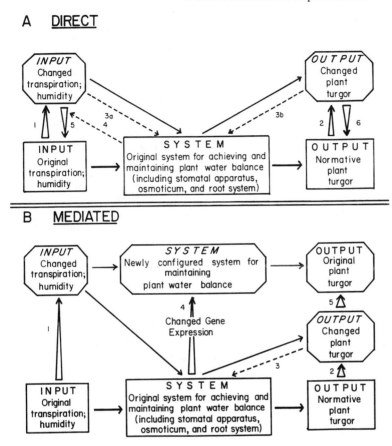

Figure 3 Model for two types of regulatory response mechanisms: A. Direct. B. Mediated. Large arrows represent conversions, solid lines show system flow, and dashed lines indicate regulatory signals.

response (Fig. 3a: regulatory signal 4) that adjusts stomatal conductance and restores transpiration to the original input range (Fig. 3a: change 5). This adjustment will also eventually restore turgor to the normative output range (Fig. 3a: change 6), allowing time for further acclimation and promoting survival.

B. Mediated Responses

Sometimes plants acclimate to the events initiated by stress via responses mediated through altered gene expression and biochemistry (Queiroz, 1983; Geiger, 1986; Fig. 3b; changes 1 and 2). A new system input and output result from the stress. The changed system output serves as a signal (Fig. 3b: regulatory signal 3) that sets in motion a mechanism mediated by changes in the system itself. Changes in gene expression (Fig. 3b: change 4) through gene activation, transcription, and translation can lead to new plant responses (Sachs and Ho, 1986). New structural or physiological characteristics create an altered system that restores the original output (Fig. 3b: change 5), even though the stress-induced changes in the input are not reversed. The successful response creates a plant with changed metabolism, biochemistry, or both, that often can function well even though the stress persists. Under water stress, for example, initiation of more extensive root growth can increase water uptake so that turgor is restored despite high transpiration in the dry air (Meyer and Boyer, 1981).

C. Mechanisms for Addressing Stress

The earliest responses are likely to be rapid, direct responses that increase the chances for immediate survival. These responses are likely to be based on the plant's current structures and physiological capabilities and, consequently, are limited in scope and ability to restore normal function, especially over the long term. For example, rapid stomatal closure in response to drought may protect a plant on the short term, but because it interferes with carbon assimilation, it cannot be maintained over a long period. Compensatory responses that are mediated by the development of new physiological capabilities and new plant morphology are more likely to be effective in restoring near-normal functioning in the face of persistent environmental stress.

It seems reasonable that with time after a stress begins, acclimation will progress from direct responses to those mediated by new physiological capabilities and, eventually, by new plant morphology (see Fig. 2), but these ideas need to be verified.

IV. Allocation and Acclimation

Allocation plays several important roles in the integration of a plant's response to stress, including maintaining phloem translocation and channeling carbon to provide chemicals involved in the responses. This carbon comes not only from allocation of recently assimilated carbon but also from mobilization of previously stored reserves.

A. Supplying Carbon for Communication by Translocation

Balanced allocation of recently fixed carbon between export and reserves supplies carbon compounds for maintaining both immediate and delayed phloem translocation (Fondy *et al.*, 1989; Servaites *et al.*, 1989a,b), thereby providing a vehicle of communication among regions of the plant. In some cases, allocation also provides compounds, such as growth regulators or metabolites, that communicate information related to the presence of the stress. In water-stressed plants, cytokinin or abscisic acid (ABA) may serve this purpose.

B. Supplying Recently Assimilated Carbon for Stress Responses

In some cases, allocation of assimilate for export actually provides the carbon needed to produce a stress response. Carbon allocated to phloem translocation may be delivered to distant points as part of the response — for example, to fuel additional root growth under water stress. In other situations, recently fixed carbon may be allocated in the photosynthesizing organs to carry out the stress responses. For instance, when the relative water content of an exporting leaf falls, carbon may be allocated to synthesis of osmotic agents in the leaf (Fox and Geiger, 1986). Carbon from starch, which contributes little to osmotic potential, is allocated to produce osmotically active sucrose. The energy required to produce one molecule of sucrose from starch is minimal, equivalent to two ATP molecules at most. Depending upon the fate of the inorganic pyrophosphate generated during this conversion, the energy requirement may be even smaller.

C. Reserve Carbon Supplies

A plant draws daily on its carbon reserves, both short and long term. Because acclimation to stress often requires additional carbon, stress increases the demand on reserve carbon, especially if photosynthesis has been inhibited. Because stress can occur at any time, allocation must provide carbon reserves throughout the plant's life. Allocation not only contributes carbon to several forms of reserves within exporting leaves but also provides for export to storage sites throughout the plant. These storage sites can be classified according to the time when the assimilate reserves are mobilized.

1. Daily Reserves Reserve carbon is needed daily, even in the absence of stress, for growth and metabolism. A variety of adaptations enable plants to maintain a steady carbon supply for distribution during the daily dark period as well as to supplement the transition hours from day to night and night to day. Allocation of newly fixed carbon to daily reserves within exporting leaves maintains this supply. Some plants, including barley (Gordon *et al.*, 1982) and spinach (Robinson, 1984), store sucrose throughout much of the day; other plants, including sugar beet (Fondy and Geiger, 1982) and garden bean (Fondy *et al.*, 1989), store mostly starch. Leaves of the latter group of plants regulate the mobilization of starch carbon to complement newly fixed carbon and thus maintain sucrose synthesis during both transition periods (Fondy *et al.*, 1989; Servaites *et al.*, 1989a). In contrast, spinach plants use stored sucrose rather than starch to supply the carbon needed to maintain steady export of sucrose during the evening transition period (Servaites *et al.,* 1989b).

Regulation of triose phosphate allocation between starch and sucrose synthesis provides daily reserves sufficient to maintain sucrose synthesis and uninterrupted translocation during the dark period (Fondy and Geiger, 1982). An important factor in regulating the rate of starch accumulation is photoperiod. Both the length of the photosynthetic period (Chatterton and Silvius, 1979) and the interaction of light with an endogenous circadian rhythm (Britz *et al.,* 1985) appear to be involved in setting the starch accumulation rate in relation to day length.

The studies of Chatterton and Silvius (1979) support the influence of photoperiod duration on both starch synthesis and degradation rates. Data from a more recent study involving a starch-deficient mutant support the view that the rate of starch breakdown during periods of low or no photosynthesis is regulated according to the rate at which recently fixed carbon is allocated to starch (Lin *et al.,* 1988). Although the activity of the enzyme ADP-glucose pyrophosphorylase in this starch-deficient *Arabidopsis* mutant is only 5% that of the wild type, starch synthesis rate is 40% that of the wild type. Starch degradation at night also occurs at 40% of the rate in the wild type, indicating that this process is somehow regulated to match the low level of starch or its slower synthesis in the mutant.

2. Long-term Reserves   Long-term reserves supply carbon that can allow a plant to respond to stresses during a given developmental stage, at the beginning of the next season's growth, or on germination of the next generation of plants. Daily and long-term carbon reserves enable plants to acclimate to stress by maintaining overall fitness, maintaining phloem translocation as a means of communication and integration, and, especially, serving as a buffer in the event of a stress that slows or stops photosynthesis.

3. Buffer Reserves   Carbon stores, located in source organs, along the translocation path, and in various sinks, serve as buffers by providing a temporary supply of carbon that can be mobilized whenever the supply of photoassimilated carbon becomes limited. To the extent that these reserves are available for mobilization, they can negate the effects of rapid decreases in carbon supply. For instance, when export from a leaf is drastically slowed by cooling a region of stem or petiole, import is maintained by mobilization of buffer reserves (Swanson and Geiger, 1967; Minchin and Thorpe, 1987).

In exporting leaves, buffer reserves include sucrose located primarily in vacuoles. Sucrose in the vacuoles of sugar beet source leaves, and likely in leaves of other species as well, appears to be in equilibrium with cytoplasmic sucrose and thus can enter the cytoplasmic pool and contribute to export when cytosolic sucrose concentration falls (Geiger *et al.,* 1983).

Starch in the mesophyll also serves as a buffer reserve. Franceschi and Giaquinta (1983b) observed that starch in the first palisade layer serves as a daily reserve, whereas starch in the other layers is only significantly mobilized during times of increased use, such as seed development. Starch in sugar beet leaves also functions as a buffer reserve under some circumstances, such as during low leaf photosynthesis (Fox and Geiger, 1984).

In the leaves of some legumes, a specialized layer of cells called paraveinal mesophyll forms a network connecting the minor veins at the level of the phloem with the spongy and palisade mesophyll cells. Because of the position of this tissue, products of photosynthesis must pass through it on the way to the minor veins, making it ideal for temporary storage (Franceschi and Giaquinta, 1983b), but the extent of this function is not fully known. During leaf development, the vacuoles of these cells accumulate a glycoprotein similar to seed storage protein (Franceschi and Giaquinta, 1983a), which appears to act as an organic nitrogen reserve at times of peak use. Several weeks after anthesis in soybean, this long-term reserve is mobilized to provide amino acids for seed development.

A proportion of the sugar translocated in phloem is stored along the translocation path and reenters the sieve elements when translocation decreases (Swanson and Geiger, 1967; Geiger *et al.*, 1969; Minchin and Thorpe, 1987). Stored carbon in a variety of sink organs can also function as buffer reserves.

V. Allocation and the Responses to Specific Stresses

Although the molecular and cellular mechanisms involved in stress responses have been studied in detail, not enough attention has been given to mechanisms for communicating the presence of stress and integrating whole-plant responses to stress. As Schulze (1986) has noted for water stress, the response of individual organs has been described at the cellular level, but our understanding of developmental and regulatory processes in the whole plant rests largely on theoretical models. Much of the problem stems from the complexity of the appropriate experiments. To observe the integration of responses at the whole-plant level with a specific cellular process such as carbon allocation, data for the cellular process must be collected without disrupting normal whole-plant functioning (Geiger, 1987). The difficulty is compounded when one attempts to look at responses to multiple stresses.

To understand how carbon allocation is involved in specific stress responses, we must consider certain critical elements: (1) the timing of the response, (2) the involvement of direct or mediated mechanisms of acclimation or both, (3) the role of translocation in communication and integration of responses in the whole plant, and (4) the contributions of carbon allocation to these responses. These aspects are especially important for understanding how stress responses are integrated.

A. Water Stress

1. Direct Responses   Responses to water stress differ according to how soon they begin after the stress itself begins (see Fig. 2). The initial effect of water stress on photosynthesis appears to be one of lowering stomatal conductance in response to low atmospheric humidity (Schulze, 1986). Other direct and relatively rapid responses include a decrease in cell turgor and an increase in apoplast water tension. Continuity of water in the xylem and cell wall allows this tension to be communicated quickly throughout the plant, and this in turn decreases turgor throughout the plant. Decreased turgor may cause cessation of growth and may affect the closure of stomata (Schulze, 1986; refs. cited in Turner, 1986); such responses may allow plants to escape sudden, irreversible damage to lessening water loss and restoring cell turgor. Carbon allocation is probably not involved in these responses.

Initial direct responses may be unable to prevent some functional disruption and may even involve processes that are detrimental to usual plant functions, as when photosynthesis is reduced by stomatal closure. In natural plant communities, many of the acclimations, such as changes in plant water status, appear to be more important for survival than for high productivity (Turner, 1986). Changes in water stress can initiate further responses, which are mediated by gene expression or biochemical change and may be needed to aid survival or even restore near-normal functioning (Fig. 3). For example, promotion of root growth allows more water uptake, permits stomata to reopen, and restores the rate of photosynthesis. These responses depend on changes in gene expression (Sachs and Ho, 1986) or enzyme activity (Hanson and Hitz, 1982) to produce new morphology or processes, or to increase levels of processes; they are likely to involve changes in the allocation of newly fixed or stored carbon. We must be cautious in interpreting responses because not all of these metabolic changes result from acclimations to stress but can result from metabolic disruption as well (Hanson and Hitz, 1982). These authors give a series of observations that can be used to determine the extent to which water stress effects are valuable for acclimation.

2. Mediated Responses   Several acclimations to water stress appear to be mediated responses in which changed gene expression alters metabolism and plant structure. One strategy for acclimating to water stress is the development of dehydration tolerance (Kramer, 1980; Turner, 1986). Plants that acclimate this way allocate carbon to form osmotic agents that help tissues endure low water status and maintain turgor. Carbon allocation appears to respond primarily to changes in turgor, or perhaps solute potential, rather than to changes in water potential, which often shows much wider variations. Osmotic adjustment occurs by synthesis of organic anions, soluble carbohydrates, amino acids, and quaternary ammonium compounds (Hanson and Hitz, 1982). These osmotic agents, including betaine, proline, and sucrose, lower solute and water potential and promote water uptake by tissues while protecting proteins, membranes, or other cellular components from the adverse effects of dehydration (Hsiao, 1973; Turner *et al.*, 1978; Hanson and Hitz, 1982; Fox & Geiger, 1986). Decreasing the osmotic potential of tissues maintains water uptake, lessens turgor reduction, and makes the decline in water potential more gradual. In a number of plants, osmotic regulation occurs largely by increased foliar synthesis of proline from glutamine (Hanson and Hitz, 1982), by both increasing activation of regulatory enzymes and enzyme synthesis. In halophytes such as beet and spinach, betaine is synthesized and accumulates in response to drought or salt stress (Hanson and Hitz, 1982).

Another strategy for acclimating to drought is the postponement of dehydration: the plant develops the ability to maintain high water status in its tissues by lessening the loss and increasing the uptake of water (Kramer, 1980; Turner, 1986). Postponing dehydration involves changes that may lead to a loss of productivity, although the decrease will likely be less than it would be without acclimation. Plants that acclimate in this manner partition more carbon to roots, thereby increasing the density and depth of root growth, which increases water uptake. Allocating carbon for export maintains translocation, and the increased carbon partitioning to the roots allows osmotic adjustment to occur, resulting in continued root growth at lower water potentials (Sharp and Davies, 1979). Osmotic adjustment in shoots also increases the amount of water that can be extracted from the soil.

For example, soybean seedlings water-stressed by transfer to a medium with low water potential show decreased hypocotyl growth, accumulation of solute in the hypocotyl, and increased root growth (Meyer and Boyer, 1981). Allocation of cotyledonary reserves for export allows translocation to continue undiminished, but partitioning among sinks changes markedly. This treatment, which decreases stem growth in seedlings to 30% of the control rate, causes parallel changes in polysome status and polysomal RNA (Mason *et al.*, 1988). Water deficits that are severe enough to inhibit growth also rapidly inhibit protein synthesis by the disaggregation of polysomes (Mason *et al.*, 1988, and references therein). Under water stress, decreases in polysomes and polysomal RNA in elongating tissue are similar to those that occur with tissue maturation; in already-mature tissues, drought induces novel mRNAs. Gene regulation appears to be involved both in prolonging growth inhibition in certain tissues and in recovering growth potential in others (Mason *et al.*, 1988).

Plants that acclimate to drought by maintaining high water status in their tissues also need to reduce water loss. Loss of cell turgor causes ABA to enter the apoplast and move to the guard cells (Cornish and Zeevaart, 1985). This ABA changes the regulation of CO_2-induced stomatal opening to give lower stomatal conductance (Raschke, 1975; Walton, 1980). Subsequently, allocation of leaf carbon helps lower water loss through accumulation of additional ABA and by decreasing growth in leaf area. ABA synthesis can be regarded as a stress integrator because ABA may be carried over from earlier episodes of stress and serve as a "memory" of recent conditions (Schulze, 1986).

Reduction in turgor in the leaves of *Pisum sativum* appears to bring about a change in the levels of at least 13 poly(A)$^+$ RNAs, some of which appear within 30 minutes, about the time that ABA levels start to rise (Guerrero and Mullet, 1988). These authors concluded that the new poly(A)$^+$ RNAs may be involved not only in the synthesis of ABA and osmotic agents, but also in growth inhibition. Factors affecting growth, such as wall extensibility or hydraulic conductivity, are also thought to change in response to plant growth regulators. The role of gene expression in these and other aspects of the regulation of leaf growth is not known.

3. Translocation   Phloem translocation is important for coordinating cellular responses to drought at the whole-plant level. Allocation of carbon to export continues during chronic water stress only to the extent that the carbon supply can be maintained by photosynthesis or mobilization of leaf carbon reserves (references cited in Hanson and Hitz, 1982). Maintenance of export, which is promoted by osmotic adjustment in the leaves, guards against carbon starvation in the rest of the plant and permits acclimations such as altered morphology, osmotic adjustment, and desiccation tolerance in tissues throughout the plant. Increased solute levels in the root tips help preserve water uptake from a drying soil and thereby maintain cytokinin synthesis (Turner, 1986). Osmotic agents such as proline and betaine are also translocated throughout the plant. Betaine is reported to be distributed mainly to young leaves and storage roots by phloem translocation (Hanson and Hitz, 1982).

Phloem translocation moves materials that convey information from sources to sinks, and xylem transport may complete the communication loop by conveying information from sink to source. Roots appear to serve as sensors of soil water deficits by means of a reduction in cytokinin production that occurs as the soil dries (Davies *et al.*, 1986; Gollan *et al.*, 1986; Schulze, 1986). Cytokinin, produced by the roots and conducted to the shoots, prevents stomata from closing under the action of ABA and helps to keep stomatal conductance as high as possible given the soil water content (Turner, 1986). Maintaining translocation from leaves supplies carbon for osmotic regulation, which helps sustain root turgor and cytokinin formation.

B. Shortened Photosynthetic Period

When plants are stressed by low irradiance, their carbon status declines; they acclimate through a number of morphological and physiological adjustments (Bunce *et al.*, 1977; Jurik *et al.*, 1979). Similarly, shortened photosynthetic period also decreases the total amount of fixed carbon available daily and appears to elicit the same set of acclimation responses (Chabot *et al.*, 1979, and references therein). Chabot *et al.* (1979) consider that plants under low light and those under shortened days are both responding to the total quanta received during the light period, i.e., to the integrated daily irradiance. Their data do not seem to rule out regulation in response to some other factor related to integrated daily irradiance, such as the level of fixed carbon in leaves or the total amount of photosynthesis.

Initially, the stress of decreased total daily irradiance lowers currently available as well as daily reserve carbon supplies (Jablonski and Geiger, 1987). Consequently, the leaf enters the dark period with less starch than before the shift. In response, the proportion of newly fixed carbon allocated to starch increases and appears to adjust to ensure a reserve related to the carbon requirements of the dark period (Chatterton and Silvius, 1979; Fondy and Geiger, 1980, 1982). Within two days, the daily rate of starch accumulation from recently fixed carbon increases 1.5- to 2-fold (Jablonski and Geiger, 1987). This response, which is similar to that observed in several species, including soybean (Chatterton and Silvius, 1979), illustrates the gradual change in carbon allocation that occurs in response to shortened day length.

Gradual changes in the morphology of the rapidly growing root and shoot accompany the change in carbon allocation in photosynthesizing leaves (Jablonski and Geiger, 1987). During leaf development, there is a positive feedback connection between current photosynthetic carbon assimilation and the investment of this assimilated carbon in leaf growth. If the proportion of carbon that is partitioned among the various plant parts does not change, the rate of leaf-area growth in plants shifted to lower irradiance or shorter days will decline in proportion to the decrease in total daily net carbon fixed (Geiger and Shieh, 1988). But, in fact, over several weeks partitioning and plant morphology do change in response to shortened daily photosynthetic period (Jablonski and Geiger, 1987). Leaf area per unit dry weight and shoot:root ratio increase, and less structural material per unit area accumulates in the leaves. As a result, growth in leaf area per plant remains nearly undiminished, and the decrease in total daily net carbon assimilation is considerably lessened.

One might expect that, as with water stress, early, direct effects of shortened photosynthetic period on carbon status could serve as signals triggering additional responses, particularly those mediated by gene expression or biochemical change. It is not yet clear if the depletion of starch under lengthened nights does generate mediated acclimations; this point needs further study.

As the photosynthetic period is shortened, a larger proportion of the carbon fixed is exported at night. Comparison of dry weight accumulated during day with that during the night indicates that the ratio of shoot growth to root growth is higher at night (Goodall, 1946; Bunce, 1978; Huber, 1983). If this is true, increasing the proportion of export that occurs at night will increase the shoot : root ratio. Huber's study identified a close association between an increased allocation of recently fixed carbon to starch and an increased ratio of carbon partitioned to shoot growth. The responses to shortened days of leaf growth and carbon allocation to starch, reported by Jablonski and Geiger (1987), are consistent with this mechanism.

In sum, acclimation to shortened photosynthetic period lessens the decrease in total daily net carbon assimilation that would be expected to result from the positive feedback loop between leaf growth and total daily net carbon fixation (Jablonski and Geiger, 1987). Instead of sustaining the original pattern of growth, altered allocation of recently fixed carbon appears to maintain the growth rate of leaf area and thereby multiply the effect of the carbon invested. Plants acclimated to short days have a higher carbon fixation rate per plant than they would have without the change in morphology.

C. Multiple Stresses

Multiple stresses can induce a number of concurrent imbalances between sink and source function, which must be brought into accord to prevent a marked deterioration in plant functioning. We hypothesize that the various responses to stresses are coordinated and that plants acclimate to multiple stresses in large measure by coordinating source and sink functioning. According to this hypothesis, if any combination of individual responses moves source and sink activity sufficiently far out of balance, regulation of allocation and translocation bring them back into balance. As a consequence, plants would be less likely to acclimate to multiple stresses in a counterproductive manner.

VI. Maintaining Source – Sink Balance

Plants maintain a balance between rates of carbon fixation and export in source leaves and carbon import and use by sinks. A given stress is likely to produce responses that induce an imbalance affecting sources and sinks. For example, low soil temperature slows growth and the import of carbon into roots. Under these conditions, carbon export will likely diminish with time as a consequence of the regulation of import and partitioning among sinks. As a result, source-leaf carbon allocation and metabolism will change, directly and possibly by responses mediated by new metabolic capabilities or altered structure. The changes in allocation following decreased sink import may act as signal, especially to initiate changes in the source.

While sinks clearly depend on source functioning, the extent to which sinks themselves exercise control over sources is less clear. Although a number of studies have contributed to our understanding of sink control over source activity (references in Neales and Incoll, 1968; Geiger, 1976; Herold, 1980; Wardlaw, 1985) — including the control resulting from the sink–source imbalance created by elevated CO_2 levels (Kimball, 1983; Cure, 1985; Allen *et al.*, 1987) — much remains to be learned about this interaction and its relation to stresses. Explanations are needed to understand how the state of source–sink imbalance is detected and communicated to the source organs and ultimately to the chloroplast, what specific signal triggers the response, and how the source acclimates to restore the balance. An effective experimental framework is needed for designing experiments, testing hypotheses, and interpreting results. The experimental approach should include descriptions or observations of the types of responses to the imbalance, distinguished according to the degree to which the plants change physiologically or morphologically; the time frame of these responses; and the sequence and nature of the mechanistic elements that changed.

A. Induction of Source–Sink Imbalance

An imbalance between source and sink can result from a decrease in carbon import by sink regions, such as that induced by cooling or excision of sink organs. In the short term, photosynthesis usually does not correlate with experimentally altered growth rates of sinks (Duncan and Hesketh, 1968). Reductions in source-leaf photosynthesis occur, but usually only after one or more days after a reduction in sink growth (Mayoral *et al.*, 1985; Plaut *et al.*, 1987; Bagnall *et al.*, 1988). For instance, when peanut plants (*Arachis hypogaea* L.) were cooled to 19° C, net assimilation rate initially remained unchanged, but during the second day it began to decrease as a result of nonstomatal factors (Bagnall *et al.*, 1988). The results were similar when the sink region alone was cooled, indicating that the sinks were the primary locus of the temperature effect. The decline in photosynthetic capacity did not occur when the source–sink balance was restored by lowering the rate of photosynthesis with reduced atmospheric CO_2 concentration. It is not clear whether the balance-restoring adjustment of photosynthesis was delayed because of mediated source responses or because the sink signal needed time to develop. Information such as translocation rates, leaf carbohydrate status, and other biochemical processes would help resolve this question. The above study does show that it is beneficial to study the types and timing of physiological changes associated with an acclimation to a sink–source imbalance by means of realistic treatments that maintain the integrity of the whole plant (Geiger, 1987).

An imbalance between source and sink can also result from an increase in carbon assimilation by source regions. Initial rates of photosynthesis per leaf area are often higher in plants exposed to increased levels of atmospheric CO_2, but, surprisingly, continued exposure often results in acclimation and a lowering of photosynthesis rate (references in Kramer, 1981). Under some conditions, sinks appear unable to use this additional carbon, but the response varies with species, stages of development, and growth habit (Kramer, 1981). Aoki and Yabuki (1977) found that rates of photosynthesis in cucumber plants grown under elevated CO_2 levels initially were about twice as high as those at atmospheric level, but the rates fell to the control level or below in 5 to 15 days. Soybean plants with either high or low sink demand, based on pod number, also showed decreasing rates of photosynthesis per plant over the several weeks following transfer to elevated CO_2 (Clough *et al.*, 1981). Plants with larger sink biomass had higher photosynthesis rates than those with smaller sink mass, but photosynthesis declined with time in both. In some cases, plants acclimate to source–sink imbalance by reducing photosynthesis rather than by accumulating more assimilate in their sinks.

B. Detection and Communication of Imbalance

Communication of the presence of stresses and of the resulting source–sink imbalance enables the plant to coordinate responses at the whole-plant level. Phloem translocation is a major link in this coordination, providing communication in the form of assimilated carbon, which may include specific signal molecules moving from source to sink. By supporting both immediate and delayed translocation of assimilated carbon, allocation maintains communication and integration of responses to stress.

A possible means for communicating reduced sink growth is lessened synthesis and transport of a growth regulator from sink to source, such as Davies *et al.*, (1986) proposed as a signal of root water deficit. In animals, hormonal regulation of carbohydrate metabolism in response to supply and demand occurs by a complex reaction cascade that activates synthesis or mobilization of storage carbohydrate to satisfy metabolic needs (Stryer, 1988). The cascade involves hormone binding to receptors on membranes of target cells, activation of adenylate cyclase and protein kinase, and enzyme phosphorylation. While the reaction cascade that coordinates carbon supply and demand in animal cells is understood well enough for it to serve as a paradigm for biochemical control systems, an analogous system in plants, if indeed it exists, has not been demonstrated. Some aspects of the control of plant carbohydrate metabolism—such as regulation of fructose bisphosphatase by the regulatory metabolite, fructose 2,6-bisphosphate (Stitt *et al.*, 1985), and of sucrose phosphate synthase activity by reversible phosphorylation (Huber *et al.*, 1989)—are very similar to parts of the reaction cascade system in animals. Nevertheless, other parts of the system and more details about the regulation of sucrose synthesis need to be identified before we can determine if an analogous regulation is present in plant cells and whether it is involved in slowing photosynthesis under conditions of source–sink imbalance.

Because slowed carbon flux into sinks likely affects sources, carbohydrate accumulation in the source could signal slowing of import into sinks. The hypothesis stating that "the accumulation of solutes in an illuminated leaf may be responsible for the reduction in the net photosynthesis rate of the leaf" was formulated by Boussingault in 1868 and rephrased by Neales and Incoll (1968) 100 years later. Accumulated carbohydrate could have a direct inhibitory effect on photosynthesis, for example, if starch grains shaded thylakoids or if accumulated sucrose directly inhibited its synthesis (Neales and Incoll, 1968; Herold, 1980). But carbohydrates do accumulate to high levels in leaves at times without any apparent inhibitory effect on photosynthesis. For example, Plaut *et al.* (1987) found that removing sinks from a number of plant species increased starch synthesis about two-fold, but in only four of the seven species did photosynthetic rate decline. In cucumber plants, Mayoral *et al.* (1985) observed a long-term decrease in carbon fixation rate three to five days after import into sinks was decreased. A transient slowing of carbon fixation several hours after treatment resulted from stomatal closure. Although starch levels were elevated and were even higher when CO_2 was raised, they apparently had no immediate effect on photosynthesis.

While accumulated carbohydrate itself may not be the signal, the changes in allocation in a source leaf that are associated with slowed export could signal the imbalance between the rates of carbon fixation and use by sinks. Changes in the rate at which carbon moves through the pathways for synthesis of sucrose or other major export species may affect metabolism and so produce subtle changes in metabolite levels or in the ratio between reactant-product pairs. For instance, slowing the flux of carbon through sucrose synthesis may decrease the ratio of free to esterified phosphate in the cytosol (Foyer, 1987; Gerhardt *et al.*, 1987). The low cytosolic inorganic phosphate level could in turn slow the exchange of this inorganic phosphate for chloroplast triose phosphate and thereby reduce the rate of sucrose synthesis. High levels of triose phosphate in the chloroplast appear to favor the synthesis of starch. If starch synthesis becomes too slow to recycle inorganic phosphate within the chloroplast, then inorganic phosphate levels could fall such that photosynthetic rate would decline (Sivak and Walker, 1986).

C. Acclimation to Restore Balance

Early responses to stress are restricted to the plant's current structures and physiological capabilities. Later, acclimation can also call on new metabolic and structural capabilities conferred by altered gene expression. The first responses aid immediate survival, whereas the later ones provide some degree of resistance to the stress that caused the imbalance (Larcher, 1987). Following the induction of a source–sink imbalance, plants are often able to maintain their photosynthetic rates unchanged for one or more days. In some cases, however, particularly when leaves are subjected to light and CO_2 levels higher than those present during growth, photosynthesis is progressively reduced following the manipulation. Although the mechanism is not clear, the decline in photosynthetic rate could result because the difference between photosynthesis and sucrose use exceeds the plant's capacity to respond to the imbalance while still maintaining undiminished carbon fixation. Whether or not carbon fixation slows immediately, the changes in allocation and carbon metabolism that accompany acclimation may require a reduction in carbon fixation should the imbalance persist over a long period.

When photosynthetic rate does change, some aspects of the photosynthetic machinery may be more pliant than others. This is demonstrated by the fact that altering nutrient and light levels during growth has considerable effects on the total amount of protein synthesized in a leaf but much less effect on the proportion of the total allocated to rubisco (Evans, 1983; Torisky and Servaites, 1984) and thylakoid proteins (Evans, 1987). Hence, changes in photosynthesis in response to stress need not be accompanied by changes in rubisco or thylakoid protein levels but depend on more subtle aspects of regulation. For example, Sage *et al.* (1989) observed that in plants grown at high CO_2 levels, leaf rubisco concentration decreased only slightly, whereas rubisco activation was rapidly and greatly reduced. Although the photosynthetic response to CO_2 enrichment varied considerably among species, all exhibited reduced rubisco activation.

VII. Conclusions

A variety of synthesis, transport, and regulatory processes enter into the allocation of both recently assimilated and newly mobilized stored carbon. Many of these processes play key roles in the export of assimilated carbon and its distribution among sink regions. Part of the carbon available in source and sink regions is allocated for acclimation, which restores and maintains homeostasis under environmental stress. Another part, allocated for phloem translocation, furnishes a vehicle of communication among plant regions, providing an important means of integrating plant responses to stress. Chemicals that may be involved in these stress responses include growth regulators such as ABA and metabolites such as betaine. Because stress can occur at any time, some newly assimilated carbon is allocated for storage and subsequent use throughout a plant's life, either as needed or as daily and long-term reserves.

Early responses to stress are likely to be rapid, direct ones that help ensure survival. Contingent on the plant's current structures and physiological capabilities, these responses are restricted in their ability to restore normal function. Later, acclimation mediated by altered gene expression enables the plant to develop new physiological capabilities and plant morphology, which may restore a plant to near-normal functioning even while the stress persists.

Under water stress, for example, direct responses such as stomatal closure initially may inhibit photosynthesis. Subsequent responses, mediated by altered gene expression, can result in changes in allocation of newly fixed or stored carbon, synthesis of osmotic agents, and promotion of root growth, which leads to greater uptake of water, reopens stomates, and restores the previous rate of photosynthesis. Phloem translocation provides communication from sources to sinks, while xylem transport can complete the loop, carrying materials such as growth regulators from sinks to sources.

Multiple stresses may induce a number of concurrent imbalances between sink and source functioning. Regulation of allocation and translocation are important means to restore balance so that acclimation does not become counterproductive. The extent to which sinks exert a controlling effect on source function and the role of this mechanism in maintaining a balance between rates of carbon fixation and sink import are not clear. Explanations are needed to understand how source–sink imbalance is detected and communicated ultimately to the chloroplast, as well as how acclimation is triggered and how balance is restored. While a traditional view favors the accumulation of carbohydrate as the sign of imbalance, the actual signal may be allied with changes in allocation associated with slowed export.

Plants are often able to maintain their photosynthesis rates unchanged for one or more days following the onset of source–sink imbalance. If an eventual decline in photosynthetic rate occurs, it probably does so because the difference between rates of photosynthesis and sucrose use exceeds the plant's capacity to respond without diminishing carbon fixation. As the plant acclimates, some aspects of the photosynthetic machinery may be more pliant to change than others. For example, whereas treatments may change the total amount of protein synthesized in a leaf, they have much less effect on the proportion of the total allocated to rubisco. Acclimation appears to involve more subtle aspects of regulation, such as changes in rubisco activation.

Acknowledgments

Assistance was provided by grants from the National Science Foundation (DMB-83031957 and DCB-8816970) and the U.S. Department of Agriculture (87-CRCR-1-2486). Discussions with Peter Minchin are gratefully acknowledged.

References

Allen Jr., L. H., Boote, K. J., Jones, J. W., Valle, R. R., Acock, B., Rogers, H. H., and Dahlman, R. C. (1987). Response of vegetation to rising carbon dioxide: Photosynthesis, biomass, and seed yield of soybeans. *Global Biogeochem. Cycles* 1: 1–14.

Aoki, M., and Yabuki, K. (1977). Studies on the carbon dioxide enrichment for plant growth. VII. Changes in dry matter production and photosynthetic rate of cucumber during carbon dioxide enrichment. *Agric. Meteorol.* 18: 475–485.

Bagnall, D. J., King, R. W., and Farquhar, G. D. (1988). Temperature-dependent feedback inhibition of photosynthesis in peanut. *Planta* 175: 348–354.

Boyer, J. S. (1982). Plant productivity and environment. *Science* 218: 443–448.

Britz, S. J., Hungerford, W. E., and Lee, D. R. (1985). Photosynthate partitioning into *Digitaria decumbens* leaf starch varies rhythmically with respect to the duration of prior incubation in continuous dim light. *Photochem. Photobiol.* 42: 741–744.

Bunce, J. A. (1978). Interrelationships of diurnal expansion rates and carbohydrate accumulation and movement in soya beans. *Ann. Bot.* 42: 1463–1466.

Bunce, J. A., Patterson, D. T., Peet, M. M., and Alberte, R. S. (1977). Light acclimation during and after leaf expansion in soybean. *Plant Physiol.* 60: 255–258.

Caspar, T., Huber, S. C., and Somerville, C. (1985). Alterations in growth, photosynthesis, and respiration in a starchless mutant of *Arabidopsis thaliana* (L.) deficient in chloroplast phosphoglucomutase activity. *Plant Physiol.* 79: 11–17.

Chabot, B. F., Jurik, T. W., and Chabot, J. F. (1979). Influence of instantaneous and integrated light-flux density on leaf anatomy and photosynthesis. *Am. J. Bot.* 66: 940–945.

Chatterton, N. J., and Silvius, J. E. (1979). Photosynthate partitioning into starch in soybean leaves. I. Effects of photoperiod versus photosynthetic duration. *Plant Physiol.* 64: 749–753.

Clough, J. M., Peet, M. M., and Kramer, P. J. (1981). Effects of high atmospheric CO_2 and sink size on rates of photosynthesis of a soybean cultivar. *Plant Physiol.* 67: 1007–1010.

Cornish, K., and Zeevaart, J. A. D. (1985). Movement of abscisic acid into the apoplast in response to water stress in *Xanthium strumarium* L. *Plant Physiol.* 78: 623–626.

Cure, J. D. (1985). Carbon dioxide doubling responses: A crop survey. U. S. Dept. Energy ER 238: 101–116.

Davies, W. J., Metcalfe, J., Lodge, T. A., and da Costa, A. R. (1986). Plant growth substances and the regulation of growth under drought. *Aust. J. Plant Physiol.* 13: 105–125.

Duncan, W. G., and Hesketh, J. D. (1968). Net photosynthesis rates, relative growth rates, and leaf numbers of 22 races of maize grown at 8 temperatures. *Crop Sci.* 8: 670–674.

Evans, J. R. (1983). Nitrogen and photosynthesis in the flag leaf of wheat (*Triticum aestivum* L.). *Plant Physiol.* 72: 297–302.

Evans, J. R. (1987). The relationship between electron transport components and photosynthetic capacity in pea leaves grown at different irradiances. *Aust. J. Plant Physiol.* 14: 157–170.

Fondy, B. R., and Geiger, D. R. (1980). Effect of rapid changes in sink–source ratio on export and distribution of products of photosynthesis in leaves of *Beta vulgaris* L. and *Phaseolus vulgaris* L. *Plant Physiol.* 66: 945–949.

Fondy, B. R., and Geiger, D. R. (1982). Diurnal pattern of translocation and carbohydrate metabolism in source leaves of *Beta vulgaris* L. *Plant Physiol.* 70: 671–676.

Fondy, B. R., Geiger, D. R., and Servaites, J. C. (1989). Photosynthesis, carbohydrate metabolism, and export in *Beta vulgaris* L. and *Phaseolus vulgaris* L. during square and sinusoidal light regimes. *Plant Physiol.* 89: 396–402.

Fox, T. C., and Geiger, D. R. (1984). Effects of decreased net carbon exchange on carbohydrate metabolism in sugar beet source leaves. *Plant Physiol.* 76: 763–768.

Fox, T. C., and Geiger, D. R. (1986). Osmotic response of sugar beet source leaves at CO_2 compensation point. *Plant Physiol.* 80: 239–241.

Foyer, C. H. (1987). The basis for source–sink interaction in leaves. *Plant Physiol. Biochem.* 25: 649–657.

Franceschi, V. R., and Giaquinta, R. T. (1983a). The paraveinal mesophyll of soybean leaves in relation to assimilate transfer and compartmentation. I. Ultrastructure and histochemistry during vegetative development. *Planta* 157: 411–421.

Franceschi, V. R., and Giaquinta, R. T. (1983b). The paraveinal mesophyll of soybean leaves in relation to assimilate transfer and compartmentation. II. Structural, metabolic and compartmental changes during reproductive growth. *Planta* 157: 422–431.

Geiger, D. R. (1976). Effects of translocation and assimilate demand on photosynthesis. *Can. J. Bot.* 54: 2337–2345.

Geiger, D. R. (1986). Processes affecting carbon allocation and partitioning among sinks. *In* "Plant Biology" Vol 1: Phloem Transport (J. Cronshaw, W. J. Lucas, and R. T. Giaquinta, eds.), pp. 375–388. Alan R. Liss, New York.

Geiger, D. R. (1987). Understanding interactions of source and sink regions of plants. *Plant Physiol. Biochem.* 25: 659–666.

Geiger, D. R., and Bestman, H. (1990). Self-limitation of herbicide mobility by phytotoxic action. *Weed Sci.* 38: 324–329.

Geiger, D. R., and Fondy, B. R. (1980). Phloem loading and unloading: pathways and mechanisms. *What's New in Plant Physiol.* 11: 25–28.

Geiger, D. R., and Shieh, W.-J. (1988). Analyzing partitioning of recently fixed and reserve carbon in reproductive *Phaseolus vulgaris* plants. *Plant, Cell Environ.* 11: 777–783.

Geiger, D. R., Saunders, M. A., and Cataldo, D. A. (1969). Translocation and accumulation of translocate in the sugar beet petiole. *Plant Physiol.* 44: 1657–1665.

Geiger, D. R., Ploeger, B. J., Fox, T. C., and Fondy, B. R. (1983). Source of sucrose translocated from illuminated sugar beet source leaves. *Plant Physiol.* 72: 964–970.

Gerhardt, R., Stitt, M., and Heldt, H. W. (1987). Subcellular metabolite levels in spinach leaves. *Plant Physiol.* 83: 399–407.

Gifford, R. M., Thorne, J. H., Hitz, W. D., and Giaquinta, R. T. (1984). Crop productivity and photoassimilate partitioning. *Science* 225: 801–808.

Gollan, T., Passioura, J. B., and Munns, R. (1986). Soil water status affects the stomatal conductance of fully turgid wheat and sunflower leaves. *Aust. J. Plant Physiol.* 13: 459–464.

Goodall, D. W. (1946). The distribution of weight changes in the young tomato plant. II. Changes in dry weight of separated organs and translocation rates. *Ann. Bot.* 40: 305–338.

Gordon, A. J., Ryle, G. J. A., Mitchell, D. F., and Powell, C. E. (1982). The dynamics of carbon supply from leaves of barley plants grown in long or short days. *J. Exp. Bot.* 33: 241–250.

Guerrero, F. D., and Mullet, J. E. (1988). Reduction of turgor induces rapid changes in leaf translatable RNA. *Plant Physiol.* 88: 401–408.

Hanson, D. R., and Hitz, W. D. (1982). Metabolic responses of mesophytes to plant water deficits. *Annu. Rev. Plant Physiol.* 33: 163–203.

Hanson, D. R., and McHale, N. A. (1988). A starchless mutant of *Nicotiana sylvestris* containing a modified plastid phosphoglucomutase. *Plant Physiol.* 88: 838–845.

Herold, A. (1980). Regulation of photosynthesis by sink activity — the missing link. *New Phytol.* 86: 131–144.

Hsiao, T. C. (1973). Plant responses to water stress. *Annu. Rev. Plant Physiol.* 24: 519–570.

Huber, S. C. (1983). Relation between photosynthetic starch formation and dry-weight partitioning between the shoot and root. *Can. J. Bot.* 61: 2709–2716.

Huber, J. L. A., Huber, S. C., and Nielson, T. H. (1989). Protein phosphorylation as a mechanism of regulation of sucrose phosphate synthase activity. *Arch. Biochem. Biophys.* 270: 681–690.

Jablonski, L. M., and Geiger, D. R. (1987). Responses of sugar beet plant morphology and carbon distribution to shortened days. *Plant Physiol. Biochem.* 25: 787–796.

Jones, T. W. A., Gottlieb, L. D., and Pichersky, E. (1986). Reduced enzyme activity and starch level in an induced mutant of chloroplast phosphoglucose isomerase. *Plant Physiol.* 81: 367–371.

Jurik, T. W., Chabot, J. F., and Chabot, B. F. (1979). Ontogeny of photosynthetic performance in *Fragaria virginiana* under changing light regimes. *Plant Physiol.* 63: 542–547.

Kimball, B. A. (1983). Carbon dioxide and agricultural yield: An assemblage and analysis of 430 prior observations. *Crop Sci.* 75: 779–788.

Kramer, P. J. (1980). Drought stress, and the origin of adaptations. *In* "Adaptation of Plants to Water and High Temperature Stress" (N. C. Turner and P. J. Kramer, eds), pp. 7–20. Wiley, New York.

Kramer, P. J. (1981). Carbon dioxide concentration, photosynthesis, and dry matter production. *Bioscience* 31: 29–33.

Larcher, W. (1987). Stress bei Pflanzen. *Naturwissenschaften* 74: 158–167.

Lin, T.-P., Caspar, T., Somerville, C. R., and Preiss, J. (1988). A starch deficient mutant of *Arabidopsis thaliana* with low ADP-glucose pyrophosphorylase activity lacks one of the two subunits of the enzyme. *Plant Physiol.* 88: 1175–1181.

Mason, H. S., Mullet, J. E., and Boyer, J. S. (1988). Polysomes, messenger RNA, and growth in soybean stems during development and water deficit. *Plant Physiol.* 86: 725–733.

Mayoral, M. L., Plaut, Z., and Reinhold, L. (1985). Effect of translocation-hindering procedures on source leaf photosynthesis in cucumber. *Plant Physiol.* 77: 712–717.

Meyer, R. F., and Boyer, J. S. (1981). Osmoregulation, solute distribution, and growth in soybean seedlings having low water potentials. *Planta* 151: 482–489.

Minchin, P. E. H., and Thorpe, M. R. (1987). Measurement of unloading and reloading of photo-assimilate within the stem of bean. *J. Exp. Bot.* 38: 211–220.

Neales, T. F., and Incoll, L. D. (1968). The control of leaf photosynthesis rate by the level of assimilate concentration in the leaf: A review of the hypothesis. *Bot. Rev.* 34: 107–125.

Plaut, Z., Mayoral, M. L., and Reinhold, L. (1987). Effect of altered sink:source ratio on photosynthetic metabolism of source leaves. *Plant Physiol.* 85: 786–791.

Queiroz, O. (1983). Interactions between external and internal factors affecting the operation of phosphoenolpyruvate carboxylase. *Physiol. Veg.* 21: 963–975.

Raschke, K. (1975). Stomatal action. *Annu. Rev. Plant Physiol.* 26: 305–340.

Robinson, J. M. (1984). Photosynthetic carbon metabolism in leaves and isolated chloroplasts from spinach plants grown under short and intermediate photosynthetic periods. *Plant Physiol.* 75: 397–409.

Sachs, M. M., and Ho, T.-H. D. (1986). Alteration of gene expression during environmental stress in plants. *Annu. Rev. Plant Physiol.* 37: 363–376.

Sage, R. F., Sharkey, T. D., and Seemann, J. R. (1989). Acclimation of photosynthesis to elevated CO_2 in five C_3 species. *Plant Physiol.* 89: 590–596.

Schulze, E.-D. (1986). Whole-plant responses to drought. *Aust. J. Plant Physiol.* 13: 127–141.

Servaites, J. C., Geiger, D. R., Tucci, M. A., and Fondy, B. R. (1989a). Leaf carbon metabolism and metabolite levels during a period of sinusoidal light. *Plant Physiol.* 89: 403–408.

Servaites, J. C., Fondy, B. R., Li, B., and Geiger, D. R. (1989b). Sources of carbon for export from spinach leaves throughout the day. *Plant Physiol.* 90: 1168–1174.

Sharkey, T. D. (1985). O_2-insensitive photosynthesis in C_3 plants: Its occurrence and a possible explanation. *Plant Physiol.* 78: 71–75.

Sharp, R. E., and Davies, W. L. (1979). Solute regulation and growth by roots and shoots of water-stressed maize plants. *Planta* 147: 43–49.

Sivak, M. N., and Walker, D. A. (1986). Photosynthesis in vivo can be limited by phosphate supply. *New Phytol.* 102: 499–512.

Stitt, M., Herzog, B., and Heldt, H. W. (1985). Control of photosynthetic sucrose synthesis by fructose 2,6-bisphosphate. V. Modulation of the spinach leaf cytosolic fructose 1,6-bisphosphatase in vitro by substrate, products, pH, magnesium, fructose 2,6- bisphosphate, adenosine monophosphate, and dihydroxyacetone phosphate. *Plant Physiol.* 79: 590–598.

Stitt, M. (1986). Regulation of photosynthetic sucrose synthesis: Integration, adaptation, and limits. *In* "Plant Biology" Vol 1: Phloem Transport (J. Cronshaw, W. J. Lucas, and R. T. Giaquinta, eds.), pp. 331–347. Alan R. Liss, New York.

Stryer, L. (1988). "Biochemistry," pp. 449–468. Freeman, New York.

Sung, S.-J. S., Xu, D.-P., Galloway, C. M., and Black, C. C. (1988). A reassessment of glycolysis and gluconeogenesis in higher plants. *What's New in Plant Physiol.* 72: 650–654.

Swanson, C. A., and Geiger, D. R. (1967). Time course of low-temperature inhibition of sucrose translocation in sugar beets. *Plant Physiol.* 42: 751–756.

Torisky, R. S., and Servaites, J. C. (1984). Effect of irradiance during growth of *Glycine max* on photosynthetic capacity and percent activation of ribulose 1,5-bisphosphate carboxylase. *Photosynth. Res.* 5: 251–261.

Turner, N. C. (1986). Adaptation to water deficits: A changing perspective. *Aust. J. Plant Physiol.* 13: 175–190.

Turner, N. C., Begg, J. E., and Tonnet, M. L. (1978). Osmotic adjustment of sorghum and sunflower crops in response to water deficits and its influence on the water potential at which stomata close. *Aust. J. Plant Physiol.* 5: 597–608.

Walker, D. A., and Herold, A. (1977). Can the chloroplast support photosynthesis unaided? *In* "Photosynthetic Organelles: Structure and Function" *Plant Cell Physiol.* Special Issue, pp. 295–310.

Walton, D. C. (1980). Biochemistry and physiology of abscisic acid. *Annu. Rev. Plant Physiol.* 31: 453–489.

Wardlaw, I. F. (1985). The regulation of photosynthesis rate by sink demand. *In* "Regulation of Sources and Sinks in Crop Plants" (B. Jeffcoat, A. F. Hawkins, A. D. Stead, eds.), pp. 145–162. Monogr. 12, British Plant Growth Regulator Group, Bristol.

6

Partitioning Response of Plants to Stress

H. A. Mooney   W. E. Winner

I. Introduction

Plants have evolved an enormous variety of forms and shapes, and biologists have spent considerable effort relating this morphological variation to particular functional roles. This has not been an easy task for a number of reasons. First, a given plant structure or organ can serve multiple functions. Second, structures may have evolved in response to selective environmental factors that no longer exist. Mostly, though, it has been difficult to develop a general understanding of relationships between form and function because compensating features in plants, which in total may optimize success in a given habitat, may not be revealed when a single morphological trait and its function are studied in isolation.

One area of investigation into form–function has, however, been particularly revealing and unifying. This is the ratio by which plants partition dry matter to roots versus stems and leaves. An analysis of plants by partitioning ratio reduces their form to simplistic dimensions but provides a universal framework to which the greater variations in growth form can be added. Importantly, growth-form variability as encompassed by partitioning ratio can be directly, although only generally, linked to function. The basis of this linkage was first elucidated by Monsi (1968). He noted that apportionment of newly acquired dry matter to new leaves resulted in a compounding of investment, whereas dry matter directed to nonphotosynthetic tissue did not. These relationships lead to the prediction that plants with greater apportionment to leaves will grow faster than those with lower leaf apportionment, even though photosynthetic capacities may be similar. If maximization of growth is important under natural conditions, and with no other considerations, plants would be mostly leaves. Plants do not occur in isolation, however, and thus must compete for light; hence, they invest in woody structures to position photosynthetic tissue in exposed positions to compete with sun-blocking neighbors. Further, roots that acquire some of the raw materials for building a canopy must also be constructed.

The general balancing of carbon partitioning to roots versus shoots under differing resource levels has been discussed in physiological terms by Davidson (1969) and in evolutionary terms by Mooney and Gulmon (1979) and Schulze (1982). Tilman (1988) has recently put these relationships into a general framework, considering also partitioning to stems. Tilman asserts that competition for light and nitrogen are the two primary selective factors acting on the partitioning of dry matter. An inverse relationship often exists between the availability of these resources, particularly along successional gradients. Hence, in closed sites plants preferentially allocate to stem and leaf, and in open sites to roots. Tilman thus provides an environmental matrix for predictions of specific partitioning ratios. Competition generally results in a convergence of partitioning types in comparable habitats. Temporal and spatial variability of resources at a given site, however, produce a certain degree of divergence in structural forms.

All plants are able to modify their basic partitioning pattern to a certain degree in response to specific environmental conditions. In this chapter we focus on this plasticity rather than on the evolutionary dimension of partitioning. That is, we examine how a given genotype responds to a changing resource base, particularly to changes in multiple resources simultaneously. To do this we must first discuss what regulates biomass partitioning (see also Chapters 1 and 5).

II. Approaches to the Problem

Biomass partitioning has been considered on theoretical grounds, generally as a growth optimization process. Iwasa and Roughgarden (1984) defined an optimal allocation strategy for an annual as the maximization of reproductive output. Extending their model to the vegetative phase, they demonstrated how a balanced growth of both roots and shoots leads to an optimal allocation strategy. In a more physiological approach, but also applying Pontryagin's Maximum principle from control theory, Schulze *et al.* (1983) showed how *Vigna unguiculata* allocates carbon to leaves versus shoots in a manner that maximizes growth without inducing water stress.

A more mechanistic approach to this problem is given by Schulze and Chapin (1987), who present the view that plants regulate partitioning so that the internal pool sizes of carbon and nutrients remain constant even under a changing resource base (Fig. 1) (*cf.* Reynolds and Thornley [1982]). In their example, light limitation with other resources held constant led to a slight drop in internal carbohydrate concentrations, presenting a deviation from a set-point ratio. Plants responded to this by producing more leaves of a reduced specific weight and by reducing root growth, thereby bringing the ratio back to the set point. Under extreme resource limitation, the set point cannot be maintained by compensation. A similar mechanism would operate with water or nutrient limitation, but in such cases root growth would be stimulated, and specific leaf area would increase to maintain an internal carbon-to-nutrient balance. This general scenario is based on the

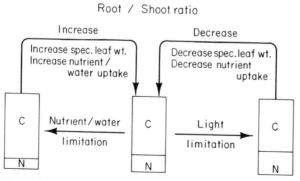

Figure 1   A model to account for regulation within a plant to balance internal resources in response to a changing external environment. [Redrawn from Schulze *et al.* (1987).]

theoretical economic argument presented by Bloom *et al.* (1985) that plants should adjust allocation so that all resources equally limit growth.

III. The Set-Point Argument

We examine here three parts of the set-point argument:

1. Plants adjust their root:shoot ratio under changing resource availabilities.
2. In doing so, they maximize growth rate.
3. The mechanism for this compensation is related to the internal ratio of carbon to nutrients.

A. Resource Response

The evidence for the directional change in root:shoot ratio dependent on either a changing light, nutrient, or water regime is overwhelming (for examples see Davidson [1969], Gulmon and Chu [1981], and Richards [1977]) and has been reviewed extensively (Novoa and Loomis, 1981; Aung, 1974; Russell, 1977; Hunt and Lloyd, 1987; Snyder and Carlson, 1984; Wilson, 1988). We will not review this information further here except in the context of multiple stresses.

B. Maximization

Apart from the theoretical treatments discussed above, few explicit tests have been done of the maximization of growth rate following allocation shifts in response to a changing resource base. Robinson and Rorison (1988) investigated the plasticity of growth-rate determinants under a series of test conditions in relation to the maximal relative growth rate obtained under all test conditions. They did not, however, determine if the adjustments made under the various treatments were optimal, in terms of growth, for each growing condition. In a different approach, Mooney *et al.* (1988) used a growth model to demonstrate that the allocation shifts that occurred in radish in response to sulfur dioxide resulted in greater growth than would have occurred if no compensatory changes had taken place (Fig. 2). Similarly, Hirose (1987) took a modeling approach to demonstrate that plasticity in partitioning or in specific leaf weight resulted in higher growth rates under changing nitrogen supply. Kachi and Rorison (1989) have also recently developed a model to predict the shoot fraction need to maximize relative growth rate at given rates of nitrogen uptake. Their experiments showed that a slow-growing species met the predicted optimality criteria, whereas a fast-growing species did not. They attribute this difference to the possibility that the fast-growing species generally encounters a competitive environment, where "over"-allocation to shoots

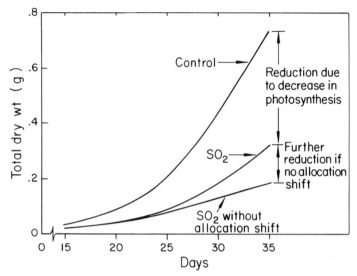

Figure 2   Yield response of radish plants subjected to SO_2 stress. Partitioning to more leaves resulted in a yield enhancement. [Redrawn from Mooney *et al.* (1988).]

may be advantageous. We need further explicit experimental tests of the optimality of plastic changes in growth parameters in response to changing resources.

C. Internal Ratio

No known mechanism explains the control of root:shoot partitioning by the ratios of carbon to nitrogen or by ratios of other nutrients. Some lines of evidence, however, indicate control points by either carbon or nitrogen concentration or of their supply rates.

1. Carbon Control   Under shortening photoperiods, partitioning to leaves is enhanced much as it is enhanced under reduced light intensity (Jablonski and Geiger, 1987). In fact, the whole adjustment syndrome occurring under low light intensity—modifications of leaf thickness, chlorophyll concentrations, and leaf duration—also occurs under short photoperiods. These modifications all minimize carbon losses that would take place without such physiological and morphological shifts. These studies thus indicate that rates of carbon flux influence allocation patterns. A reduction in flux entrains a series of events resulting in a new, higher flux capacity. As discussed below, these compensations occur no matter what factor causes the flux reduction (e.g., light intensity, photoperiod, carbon dioxide, pollutants, and so on).

Huber (1983) proposed a mechanism that could regulate root:shoot partitioning through variability in carbon flux (see also model of this mechanism by Geiger *et al.* [1985]). The amount of photosynthate that goes into starch, rather than sucrose, in the leaves during the day controls how much carbon is partitioned to roots. The greater the proportion that goes to starch, the less carbon that is exported daily to roots. The reason for this partitioning differential is that the source of carbon for root growth during the night is sucrose from starch mobilization, and this source is limited. At night, developing leaves are stronger sinks for this sucrose because of their proximity and their possibly greater sink strength due to nighttime increases in cell turgor, causing a potential for cell expansion. During the day, all sinks, both roots and shoots, receive carbon equally, although the amount is determined by how much sucrose is available. If a large proportion is put into leaf starch, the potential for a nighttime differential in partitioning is set up. What then controls starch:sugar allocation in the leaves, and how does this relate to carbon flux rates in the whole canopy? Starch synthesis rates generally increase under conditions, such as shortened photoperiod, where daily production of assimilate is reduced (Baysdorfer and Robinson, 1985, and references therein). The initial increase in leaf starch is followed by a whole-plant adjustment, as described above, that leads to greater carbon acquisition under the new prevailing conditions.

Reductions in sink strength can also lead to an increase in leaf starch content. For example, Plaut *et al.* (1987) showed a progressively greater increase in leaf starch with greater sink removal in cotton and a concomitant decrease in photosynthetic capacity. These and other results led Baysdorfer and Robinson (1985) to conclude that under carbon limitation, starch synthesis has priority over sucrose synthesis. There is now considerable evidence that the concentration of fructose 2,6-bisphosphate in the leaf serves as the control for allocation between starch and sucrose. Increased concentrations of this compound restrict sucrose synthesis and lead to starch production (Stitt, 1987). Thus, a molecular control point for carbon allocation and partitioning may exist.

2. Nitrogen Control   Vessey and Layzell (1987) indicate how nitrogen supply rate can alter carbon partitioning. Under nitrogen limitation nitrogen circulation through the plant is preferentially diverted to roots, resulting in a stimulation of root growth over shoot growth. Such a mechanism would alter partitioning before nitrogen became limiting to photosynthesis. This proposal differs from the mechanism proposed by Brouwer (1962), where absorbed nitrogen is preferentially retained by roots before circulation in the xylem–phloem system.

IV. Linking Resource Partitioning with Growth Rates

Although our understanding of the mechanisms driving allocation and partitioning is still incomplete, we can make some predictions regarding growth rates and allocation patterns. The bases of these predictions are the tight relationships among nitrogen content and carbon gain, nitrogen and allocation, and carbon gain and allocation as controllers of growth rate. We still, however, do not understand the general mechanisms integrating carbon and nitrogen metabolism in plants (Stulen, 1986).

A. Nitrogen and Photosynthesis

The relationship between nitrogen content and photosynthesis has been explored in some detail (Field and Mooney, 1986). Net photosynthesis in leaves generally increases linearly with increasing concentrations of organic nitrogen under saturating light conditions (Fig. 3). Data are limited for whole plants. Küppers *et al.* (1988), however, found that they could successfully predict whole-canopy photosynthesis from information on the age distribution of leaves and on the relationships between leaf nitrogen and photosynthesis before self-shading.

B. Within-Plant Nitrogen Distribution

Nitrogen distribution within plants follows predictable patterns. These patterns have been interpreted in part as mechanisms to maximize plant carbon gain (Field, 1983). First, nitrogen concentrations are always highest in the youngest leaves, which, by their position, receive the greatest amount of solar radiation (Mooney and Gulmon, 1982). Nitrogen content de-

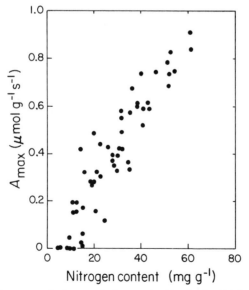

Figure 3 Light-saturated photosynthetic capacity (A_{max}) of radish leaves with differing organic nitrogen contents. [Redrawn from Küppers *et al.* (1988).]

creases with leaf age. These trends hold independent of plant nitrogen status (Fig. 4). Furthermore, leaf nitrogen content is strongly related to whole-plant nitrogen content (Greenwood *et al.*, 1986). This relationship is useful in predicting partitioning, as described below.

C. Nitrogen and Allocation

A strong linear relationship exists between total plant nitrogen concentration and the fraction of plant dry matter that is in leaf tissue (Ågren and Ingestad, 1987) (Fig. 5). This derives from the fact that the bulk of plant nitrogen resides in the plant canopy. Although whole-plant nitrogen content is not generally available from experiments, it can be predicted from leaf concentrations, as noted above.

D. Nitrogen and Growth Rate

Ågren (1985) demonstrated theoretically that plant relative growth rate is a linear function over a wide range of whole-plant internal nitrogen concentrations. Hirose (1988) cites a large number of studies verifying this prediction. He further modeled the interrelationships among the components controlling growth rate, showing how they interact to produce a linear relative growth rate over a wide range of nitrogen concentrations (Fig. 6).

E. Quantitively Linking Nitrogen Growth Controllers

As noted above, the growth of plants is a function of their photosynthetic capacity and allocation to leaf tissue. Since both of these are linked to nitrogen concentration, predictions of growth rates from nitrogen concentration alone are possible. Ågren and Ingestad (1987) have discussed these relationships formally, showing that

$$P \times S = dW/dT = aN,$$

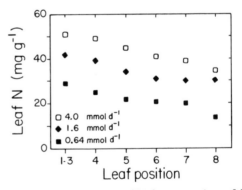

Figure 4   Leaf nitrogen contents of leaves of *Diplacus aurantiacus* of different ages grown at different nitrate supply rates. [Redrawn from Gulmon and Chu (1981).]

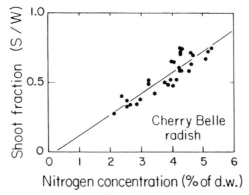

Figure 5 The fraction of shoot to total plant weight of radish plants with different nitrogen concentrations. [Redrawn from Levin and Mooney (1989).]

where P is net photosynthesis; S, shoot weight; W, total plant weight; N, nitrogen concentration; and a is a constant. Hirose (1986) has given empirical tests of these relationships.

What then controls whole-plant nitrogen concentration? Reduction in supply rate at low concentrations leads to a reduction in tissue concentrations, with roots having priority over this nitrogen (Fig. 7). With the lower

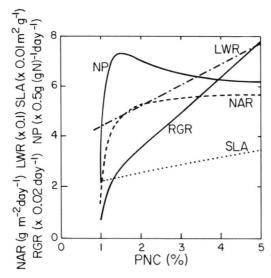

Figure 6 Modeled interrelationships among nitrogen productivity (NP), leaf weight ratio (LWR), specific leaf area (SLA), net assimilation rate (NAR), and relative growth rate (RGR). [Redrawn from Hirose (1988).]

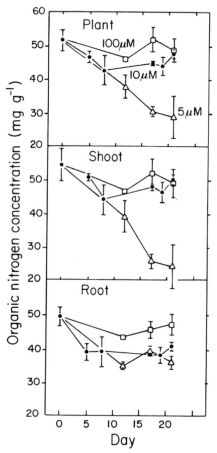

Figure 7 Repsonse of radish shoots and roots to limiting nitrate supplies. Shoots are more sensitive than roots to limiting nitrogen (Koch and Mooney, in press).

nitrogen concentration, shoot growth is restricted, and if this growth restriction exceeds the coincident decline in photosynthesis, carbon will be preferentially translocated to roots (Rufty *et al.*, 1988).

V. Conclusions

The assumption in the above concepts is that the particular partitioning configuration a plant deploys is related to optimizing growth under the available resources. It follows then that a given species, which has evolved in a given resource matrix, will have a genetically controlled partitioning schedule that ensures maximal growth for the prevailing environmental conditions. These conditions, however, change from day to day and from year to year and, hence, á certain flexibility must be built into the adjustment, an adjustment involving a great number of processes and plant structures. These adjustments take place in differing time scales varying from hours to days and weeks. In nature, demands for internal resources may conflict. We are still far from understanding what mechanism plants use to "resolve" these conflicts to maximize growth and reproductive output. We do know, however, that new conflicting demands are being imposed on plants as human activities modify natural environments. These conflicts may result in nonadaptive partitioning responses and lead to plant mortality, such as we may be seeing in forests enriched with atmospheric sources of nitrogen.

Acknowledgments

We express appreciation to Ellen Chu, Eva Pell, and Detlef Schulze for comments on this manuscript, and to the National Science Foundation and the Electrical Power Research Institute for support of our work.

References

Ågren, G. I. (1985). Theory for growth of plants derived from the nitrogen productivity concept. *Physiol. Plant.* 64: 17–28.

Ågren, G. I., and Ingestad, T. (1987). Root:shoot ratio as a balance between nitrogen productivity and photosynthesis. *Plant, Cell Environ.* 10: 579–586.

Aung, L. A. (1974). Root–shoot relationships. *In* "The Plant Root and Its Environment" (E. W. Carson, ed.), pp. 29–61. University Press of Virginia, Charlottesville.

Baysdorfer, C., and Robinson, M. J. (1985). Sucrose and starch synthesis in spinach plants grown under long and short photosynthetic periods. *Plant Physiol.* 79: 838–842.

Bloom, A. J., Chapin, F. S., III., and Mooney, H. A. (1985). Resource limitation in plants — an economic analogy. *Ann. Rev. Ecol. Syst.* 16: 363–392.

Brouwer, R. (1962). Nutritive influences on the distribution of dry matter in the plant. *Neth. J. Agric. Sci.* 10: 399–408.

Davidson, R. L. (1969). Effect of root:leaf temperature differentials on root:shoot ratios in some pasture grasses and clover. *Ann. Bot.* 33: 561–569.

Field, C. (1983). Allocating leaf nitrogen for the maximization of carbon gain: Leaf age as a control on the allocation program. *Oecologia* 56: 341–347.

Field, C., and Mooney, H. A. (1986). The photosynthesis–nitrogen relationship in wild plants. *In* "On the Economy of Plant Form and Function" (T. J. Givnish, ed.), pp. 25–55. Cambridge Univ. Press, Cambridge.

Geiger, D. R., Jablonski, L. M., and Ploeger, B. J. (1985). Significance of carbon allocation to starch in growth of *Beta vulgaris* L. *In* "Regulation of Carbon Partitioning in Photosynthetic Tissue" (R. L. Heath and J. Preiss, eds.), pp. 289–308. Waverly Press, Baltimore.

Greenwood, D. J., Neeteson, J. J., and Draycott, A. (1986). Quantitative relationships for the dependence of growth rate of arable crops on their nitrogen content, dry weight, and aerial environment. *Plant Soil* 91: 281–301.

Gulmon, S. L., and Chu, C. C. (1981). The effects of light and nitrogen on photosynthesis, leaf characteristics, and dry matter allocation in the chaparral shrub, *Diplacus aurantiacus*. *Oecologia* 49: 207–212.

Hirose, T. (1986). Nitrogen uptake and plant growth. II. An empirical model of vegetative growth and partitioning. *Ann. Bot.* 58: 487–496.

Hirose, T. (1987). A vegetative plant growth model: Adaptive significance of phenotypic plasticity in matter partitioning. *Funct. Ecol.* 1: 195–202.

Hirose, T. (1988). Modeling the relative growth rate as a function of plant nitrogen concentration. *Physiol. Plant.* 72: 185–189.

Huber, S. C. (1983). Relation between photosynthetic starch formation and dry-weight partitioning between the shoot and root. *Can. J. Bot.* 61: 2709–2716.

Hunt, R., and Lloyd, P. S. (1987). Growth and partitioning. *New Phytol.* 106 (Suppl.): 235–249.

Iwasa, Y., and Roughgarden, J. D. (1984). Shoot:root balance of plants: Optimal growth of a system with many vegetative organs. *Theor. Popul. Biol.* 25: 78–105.

Jablonski, L. M., and Geiger, D. R. (1987). Responses of sugar beet plant morphology and carbon distribution to shortened days. *Plant Physiol. Biochem.* 25: 787–796.

Kachi, N., and Rorison, I. H. (1989). Optimal partitioning between root and shoot in plants with contrasted growth rates in response to nitrogen availability and temperature. *Func. Ecol.* 3: 549–559.

Koch, G. W., and Mooney, H. A. (1991). Response of wild plants to nitrate availability: Determinants of plant nitrogen demand and sensitivity to solution nitrate concentration in wild radish. *Oecologia*, in press.

Küppers, M., Koch, G., and Mooney, H. A. (1988). Compensating effects to growth of changes in dry matter allocation in response to variation in photosynthetic characteristics induced by photoperiod, light, and nitrogen. *Aust. J. Plant Physiol.* 15: 287–298.

Levin, S. A., and Mooney, H. A. (1989). The dependence of plant root:shoot ratios on internal nitrogen concentration. *Ann. Bot.* 64: 71–75.

Monsi, M. (1968). Mathematical models of plant communities. *In* "Functioning of Terrestrial Ecosystems at the Primary Production Level" (F. Eckardt, ed.), pp. 131–149. UNESCO, Paris.

Mooney, H. A., and Gulmon, S. L. (1979). Environmental and evolutionary constraints on the photosynthetic characteristics of higher plants. *In* "Topics in Plant Population Biology" (O. T. Solbrig, S. Jain, G. B. Johnson, and P. H. Raven, eds.), pp. 316–337. Columbia Univ. Press, New York.

Mooney, H. A., and Gulmon, S. L. (1982). Constraints on leaf structure and function in reference to herbivory. *Bio Science* 32: 189–206.

Mooney, H. A., Küppers, M., Koch, G., Gorham, J., Chu, C. C., and Winner, W. E. (1988). Compensating effects to growth of carbon partitioning changes in response to SO_2-induced photosynthetic reduction in radish. *Oecologia* 75: 502–506.

Novoa, R., and Loomis, R. S. (1981). Nitrogen and plant production. *Plant Soil* 58: 177–204.

Plaut, Z., Mayoral, M. L., and Reinhold, L. (1987). Effect of altered sink:source ratio on photosynthetic metabolism of source leaves. *Plant Physiol.* 85: 786–791.

Reynolds, J. F., and Thornley, J. H. M. (1982). A shoot:root partitioning model. *Ann. Bot.* 49: 585–597.

Richards, D. (1977). Root–shoot interactions: A functional equilibrium for water uptake in peach (*Prunus persica* [L.] Batsch). *Ann. Bot.* 41: 279–281.

Robinson, D., and Rorison, I. H. (1988). Plasticity in grass species in relation to nitrogen supply. *Func. Ecol.* 2: 249–257.

Rufty, T. W., Jr., Huber, S. C., and Volk, R. J. (1988). Alterations in leaf carbohydrate metabolism in response to nitrogen stress. *Plant Physiol.* 88: 725–730.

Russell, R. S. (1977). "Plant Root Systems." McGraw-Hill, London.

Schulze, E.-D. (1982). Plant life-forms and their carbon, water, and nutrient relations. *In* "Encyclopedia of Plant Physiology, New Series" (O. L. Lange, P. S. Nobel, C. B. Osmond, and H. Ziegler, eds.), Volume 12B, p. 615–676. Springer-Verlag, Berlin.

Schulze, E.-D., and Chapin, F. S., III. (1987). Plant specialization to environments of different resource availabilities. *In* "Potentials and Limitations of Ecosystem Analysis" (E.-D. Schulze and H. Zwölfer, eds.), pp. 120–148. Springer-Verlag, Berlin.

Schulze, E.-D., Schilling, K., and Nagarajah, S. (1983). Carbohydrate partitioning in relation to whole-plant production and water use of *Vigna unguiculata* (L.) Walp. *Oecologia* 58: 169–177.

Snyder, F. W., and G. E. Carlson. (1984). Selecting for partitioning of photosynthetic products in crops. *Adv. Agron.* 37: 47–72.

Stitt, M. (1987). Fructose 2,6-bisphosphate and plant carbohydrate metabolism. *Plant Physiol.* 84: 201–204.

Stulen, I. (1986). Interactions between nitrogen and carbon metabolism in a whole-plant context. *In* "Fundamental, Ecological, and Agricultural Aspects of Nitrogen Metabolism in Higher Plants" (H. Lambers, J. J. Neeteson, and I. Stulen, eds.), pp. 261–274. Martin. Nijhoff, Dordrecht.

Tilman, D. (1988). "Plant Strategies and the Dynamics and Structure of Plant Communities." Princeton Univ. Press, Princeton, New Jersey.

Vessey, J. K., and Layzell, D. B. (1987). Regulation of assimilate partitioning in soybean: Initial effects following change in nitrate supply. *Plant Physiol.* 83: 341–348.

Wilson, J. B. (1988). A review of evidence on the control of shoot:root ratio in relation to models. *Ann. Bot.* 61: 433–449.

7

Growth Rate, Habitat Productivity, and Plant Strategy as Predictors of Stress Response

J. P. Grime   B. D. Campbell

I. Introduction

During 1950–1970 experimental studies revealed that plant species show genetically controlled differences in rates of dry-matter production. Three principles were established: (1) that many, if not all, plants grow more rapidly in productive controlled environments than in their natural habitats; (2) that plant species and populations differ in the conditions required for optimal growth; and (3) that even under their optimal conditions, some plant species do not have the potential to grow rapidly.

It was the last of these three findings that captured the attention of

Response of Plants to Multiple Stresses. Copyright © 1991 by Academic Press, Inc. All rights of reproduction in any form reserved.

ecophysiologists and prompted several authors to suggest that the potential growth rates of plants vary in relation to the productivity of the natural habitats to which they have been attuned by natural selection (Beadle, 1954, 1962; Kruckeberg, 1954; Bradshaw *et al.*, 1964; Jowett, 1964; Hackett, 1965; Clarkson, 1967; Parsons, 1968a,b; Higgs and James, 1969). This led to the expectation that studies of potential growth rate and its physiological and biochemical concomitants would help to elucidate the mechanisms controlling the species composition of productive and unproductive habitats.

Since 1970, measurements of relative growth rate have retained their value as a probe with which to explore adaptation to environment, but they have been incorporated into a more sophisticated framework (plant strategy theory) that includes other variable attributes of plants and takes account of habitat factors other than productivity. In this chapter, we explain how plant strategy theory can be used to predict differences among ecologically contrasted plants in their reactions to various stresses occurring either singly, sequentially, or coincidentally. The first step in this analysis is to examine some conspicuous exceptions to the correlation between potential growth rate and habitat productivity.

II. Growth Rate, Productivity, and Plant Niches

Grime and Hunt (1975) present the results of an extensive comparative study of plant growth rates, conducted in a standardized productive environment. These data confirm that fast-growing species tend to be associated with productive habitats; they also show that plants from unproductive sites include a high proportion that grow slowly. Closer inspection of the data, however, reveals that among the species drawn from unproductive habitats (e.g., rock outcrops, cliffs, and walls), some (e.g., *Arenaria serpyllifolia, Arabidopsis thaliana, Geranium robertianum*) have comparatively high growth rates. These are ephemeral species able to exploit local patches of bare soil exposed by summer drought, soil-slip, or biotic disturbance. This exception to the general pattern serves as a reminder that the conditions exploited by a plant refer to the productivity and temporal stability of its realized niche rather than the conditions of its habitat at large.

Further departures from the general relationship between growth rate and habitat productivity are evident in the data of Grime and Hunt (1975), specifically the slow growth rates of tall herbs such as *Anthriscus sylvestris* and *Heracleum sphondylium*, both of which occur in vegetation of high productivity. In these two plants, slow growth is related to the fact that even in the early seedling phase, captured resources are invested in an underground storage organ that provides the capital for vegetative and reproductive development in later growing seasons (Hunt and Lloyd, 1987). This example not only exposes the occasional fallibility of growth rate as an index of habitat productivity but implies a trade-off between, on the one hand, allocation to leaves and fine roots (strongly developed in the pre-reproductive phase of ephemeral species) and, on the other, investments in storage organs and arterial roots and shoots (strongly developed in robust perennials). This trade-off is even more apparent in the seedlings of trees and shrubs of fertile habitats, where diversion of photosynthate into wood and other supporting tissue often dictates a slow rate of seedling growth, yet is essential to the construction of the extensive root and shoot systems that allow long-term occupation of fertile, relatively undisturbed habitats.

III. Low Productivity, Slow Growth, and Stress Tolerance

Slow growth rates characterize a wide range of unproductive habitats and have been documented in trees, shrubs, herbs, bryophytes, and lichens (Billings and Mooney, 1968; Grime and Hunt, 1975; Chapin, 1980; Furness and Grime, 1982a,b; Callaghan and Emanuelsson, 1985). In these plants, slow growth is allied to a suite of attributes, including long life histories, delayed and intermittent reproduction, long lifespans in individual leaves, and a tendency for the tissues to be strongly defended against herbivory (Grime, 1979; Coley, 1983; Coley *et al.*, 1985); these characteristics have been collectively described as stress-tolerance (Grime, 1974).

Despite many features in common, stress tolerators exhibit wide variation in size and morphology. Individual species are usually restricted to one type of unproductive ecosystem (hot, cold, dry, acidic, calcareous, saline, shaded, or metal-contaminated), and many have been shown to possess tolerance mechanisms specific to the stresses operating in their habitats. On first inspection, this diversity of morphologies and physiologies calls into question the hypothesis that these plants are the product of a similar form of natural selection. It is, however, a key assertion of the theory of stress tolerance (Chapin, 1980; Grime, 1988) that the habitats of stress tolerators share a common underlying constraint: chronic deficiencies in major mineral nutrients (in particular, phosphorus and nitrogen). The characteristics of stress tolerators strongly suggest that they have been subjected to forms of natural selection that have promoted features (slow growth, intermittent reproduction, strong antiherbivore defenses) conducive to the retention and conservative use of captured mineral nutrients.

Chronic limitations in mineral nutrient supply impose major constraints upon the capacity of plants to capture water and assimilate carbon (Chapters 3 and 4). Experiments at high and low levels of mineral nutrient supply (Donald, 1958; Mahmoud and Grime, 1976) have established the strong interdependence of competitive abilities above- and belowground, and there is strong evidence that photosynthetic competence is often controlled by nitrogen and phosphorus supplies to the leaf (Field and Mooney, 1986; Sharkey, 1985). The generality of this phenomenon is corroborated by the data in Figure 1, which shows the positive correlation between the maximum potential growth rates of a wide range of vascular plants and the average concentrations of nitrogen and phosphorus found in the leaves of these same species in their natural habitats.

An emerging theme in plant physiological ecology is the need for integration of studies on diverse aspects of physiological specialization, especially those linking root and shoot functions. However, it must be acknowledged that a rival philosophy—that of limiting factors—remains a powerful influence.

Until recently, efforts to understand the controlling influence of environment upon the productivity and species composition of vegetation have been dominated by the concept of limiting factors (Liebig, 1840; Blackman, 1905). Because all autotrophic plants use the same set of resources, it has been convenient to classify vegetation and species according to the identity of the major resource limiting production in particular habitats. The notion of limiting factors has also given rise to a distinctive theoretical approach to studying competition, expressed recently by Newman (1973, 1983) and Tilman (1982). According to these authors, plants differ in their abilities to capture particular resources, and consequently, each site is occupied by those species best equipped to compete for the currently limiting resource.

The concept of limiting factors remains useful as a means of recognizing the measures necessary to stimulate agricultural productivity at particular sites, but it is inadequate for analyzing how vegetation functions or for predicting plant responses to stress. A more realistic theoretical framework must take into account modifying factors, such as seasonal variation in resource supply (and temperature) and interruption of growth by biotic or climatic disturbance. A convenient approach here is to harness comparative plant physiology to the concept of the "Habitat Templet" (Southwood, 1977).

IV. The Habitat Templet and Primary Plant Strategies

Plant habitats differ in the length and quality of the opportunities they provide for resource capture, growth, and reproduction. According to Southwood (1977) and Grime (1977), the matrix *habitat duration × habitat productivity* generates an array of conditions and associated types of functional specialization (primary strategies) that can form the basis for a

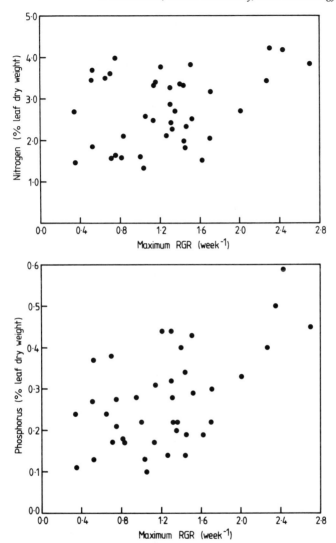

Figure 1 The relationship between maximum potential relative growth rate (RGR; data from Grime and Hunt, 1975) in the seedling phase and the average concentration of nitrogen and phosphorus in the leaf. For nitrogen, $r^2 = 0.339$; for phosphorus, $r^2 = 0.564$; in both cases d.f. = 39 and $P < 0.05$). Each determination of nutrient concentration refers to mature, nonsenescent leaves and represents the mean of populations sampled from a wide range of natural habitats distributed within a 2400-km^2 area in northern England [Data from Band and Grime (1981).]

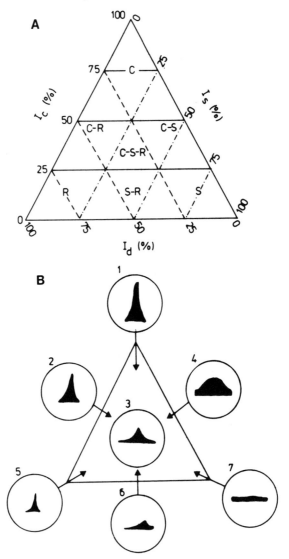

Figure 2 (A) Model describing the equilibria among competition, stress, and disturbance in vegetation and primary and secondary plant strategies. C, competitor; S, stress tolerator; R, ruderal; C–R, competitive–ruderal; S–R, stress-tolerant ruderal; C–S, stress-tolerant competitor; C–S–R, "C–S–R strategist." I_c, relative importance of competition (——); I_s, relative importance of stress (– · –); I_d, relative importance of disturbance (– – –). (B) Scheme relating strategy to pattern of seasonal change in shoot biomass. 1, competitor; 2, competitive–ruderal; 3, C–S–R strategist; 4, stress-tolerant competitor; 5, ruderal; 6, stress-tolerant ruderal; 7, stress tolerator.

universal ecological classification of plants and animals (Fig. 2). The relevance of the primary strategies to vegetation dynamics has been explored elsewhere (Grime, 1987, 1988); here we confine our attention to an aspect of strategy theory that is particularly relevant to predicting and explaining plant responses to stress. The main objective will be to use strategy theory to identify the roles of morphological plasticity and cellular acclimation, two alternative modes of stress response.

V. Morphological Plasticity and Cellular Acclimation

When plants are exposed to stress as a consequence of resource depletion or climatic fluctuation, many different responses are possible depending upon the species and the nature and severity of the stress. As Bradshaw (1965) recognized, however, stress responses can be classified into two basic types, one morphological and the other physiological. Plant strategy theory can provide a basis for predicting which of the two mechanisms is likely to operate in particular species, populations, and situations. Plant growth rates and habitat productivity are of key importance in these predictions.

To explain the linkages between strategy and stress response, we refer to Figure 2b, which depicts the patterns of seasonal change in shoot biomass associated with the full spectrum of primary strategies shown in Figure 2a. For simplicity, this diagram refers only to the patterns observed in herbaceous plants in a temperate zone situation with a sharply defined growing season, but the principles apply to any life-form or biome.

At position 1 in Figure 2b, corresponding to conditions of high productivity and low disturbance, the growing season is marked by the rapid accumulation of biomass above- and belowground. The fast-growing perennials (e.g., *Urtica dioica* and *Phalaris arundinacea*) which usually enjoy a selective advantage under these conditions, achieve their initial surge in spring growth by mobilizing reserves from storage organs; this growth surge quickly produces a dense and extensive leaf canopy and root network. As development continues, resources are rapidly intercepted by the plant and its neighbors, and local stresses arise through the development of depletion zones that expand above and below the soil surface. In this swiftly changing scenario, we predict that morphological plasticity, acting mainly but not exclusively at the time of cell differentiation, will be the dominant form of stress response. Attenuation of stems, petioles, and roots provides a mechanism of escape from the depleted sectors of the environment and, in conjunction with the short lifespans of individual leaves and roots, results in a continuous relocation of the effective leaf and root surfaces during the growing season. In theory, the high leaf and root turnover might also be expected to allow continuous modulation of leaf and root biochemistry during organ differentiation as conditions change over the growing season.

Position 5 in Figure 2b represents the brief explosive development of biomass in an ephemeral species able to exploit a productive but temporary habitat. Here again, morphological plasticity would be expected to predominate. In the vegetative phase, plasticity in root and shoot morphology will be an integral part of the mechanism of resource capture. As first elaborated by Salisbury (1942), however, there is a "reproductive imperative" in the biology of ephemerals; that is, stresses imposed by the environment or by crowding from neighbors tend to induce early flowering and to sustain reproductive allocation as a proportion of the biomass even in circumstances of extreme stunting and early mortality. Although Salisbury's observation has been confirmed by subsequent investigations (Harper and Ogden, 1970; Hickman, 1975; Boot *et al.*, 1986), more research is required to document precisely the stress cues that, in particular ephemeral species, promote early flowering. Some indication of the complications that may arise in natural habitats, where several stresses may coincide, can be gained from Figure 3, which examines flowering allocation in a population of *Poa annua* exposed to various stresses. It is evident from these data that an unpredictable situation could arise where neighbors have depleted several resources simultaneously, and conflicting signals to reproductive allocation originate from moisture stress and shading.

Under stable conditions of extremely low productivity imposed by mineral nutrient stress (Position 7 in Figure 2b), there is little seasonal change in biomass. Leaves and roots often have a functional life of several years, and resource capture is usually uncoupled from growth. Because of the slow turnover of plant parts, differentiating cells occupy a small proportion of the biomass, and morphogenetic changes are less likely to provide a viable mechanism of stress response. In these circumstances, where the same tissues experience a sequence of different climatic stresses as the seasons pass, the dominant form of stress response is likely to be cellular acclimation, whereby the functional characteristics and "hardiness" of the tissues change rapidly through biochemical adjustments of membranes and organelles. These changes are reversible and can occur extremely rapidly in certain species (Hosakawa *et al.*, 1964; Mooney and West, 1964; Strain and Chase, 1966; Bjorkman, 1968; Mooney, 1972; Oechel and Collins, 1973; Larsen and Kershaw, 1975; Taylor and Pearcy, 1976).

By confining attention to the three extreme contingencies of Figure 2b, we have illustrated the strong link between the stress responses of plants and the productivity and temporal stability of their habitats. It is important to recognize, however, that many plants exploit intermediate habitats corresponding to the central areas of the triangular model; here we may

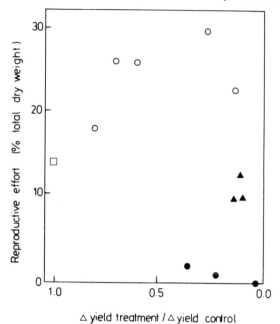

Figure 3   Comparison of the effects of various stress treatments upon reproductive effort in *Poa annua*. The intensity of each treatment is characterized by comparing the growth increment of stressed plants to that of unstressed controls over the same experimental period. □, control; O, water stress (polyethylene glycol); s, mineral nutrient stress (dilute concentrations of Rorison solution); ●, shading by neutral filters [From Smit (1980).]

expect to find morphogenetic plasticity and cellular acclimation coexisting within the same genotypes.

VI. Stress Responses and Resource Foraging

Recent studies have tested the hypothesis that plants from productive and unproductive habitats differ in morphological plasticity when exposed to resource stress. These experiments also measure to what extent the response (or lack of response) influences resource capture under stress conditions mimicking those characterizing productive and unproductive habitats. The experiments are analogous to some already applied to animals and can be legitimately described as resource foraging investigations. An account of some of these studies has been published (Grime *et al.*, 1986; Crick and Grime, 1987); here we summarize only the main methods and results.

The experiments feature partitioned containers in which each plant can be exposed to a "controlled patchiness" in its supply of light or mineral nutrients. Some treatments maintain spatially predictable patches long enough to allow morphogenetic adjustment of leaves and roots. Other treatments involve brief pulses of enrichment (24-hr mineral nutrient flushes or simulated sunflecks) localized in space, random in space and time, and too short to permit significant growth adjustments. The main conclusion drawn from these experiments is that although all the species so far examined have some capacity for concentration of leaves and roots in the resource-rich sectors of a stable patchy environment, this capacity is much more apparent in potentially fast-growing species of productive habitats. It has been clearly established in one experiment (Crick and Grime, 1987) that high rates of nitrogen capture from a stable patchy rooting environment are achieved by *Agrostis stolonifera*, a fast-growing species with a small but morphologically dynamic root system (Fig. 4).

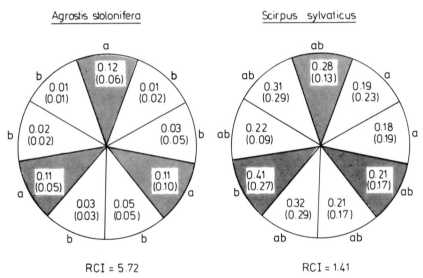

Figure 4 Comparison of the ability of two ecologically different wetland species with different potential growth rates (*Agrostis stolonifera*, rapid; *Scirpus sylvaticus*, slow) to modify allocation between different sectors of the root system when exposed to a nutritionally patchy rooting medium. Each plant was grown in uniform nutrient-sufficient solution culture for ten weeks, then exposed to patchy conditions by distributing the root system equally between nine radial compartments, three of which (▲) contained full-strength Rorison nutrient solution. The remaining compartments (△) were provided with 1/100 Rorison solution. The value in each compartment is the mean increment of root dry matter to the compartment over 27 days (95% confidence value in parentheses). In each species, compartments differing significantly ($P < 0.05$) in root increment have no letters in common. Root concentration index (RCI) was calculated by dividing the mean root increment in compartments receiving full-strength solution by the mean increment in compartments containing 1/100 solution [Redrawn from Crick and Grime (1987).]

Under the same conditions, nitrogen capture is much inferior in *Scirpus sylvaticus,* a slow-growing species with a massive root system but a relatively low nitrogen-specific absorption rate and slow root adjustment to mineral nutrient patchiness. These results are consistent with the hypothesis that rapid morphogenetic changes ("active foraging") are an integral part of the mechanism whereby fast-growing plants sustain high rates of resource capture in the rapidly changing resource mosaics created by actively competing plants. The experiment has also yielded evidence that the larger, evenly distributed root system of *S. sylvaticus* enjoys an advantage when nutrients are supplied in brief localized pulses; this accords with previous studies suggesting that the major influxes of mineral nutrients on infertile soils come from brief pulses (Gupta and Rorison, 1975).

VII. Stored Growth

So far we have heavily emphasized plant strategy theory and resource capture under stress. The purpose of this last section is to examine another dimension of plant attunement to stress by natural selection. This occurs in habitats or niches where access to resources depends upon rapid growth under climatic stress. Examples of this phenomenon are particularly obvious on shallow soils in continental climates, where the growth window between winter cold and summer desiccation may be extremely short. In deciduous woodlands in the cool temperate zone, an essentially similar niche arises between the time that the snow melts and the tree canopy closes. Both circumstances provide opportunities for high rates of photosynthesis and mineral nutrient capture in the late spring but depend upon rapid expansion of roots and shoots while it is still cold, in later winter and early spring.

Vernal geophytes provide a well-known solution to this ecological challenge, and controlled-environment studies (e.g., Hartsema, 1961) have documented the crucial requirement for warm temperatures for the preformation of leaves and flowers during the "inactive" summer period. Grime and Mowforth (1982) suggested that the essential feature of the vernal geophyte life cycle was the extent to which cell division took place during the summer developmental phase, thus circumventing the potentially limiting effect upon mitosis of low spring temperatures. According to this hypothesis, the rapid growth of the geophytes at low temperatures is mainly due to cell expansion, a process less inhibited than mitosis by low temperatures.

The geophyte growth strategy is thus associated with temporal separation of the main phases of cell division and cell expansion. It is also correlated with possession of unusually large cells (Grime and Mowforth, 1982;

Grime, 1983). This suggests that another important feature of vernal geophyte phenology is the construction of tissues that, as unexpanded cells in the summer, occupy a small volume in the densely packed interior of the bulb but have high expansion coefficients as they become vacuolated in the spring.

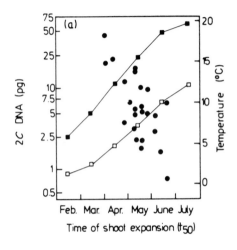

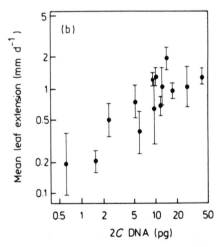

Figure 5   (a) The relationship between quantity of nuclear DNA and the time of shoot expansion in 24 plant species commonly found in the Sheffield region. Temperature is expressed as the long-term average for each month of daily minima (□) and maxima (■) in air temperature 1.5 m aboveground. [Redrawn from Grime and Mowforth (1982).] (b) The relationship between quantity of nuclear DNA and the mean rate of leaf extension from 25 March to 5 April in 14 grassland species coexisting in the same turf; vertical lines represent 95% confidence limits. [Redrawn from Grime *et al.* (1985).]

Spring growth at low temperatures is not confined to vernal geophytes, however, and it is likely that other growth forms exhibit similar mechanisms. One source of evidence here relies upon surveys of the quantity of DNA in the nucleus of various native plant cells (Bennett and Smith, 1976; Grime and Mowforth, 1982). The quantity of nuclear DNA is strongly correlated with cell size and provides a convenient identifier of large-celled

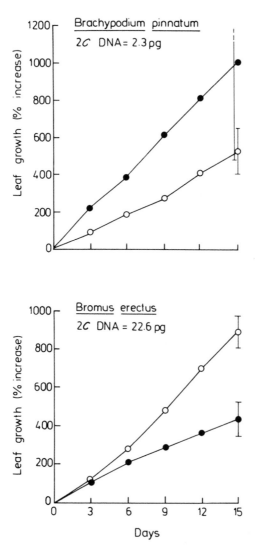

Figure 6 Comparison of leaf extension rates in two grasses with different cell size and quantity of nuclear DNA; vertical lines indicate 95% confidence limits. ●, watered before measurement; ○, water withheld for three weeks before measurement.

species (Olmo, 1983; Grime, 1983). This has prompted field investigations examining the relationship between quantity of nuclear DNA in species and their rates of growth in the spring. The results of these studies reveal that quantity of DNA is useful both as a predictor of phenology (Fig. 5a) and as an index of temporal niche differentiation within a plant community (Figure 5b).

Temporal separation of cell division and cell expansion also characterizes the response to moisture stress of some grasses and cereals, where the phenomenon has been described as "stored growth" (Salter and Goode, 1967). Recent evidence suggests that, as in the case of the geophytes, this response depends strongly upon possession of large cells with large nuclei. Figure 6 compares the potential for stored growth in two grasses that differ in quantity of nuclear DNA by measuring their rates of leaf extension after standardized exposure to experimental drought. No capacity for stored growth is apparent in *Brachypodium pinnatum,* a species with relatively little nuclear DNA and restricted in its distribution to sites with a large reservoir of subsoil moisture. In marked contrast, a sustained surge in leaf extension characterizes the resumption of growth after drought in *Bromus erectus,* a species with a relatively large quantity of nuclear DNA and distinct association with shallow soils and summer desiccation.

VIII. Conclusions

Stress sensitivity and response are related to the form, duration, intensity, and spatial/temporal distribution of the stresses experienced, but they also depend heavily upon the ecological and evolutionary history of the species or population. Stress physiology will not be well served by models that reduce botanical diversity to that familiar panmictic species, "the plant." Rather, a coherent theoretical framework for predicting and interpretating stress responses should be based upon a functional classification of plants. The concept of primary plant strategies and the study of specialization associated with cell and genome enlargement are contributions toward this typology, which is needed not only to elucidate patterns of existing adaptation to stress but to forecast the rate and direction of contemporary evolutionary responses to stresses of recent origin.

Acknowledgments

This paper has been prepared with the skilled assistance of J. M. L. Mackey, S. R. Band, and N. Ruttle. It is based upon research supported in the U.K. by the Natural Environment Research Council.

References

Al-Mufti, M. M., Sydes, C. L., Furness, S. B., Grime, J. P., and Band, S. R. (1977). A quantitative analysis of shoot phenology and dominance in herbaceous vegetation. *J. Ecol.* 65: 759–791.

Band, S. R., and Grime, J. P. (1981). Chemical composition of leaves. *In* "Annual Report 1981," pp. 6–8. Unit of Comparative Plant Ecology (NERC), University of Sheffield.

Beadle, N. C. W. (1954). Soil phosphate and the delimitation of plant communities in eastern Australia I. *Ecology* 35: 370–375.

Beadle, N. C. W. (1962). Soil phosphate and the delimitation of plant communities in eastern Australia II. *Ecology* 43: 281–288.

Bennett, M. D., and Smith, J. B. (1976). Nuclear DNA amounts in angiosperms. *Philos. Trans. R. Soc. Lond. B Biol. Sci.* 274: 227–274.

Billings, W. D., and Mooney, H. A. (1968). The ecology of arctic and alpine plants. *Biol. Rev.* 3: 277–289.

Bjorkman, O. (1968). Carboxydismutase activity in shade-adapted and sun-adapted species of higher plants. *Physiol. Plant.* 21: 1–10.

Blackman, F. F. (1905). Optima and lighting factors. *Ann. Bot.* 14: 281–295.

Boot, R., Raynal, D. J., and Grime, J. P. (1986). A comparative study of the influence of drought stress on flowering in *Urtica dioica* and *U. urens. J. Ecol.* 74: 485–495.

Bradshaw, A. D. (1965). Evolutionary significance of phenotypic plasticity in plants. *Adv. Genet.* 13: 115–155.

Bradshaw, A. D., Chadwick, M. J., Jowett, D., and Snaydon, R. W. (1964). Experimental investigations into the mineral nutrition of several grass species. IV. Nitrogen level. *J. Ecol.* 52: 665–676.

Callaghan, T. V., and Emanuelsson, U. (1985). Population structure and processes of tundra plants and vegetation. *In* "The Population Structure of Vegetation" (J. White, ed.), pp. 399–439. Dr. W. Junk, Dordrecht.

Chapin, F. S., III. (1980). The mineral nutrition of wild plants. *Ann. Rev. Ecol. Syst.* 11: 233–260.

Clarkson, D. T. (1967). Phosphorus supply and growth rate in species of *Agrostis* L. *J. Ecol.* 55: 111–118.

Coley, P. D. (1983). Herbivory and defensive characteristics of tree species in a lowland tropical forest. *Ecol. Monogr.* 53: 209–232.

Coley, P. D., Bryant, J. P., and Chapin, F. S., III. (1985). Resource availability and plant antiherbivore defence. *Science* 230: 895–899.

Crick, J. C., and Grime, J. P. (1987). Morphological plasticity and mineral nutrient capture in two herbaceous species of contrasted ecology. *New Phytol.* 107: 403–414.

Donald, C. M. (1958). The interaction of competition for light and for nutrients. *Aust. J. Agric. Res.* 9: 421–432.

Field, C., and Mooney, H. A. (1986). The photosynthesis–nitrogen relationship in wild plants. *In* "On the Economy of Plant Form and Function" (T. J. Givnish, ed.), pp. 25–55. Cambridge Univ. Press, Cambridge.

Furness, S. B., and Grime, J. P. (1982a). Growth rate and temperature responses in bryophytes. I. An investigation of *Brachythecium rutabulum. J. Ecol.* 70: 513–523.

Furness, S. B., and Grime, J. P. (1982b). Growth rate and temperature responses in bryophytes. II. A comparative study of species of contrasted ecology. *J. Ecol.* 70: 525–536.

Grime, J. P. (1974). Vegetation classification by reference to strategies. *Nature* 250: 26–31.

Grime, J. P. (1977). Evidence for the existence of three primary strategies in plants and its relevance to ecological and evolutionary theory. *Am. Nat.* 111: 1169–1194.

Grime, J. P. (1979). "Plant Strategies and Vegetation Processes." Wiley, Chichester.

158   *J. P. Grime and B. D. Campbell*

Grime, J. P. (1983). Prediction of weed and crop response to climate based upon measurements of nuclear DNA content. *Aspects Appl. Biol.* 4: 87–98.

Grime, J. P. (1987). Dominant and subordinate components of plant communities—implications for succession, stability and diversity. *In* "Colonisation, Succession, and Stability" (A. Gray, P. Edwards, and M. Crawley, eds.), pp. 413–428. Blackwell, Oxford.

Grime, J. P. (1988). The C–S–R model of primary plant strategies—origins, implications, and tests. *In* "Plant Evolutionary Biology" (L. D. Gottlieb and K. S. Jain, eds.), pp. 371–393. Chapman and Hall, London.

Grime, J. P., and Hunt, R. (1975). Relative growth rate: Its range and adaptive significance in a local flora. *J. Ecol.* 63: 393–422.

Grime, J. P., and Mowforth, M. A. (1982). Variation in genome size—an ecological interpretation. *Nature* 299: 151–153.

Grime, J. P., Shacklock, J. M. L., and Band, S. R. (1985). Nuclear DNA contents, shoot phenology, and species coexistence in a limestone grassland community. *New Phytol.* 100: 435–445.

Grime, J. P., Crick, J. C., and Rincon, E. (1986). The ecological significance of plasticity. *In* "Plasticity in Plants" (D. H. Jennings and A. J. Trewavas, eds.), pp. 5–19. Company of Biologists, Cambridge.

Gupta, P. L., and Rorison, I. H. (1975). Seasonal differences in the availability of nutrients down a podzolic profile. *J. Ecol.* 63: 521–534.

Hackett, C. (1965). Ecological aspects of the nutrition of *Deschampsia flexuosa* (L.) Trin. II. The effects of Al, Ca, Fe, K, Mn, N, P, and pH on the growth of seedlings and established plants. *J. Ecol.* 53: 315–333.

Harper, J. L., and Ogden, J. (1970). The reproductive strategy of higher plants. I. The concept of strategy with special reference to *Senecio vulgaris* L. *J. Ecol.* 58: 681–698.

Hartsema, A. M. (1961). Influence of temperature on flower formation and flowering of bulbous and tuberous plants. *Handb. Pflanzen-physiol.* 16: 123–167.

Hickman, J. C. (1975). Environmental unpredictability and plastic energy allocation strategies in the annual *Polygonum cascadense* (Polygonaceae). *J. Ecol.* 63: 689–701.

Higgs, D. E. B., and James, D. B. (1969). Comparative studies in the biology of upland grasses. I. Rate of dry-matter production and its control in four grass species. *J. Ecol.* 57: 553–563.

Hosakawa, T., Odani, H., and Tagawa, H. (1964). Causality of the distribution of corticolous species in forests with special reference to the physiological approach. *Bryologist* 67: 396–411.

Hunt, R., and Lloyd, P. S. (1987). Growth and partitioning. *New Phytol.* 103 (Suppl.): 235–250.

Jowett, D. (1964). Population studies on lead-tolerant *Agrostis tenuis*. *Evolution* 18: 70–80.

Kruckeberg, A. R. (1954). The ecology of serpentine soils. III. Plant species in relation to serpentine soils. *Ecology* 35: 267–274.

Larsen, D. W., and Kershaw, K. A. (1975). Acclimation in arctic lichens. *Nature* 254: 421–423.

Liebig, J. (1840). "Chemistry and Its Application to Agriculture and Physiology." Taylor and Walton, London.

Mahmoud, A., and Grime, J. P. (1976). An analysis of competitive ability in three perennial grasses. *New Phytol.* 77: 431–435.

Mooney, H. A. (1972). The carbon balance of plants. *Ann. Rev. Ecol. Syst.* 3: 315–346.

Mooney, H. A., and West, M. (1964). Photosynthetic acclimation of plants of diverse origin. *Am. J. Bot.* 51: 825–827.

Newman, E. I. (1973). Competition and diversity in herbaceous vegetation. *Nature* 244: 310.

Newman, E. I. (1983). Interactions between plants. *In* "Encyclopedia of Plant Physiology, New Series" (O. L. Lange, P. S. Nobel, C. B. Osmond, and H. Ziegler, eds.), Vol. 12C. Springer-Verlag, Berlin.

Oechel, W. D., and Collins, N. J. (1973). Seasonal patterns of CO_2 exchange in bryophytes at Barrow, Alaska. *In* "Primary Production and Production Processes" (L. C. Bliss and F. E. Wielogolaski, eds.) Wenner-Gren Center, Stockholm.

Olmo, E. (1983). Nucleotype and cell size in vertebrates: A review. *Basic Appl. Histochem.* 27: 227–256.

Parsons, R. F. (1968a). The significance of growth-rate comparisons for plant ecology. *Am. Nat.* 102: 595–597.

Parsons, R. F. (1968b). Ecological aspects of the growth and mineral nutrition of three mallee species of Eucalyptus. *Oecol. Plant.* 3: 121–136.

Salisbury, E. J. (1942). "The Reproductive Capacity of Plants." Bell, London.

Salter, P. J., and Goode, J. E. (1967). Crop responses to water at different stages of growth. *Commonw. Agric. Bur. Res. Rev.* 2.

Sharkey, T. D. (1985). Photosynthesis in intact leaves of C_3 plants: Physics, physiology, and rate limitations. *Bot. Rev.* 51: 53.

Smit, P. T. (1980). Phenotypic plasticity of four grass species under water, light, and nutrient stress. BSc. thesis, University of Utrecht.

Southwood, T. R. E. (1977). Habitat, the templet for ecological strategies? *J. Anim. Ecol.* 46: 337–365.

Strain, B. R., and Chase, V. C. (1966). Effect of past and prevailing temperatures on the carbon dioxide exchange capacities of some woody desert perennials. *Ecology* 47: 1043–1045.

Taylor, R. J., and Pearcy, R. W. (1976). Seasonal patterns in the CO_2 exchange characteristics of understorey plants from a deciduous forest. *Can. J. Bot.* 54: 1094–1103.

Tilman, D. (1982). "Resource Competition and Community Structure." Princeton Univ. Press, Princeton, New Jersey.

8

Stress Effects on Plant Reproduction

N. R. Chiariello   **S. L. Gulmon**

I. Introduction

All of the environmental factors affecting a plant's carbon fixation and nutrient uptake also affect plant reproduction. Reproductive organs are constructed from resources either recently acquired or previously stored by the vegetative parts. Therefore, any environmental stress that affects vegetative parts ultimately affects reproductive yield as well. In some species, reproductive structures themselves exhibit significant photosynthetic activity. Stresses that reduce carbon fixation may directly alter the reproductive yield of these species. Despite this coupling between vegetative and reproductive growth, reproduction is often more sensitive to stress than is vegetative growth. This sensitivity arises from the precarious choreography of reproductive development and from the frequent reliance on external pollen vectors. From the time of meristem differentiation to seed filling, even moderate stresses may terminate reproductive development by disabling pollen vectors or by causing sterility, asynchrony in pollen and ovule production, or abortion of reproductive structures. During the vegetative phase, the same stresses are more likely to slow or arrest growth, rather than terminate it completely.

This chapter emphasizes these two aspects of the relationship between reproductive yield and vegetative processes: (1) the coupling between reproduction and growth that is inherent in a plant's modular, open growth form and (2) the distinctive sensitivity of reproduction to stress. Where possible, we have included both wild plants and crops in our discussion to provide a broad picture of the response of reproduction to stress. Despite their similarity at the level of the individual plant, differences between wild and cultivated plants in genetic variability and in habitat variability within populations result in distinct patterns of yield compensation in response to stress.

II. Windows of Sensitivity

Many developmental stages contribute to a plant's production of seeds. These include the differentiation of meristems as reproductive rather than vegetative; the development of an inflorescence, flowers, ovaries, and ovules within each ovary; sporogenesis; gametogenesis; fertilization; and the filling and maturation of fertilized seeds. These stages of development are windows of varying sensitivity to environmental stress and can be viewed as multiple control points for yield determination (Fig. 1). Each controls a component of yield. For example, the level of stress at the time of meristem differentiation determines the number of inflorescences per plant and flowers per inflorescence. Stress during seed fill affects individual seed mass and the fraction of seeds aborted. Because reproductive

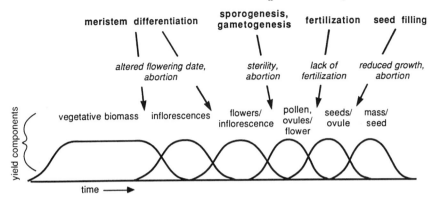

Figure 1 Yield components in relation to critical developmental–physiological processes that are sensitive to stress. Yield components are represented schematically along a time axis showing duration, intensity, and overlap of yield determination stages, which are functions of growth form and inflorescence type. Arrows indicate control points for developmental processes (indicated in bold); stress effects (italics) moderate these control points. [Adapted from Adams (1967).]

development is a linear progression of stages, stress at any point can affect processes occurring later. As we discuss in section VII,C, however, it is also possible for some control points to compensate for earlier stresses.

A. Water Stress

Water stress can influence every stage of reproduction, beginning with the switch to flowering. Water stress during vegetative growth may render plants insensitive to cues for meristem differentiation, such as inductive photoperiods. In photoperiodically controlled, single-cycle species (e.g., *Xanthium strumarium, Lolium temulentum,* and *Pharbitis nil*), flowering may be suppressed or eliminated if the single inductive photoperiod coincides with osmotic stress (Aspinall and Hussain, 1970). This effect may be mediated by stress effects on translocation from leaves to stem apices.

Other species have quite different responses of flowering time to water stress. Flowering in sunflower cultivars is a function of plant age and responds little to water stress, whether brief (Yegappan *et al.,* 1980) or prolonged (Marc and Palmer, 1976). In the tropical pasture legume *Macroptilium atropurpureum* cv. Siratro, moderate water stress (followed by rewatering) stimulates flowering relative to unstressed controls (Kowithayakorn and Humphreys, 1987). This range of stress effects on flowering time illustrates that "stress" must be defined relative to optimal conditions that are unique to a species or genotype.

Once the switch to reproduction is made, water stress can disrupt many of the phenological events necessary for maximum seed yield. Dioecious and monoecious species appear to be particularly vulnerable. In maize, water stress beginning at anthesis may delay silking relative to pollen release, disrupting the synchrony of events necessary for pollination (Herrero and Johnson, 1981). This may be a major cause of yield reductions in the field (Du Plessis and Dijkhuis, 1967). Water stress can also cause abnormal development in maize. Drought stress during embryo sac formation may cause female sterility through embryo sac abortion (Moss and Downey, 1971). Male sterility may result when drought stress occurs during male meiosis in the tassel (Downey and Miller, 1971). To various degrees, other grain crops share these periods of sensitivity (Begg and Turner, 1976; Boyer and McPerson, 1975). Sporogenesis is a particularly vulnerable stage for barley (Aspinall *et al.*, 1964) and other grains (Evans and Wardlaw, 1976).

B. Atmospheric Pollutants and Acid Precipitation

Developing reproductive structures may also be sensitive to atmospheric pollutants and acid precipitation, which can alter stigma chemistry and pollen germinability or viability. In vitro studies have found that SO_2 reduced pollen germination, pollen tube growth, or both in several tree species (Karnosky and Stairs, 1974) and a variety of herbaceous plants (Ma and Khan, 1976; Masaru *et al.*, 1976; Varshney and Varshney, 1981). Chromosome aberrations have been associated with SO_2 exposure in pollen-tube cultures of *Tradescantia* (Ma *et al.*, 1973).

Certain breeding strategies of plants make reproductive structures particularly vulnerable. Outcrossing species that rely on wind or insects to carry pollen typically have tall or prominent inflorescences. Their reproductive tissues are higher in the wind profile than their leaves and may experience greater exposure to airborne pollutants. Moreover, these tissues generally lack the degree of control over exposure afforded by the leaf cuticle and stomata. In wind-pollinated plants, reproductive tissues may receive especially high exposure because their breeding system tends to minimize the boundary layer of still air surrounding these structures. Pollen germination in red and white pine may be reduced by ambient pollution levels below the threshold for apparent leaf injury (Houston and Dochinger, 1977). In corn, pollen germination is lower on silks exposed to simulated rain of pH 4.6, 3.6, and 2.6 relative to silks exposed to simulated rain of pH 5.6 (Wertheim and Craker, 1988).

The sensitivity of reproductive tissues to air pollution may make the timing of exposure important in determining effects on seed yield. Yield of wheat is more sensitive to SO_2 exposure during flowering than other stages (Godzik and Krupa, 1982). Fluorides affect the growth of the strawberry receptacle, but only if exposure occurs during the few days between anthesis and fertilization and prevents the development of normal, fertile achenes (Bonte, 1982).

C. Low-Temperature Stress

Low-temperature stress can also cause massive crop losses because of the sensitivity of reproductive structures. Temperate deciduous trees are particularly at risk during spring deacclimation from their dormant winter state. Orchard yields throughout temperate regions are vulnerable to late winter frosts that can decimate developing flowers yet spare the woody plant. Citrus crops are notoriously sensitive to winter injury; extensive losses have been caused by single frosts.

The well-studied genus *Prunus* provides a good example of phenological windows of sensitivity and differential cold-hardiness among plant tissues. Before petal tips emerge from the calyx, freezing injury is prevented first by supercooling, and then by extracellular freezing. Overwintering flower buds supercool during low-temperature stress, enabling them to survive temperatures as low as $-23°$ C (Andrews *et al.*, 1983; Durner and Gianfagna, 1988) or $-27°$ C (Quamme, 1978). Supercooling involves water movement from the flower primordium to the bud scales and bud axis (Quamme and Gusta, 1987), which freeze near $0°$ C. This mechanism protects flowers until early bud swell, when the ice nucleation temperature rises in buds. Frost tolerance is lost when petal tips emerge, leaving the flower and fruit population at risk to late winter frosts (Andrews *et al.*, 1983).

Herbaceous crops tend to be more sensitive to cold than woody plants. Tomatoes *(Lycopersicon esculentum)* are very sensitive to low temperatures, especially during sporogenesis and gametogenesis. Repeated night temperatures below $10°$ C result in infertile pollen grains, particularly if the exposure occurs 11 and 6 days before anthesis (Patterson *et al.*, 1987).

We have chosen these examples to show that pronounced (but not unexpected) stresses may damage a fruit or seed crop because of periods of particular sensitivity during reproductive development. Of course, not all species show greater sensitivity of reproductive structures to stress. In some cases, the reproductive structures are more resilient. Freezing sufficient to damage 80% of the leaves of soybean plants during various stages of pod fill may spare the pods almost completely (Saliba *et al.*, 1982). Also, the relative sensitivity of reproductive and vegetative structures may change with time. For example, in July, before kernel development in pecans, water stress is more likely to cause fruit abortion than leaf abscission, but in September, during rapid kernel development, water stress is more likely to cause leaf abscission (Sparks, 1989).

III. Resource Limitation of Reproduction

When stress is less severe or occurs outside periods of developmental sensitivity, the stress is generally mediated through resources and affects plant growth. The stress may itself be a limited resource (e.g., low nitrogen) or some factor that reduces resource availability (low-temperature limitation of nitrogen uptake). Competition reduces resource availability, and herbivory removes resources from plants. Because all of these stresses are mediated through resources, we view resource availability as the most general limitation to reproduction of wild plants.

The modular, usually indeterminate growth form of plants results in strong coupling between vegetative and reproductive growth. The implication of this coupling for stress responses is inherent in even the simplest models of plant growth which consider the whole plant a single module (Fig. 2). During vegetative growth, biomass accumulation is roughly exponential. At the switch to reproduction, the vegetative portion remains fixed, and new biomass accumulates in reproductive structures. This gives rise to a phase of linear reproductive growth at a rate proportional to two terms: the vegetative biomass (which is determined by the degree of resource limitation during the vegetative phase) and resource availability

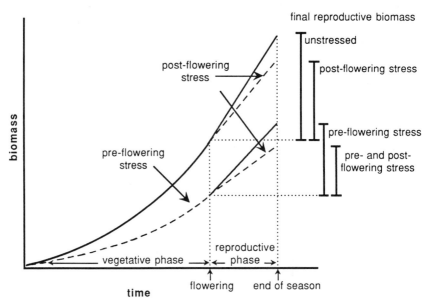

Figure 2 Simulated biomass growth of a plant that is unstressed (solid line) or stressed (dashed line) during the vegetative and/or flowering stage. Stress is assumed to be mediated by resources and to affect the unit rate of biomass increase. During exponential vegetative growth, vegetative biomass (V) follows the equation $V = se^{rt}$, where s is the seed weight at germination, r is the rate of biomass growth per unit vegetative biomass and t is the time since germination. At flowering, all new biomass is put into reproduction, resulting in linear growth. Reproductive biomass (R) follows the equation $R = V_f r t_f$, where V_f is the vegetative biomass at flowering, and t_f is time past flowering.

during the reproductive phase. According to this model, the final reproductive biomass is proportional to these two terms, plus a third: the duration of the reproductive phase. These terms may be considered the physiological components of reproductive yield, as opposed to the developmental components determined through meristem differentiation, gametogenesis, and fertilization (Fig. 1).

Declines in absolute reproductive yield under stress may be accompanied and/or caused by changes in the relative amount of biomass in reproductive structures, or reproductive partitioning. In wild plants, reproductive effort (the amount of final reproductive biomass relative to vegetative or total biomass) has been viewed as one aspect of a plant's strategy for responding to its environment. For seed crops, harvest index (the ratio of economic yield to aboveground biomass at harvest) excludes roots but is otherwise comparable to the ecological concept of reproductive effort. The components of yield that determine absolute reproductive yield also determine reproductive partitioning, an aspect of whole-plant responses to stress that we discuss in the following two sections.

IV. Tolerance and Avoidance of Environmental Stress

If a stress is predictable over the lifespan of individuals, then plant responses may become genetically fixed and appear as ecotypic or clinal variation across environments. We will not attempt to review all the possibilities but present a few examples where differentiation in physiology, morphology, or life history occurs along a stress gradient.

The concepts of avoidance and tolerance provide a useful dichotomy for evaluating reproductive responses to stress. Tolerance refers to adaptations that increase reproductive output under stress conditions. *Prunus* and *Lycopersicon* again provide examples. Extracellular freezing protects tissues from freezing damage at temperatures below the limits of supercooling and generally replaces supercooling in *Prunus* species of the northern boreal forests (Burke and Stushnoff, 1979). In contrast to the domestic tomato, *Lycopersicon esculentum*, which is very cold sensitive, *Lycopersicon hirsutum*, a wild tomato from high altitudes, produces normal pollen at temperatures below 10° C (Patterson *et al.*, 1987). Tolerance may include morphological adaptations, such as floral pubescence in high-altitude species, which conserves heat in developing reproductive organs (Miller, 1986). Similarly, the drought-sensitive pollen of *Cucurbita pepo* L. is sheltered from dehydration injury by the humid microclimate created by the corolla (Gay *et al.*, 1987). Frequently, multiple mechanisms are involved in tolerance. *Poa alpina* populations from different altitudes vary in temperature responses of days to flowering, number of panicles, florets per panicle, and seeds per panicle, which all act to maximize potential reproduction at the altitude of origin (Hermesh and Acharya, 1987).

Because all plants require the same resources for growth and reproduction, it is possible to compare habitats in terms of the degree to which the physical environment limits plant growth and ask whether plants of extreme environments exhibit characteristic strategies. Grime (1979) proposed that in unfavorable environments with a low frequency of disturbance, a "stress-tolerator" strategy predominates, typified by perennials with short stature and high partitioning to belowground structures. In contrast, annuals and high reproductive partitioning characterize favorable environments subject to frequent disturbance.

While general trends in adaptation may be recognized this way, divergent strategies coexist in many environments, such as ephemeral annuals and long-lived perennials in warm deserts. From this, it would appear that the degree of stress in environments must be compared from the perspective of specific taxa or life-history groups. When this narrower focus is taken, it appears that much of the between-habitat variation in reproductive partitioning is plastic. Detecting small genotypic differences in partitioning is complicated by the difficulties of quantifying all partitioning to reproduction, such as pollen, nectar, and other investments. This makes it difficult to test general predictions about strategies of stress responses.

A clearer pattern can be seen in species differentiated into life-history ecotypes along a stress axis. Some monocarpic species, such as *Verbascum thapsus* (Reinartz, 1984), *Capsella bursa-pastoris* (Neuffer and Hurka, 1986), and *Cynoglossum officinale* (DeJong *et al.*, 1986), have populations that are biennial or triennial in areas where the growing season is short or resources are scarce but have annual populations in more favorable environments.

As we have noted for deserts, however, many species survive in unfavorable environments through avoidance of stress, i.e., adjusting sensitive reproductive stages to periods of reduced stress. For example, Mediterranean-climate vegetation shows phenological shifts as a function of altitude (Reader, 1984). At high altitude, flowering is constrained to the warmest part of the summer, probably in part to maximize pollinator activity and because water stress is ameliorated by cooler temperatures. At low elevation, with temperature less limiting, flowering occurs throughout the year but peaks in late spring before summer drought begins. Avoidance may also take the form of early spring flowering by ephemerals in deciduous forest understory; the extreme of this strategy is production of preformed floral buds during the previous season (Grandtner and Gervais, 1985). On a population scale, *Impatiens* exhibits significant interpopulation variation in both timing and duration of flowering, and this is related to duration of the growing season in different habitats (Simpson *et al.*, 1985).

The dichotomy between tolerance and avoidance is not strict. The strategy of avoidance along one environmental axis, such as temperature, light, or water supply, may often entail increased tolerance along another, covarying axis. Moreover, increased tolerance generally results in a trade-off where reduced stress at one axis position results in increased stress at the old position. This trade-off appears even in species (e.g., *Silene*) adapted to pollutants such as SO_2. Plants from both a rural, unpolluted habitat and a polluted habitat produce significantly smaller seeds when grown in the reciprocal environment (Ernst *et al.*, 1985). Again, this observation illustrates that stress is always relative to a set of optimal conditions unique to a genotype.

V. Plasticity of Relative and Total Yield

Because much of the interpopulation variation in reproductive partitioning appears to be plastic, it is useful to examine trends in plastic responses and their implications for reproductive yield under stress. In many species, the magnitude of the partitioning response suggests it is a major determinant of yield.

A. Plastic Responses in Wild Plants

In species capable of vegetative reproduction, one aspect of stress response is the balance between sexual and vegetative reproduction. In favorable sites, it appears that increased plant density is associated with greater partitioning to sexual reproduction at the expense of vegetative proliferation (e.g., Abrahamson, 1975; Douglas, 1981; Ohlson, 1986). Since increased density results in fewer resources available per plant, the shift toward greater sexual reproduction may be considered a response to increased stress. In an evolutionary context, however, this response may not serve to ameliorate or compensate for stress but rather to favor seed dispersal to a new site.

Like crowding, the stress of a shorter growing season appears to result in a plastic response that increases partitioning to sexual reproduction. This plastic response occurs even in species that show an evolutionary trend toward greater vegetative reproduction with decreasing season length. For example, in species spanning a range of altitudes or latitudes, populations at more extreme latitudes and higher altitudes may have a genetic predisposition toward vegetative reproduction (Harris, 1970; McNaughton, 1966; Mooney and Billings, 1961; but see also Douglas, 1981, for an exception). In the first two studies, however, increased altitude or cooler temperatures resulted in plastic responses of reduced proportional vegetative reproduction in a population, regardless of origin.

Studies on nonclonal species have examined reproductive partitioning in a wide variety of perennials and annuals. These studies suggest that in polycarpic perennials, and some annuals, reproductive partitioning generally decreases with resource limitation (Fig. 3). This applies to all major resources: light, water, minerals, or a combination, as, for example, in a successional sequence (Andel and Vera, 1977; Cartica and Quinn, 1982; Foulds, 1978; Haase, 1986; Lee and Bazzaz, 1980, 1982; Roos and Quinn, 1977; Willson and Price, 1980; Wyatt, 1981). Annuals and perennials that tend to act as annuals under stress may show different trends. In *Senecio sylvaticus* (Andel and Vera, 1977) and *Senecio vulgaris* (Fenner, 1986), both short-lived annuals of transient habitats, reproductive output was independent of mineral nutrition. Hickman (1975) found that reproductive output increased with water stress in *Polygonum cascadense,* a species characteristic of depauperate annual vegetation in the Cascade Mountains. Severe drought also resulted in a large increase in proportional reproductive output in *Trifolium repens,* a perennial normally found in moist habitats (Foulds, 1978).

From these examples, it appears that stress tends to decrease partitioning to reproductive output except in situations where survival probability is low and the options for reproduction are now or never. Polycarps and monocarps thus tend to have different responses. But there are numerous crossovers and exceptions, such as facultative perennials that act as annuals

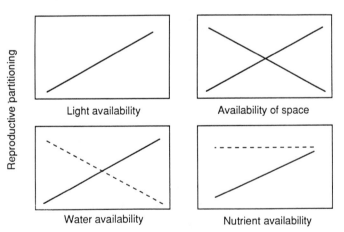

Figure 3 Plastic response of reproductive partitioning, or the fractional biomass in reproduction, to availability of resources aboveground (light), belowground (water and nutrients), or both (space, i.e., area per plant). Trends are based on a variety of studies of wild plants (see text). Solid lines reflect trends observed for most perennials and some annuals. Dashed lines reflect some annuals and perennials that act like annuals under stress. A positive relationship between reproductive partitioning and light (photon flux density) appears to hold for all cases studied, whereas species differ in response to belowground resources.

under stress. Coupled shifts in life history and biomass partitioning appear to be common in deserts with highly unpredictable precipitation (Beatley, 1970; Royce and Cunningham, 1982).

B. Constraints on Optimal Partitioning Responses

Given the variety of partitioning responses to stress reported for both annuals and perennials, it is useful to consider how relative reproductive yield relates to absolute yield. For annuals, we can return to the simple growth model of Figure 2 and ask how the partitioning pattern should change to maximize seed set under stress. If we assume that the plant is governed by the physiological components of yield identified in Figure 2 (Section III) and that flowering date is completely plastic, this model predicts that an annual's reproductive effort should remain constant in response to stress. In the simplest case, the reproductive effort should be 0.5 for maximizing yield (Fig. 4). High-yielding cultivars of seed crops often have a harvest index approximating this value; many wild annuals do also.

To maintain an optimal reproductive effort (for maximum seed set), a plant should respond to stress by flowering slightly early. This applies to the stress of chronically reduced resource availability or shortened growing season. Although all plants are far more complex than this model, it illustrates an important point. Whether stressed or not, an annual's seed yield is maximized at a reproductive effort near 0.5. To the extent that real

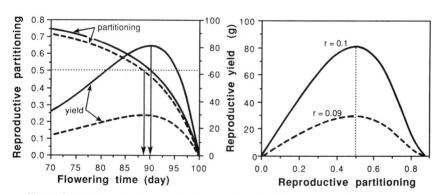

Figure 4 Relationship between absolute and fractional reproductive yield at the season's end based on the simulations in Figure 2 and with $r = 0.1$ (unstressed, solid line) or $r = 0.09$ (stressed, dashed line). Variation in final reproductive yield and partitioning results from changes in flowering date within a season fixed at 100 days. Left: The optimal flowering date for maximum seed set is indicated for each plant by a vertical arrow. The stressed plant has maximum yield when it flowers slightly earlier than the unstressed. Maximum yield coincides with reproductive partitioning (reproductive biomass/total biomass) of 0.5. Right: Sensitivity of final reproductive yield to partitioning when it varies over the entire range possible, resulting from flowering at all possible times in the growing season.

plants differ from the model's assumptions (e.g., the complete switch in allocation at flowering), the optimal reproductive effort for a species may depart from 0.5 but should stay constant in response to resource stress. Increased reproductive partitioning in response to stress is not, in itself, a compensatory mechanism because it may result from reducing vegetative biomass below that necessary for producing maximum reproductive biomass. Studies of stress effects on partitioning should therefore consider both the relative and absolute reproductive biomass.

Plant growth form, phenological controls, and taxonomic relationships may be interacting constraints on biomass partitioning and its response to stress. Their effects are evident in the size dependence of partitioning in many species. At maturity, vegetative and reproductive biomass in annuals are usually linearly related, but species differ in both the slope and y-intercept of the linear fit (Samson and Werk, 1986). The linear model can extrapolate to a positive, zero, or negative intercept, in which case reproductive partitioning will increase, remain the same, or decrease as stress reduces total plant biomass (Fig. 5). The linear fit is not valid over the entire range of extrapolation, especially for species with a positive y-intercept, but can be used to determine whether reproductive effort is sensitive to the size (biomass) distribution of a population or a sample taken from it.

1. Growth Form Aspects of plant growth form that may affect the y-intercept include the threshold size for reproduction, the structure and placement of inflorescences, and the indeterminacy of growth. There are many

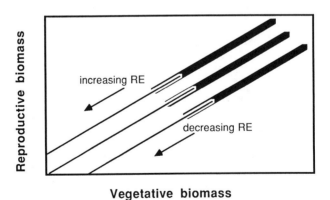

Vegetative biomass

Figure 5 Size dependence of reproductive effort for species with linear partitioning between vegetative and reproductive biomass. The observed relationship between vegetative and reproductive biomass (thick lines) can extrapolate (thin lines) to a negative, zero, or positive intercept. Whenever the regression has an intercept other than zero, reproductive effort differs at each point on the line, and the value determined for a population depends on the size distribution of the sample. [Redrawn from Samson and Werk (1986).]

possible morphological relationships between vegetative and reproductive growth, ranging from indeterminate growth with complete coupling to determinate, scapose inflorescences (Fig. 6). The importance of inflorescence morphology is that it determines the developmental patterns of the components of yield. The potential mechanisms of compensation may be related to inflorescence type and to the timing and type of stress.

2. Phenology Controls on the transition from vegetative growth to flowering may also act as constraints on partitioning responses. Where plant growth is seasonally limited by cold winter temperatures or by summer drought and heat, phenology is often regulated by environmental cues, such as day length, so that plants can reproduce, enter dormancy, or both before a prolonged stress. When stress occurs during a growing season, phenological responses can either increase or decrease reproductive output. If a plant under short-term stress can delay flowering until conditions are more favorable for reproductive development, it may achieve a higher reproductive output than a less plastic plant (e.g., late-flowering pearl millet: Mahalakshmi and Bidinger, 1985). On the other hand, phenological plasticity can disrupt the coordination of steps necessary for successful

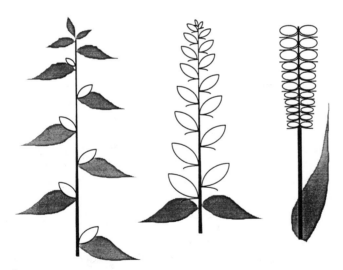

Figure 6   Examples of different degrees of coupling between vegetative and reproductive growth resulting from inflorescence morphology. Where solitary flowers occur in the axil of each leaf (left), as in *Impatiens,* vegetative and reproductive growth occur together. In determinate, scapose inflorescences (right), as in sorghum, grain number is determined over a much shorter period with less overlap between vegetative and reproductive growth. Intermediate degrees of coupling and determination of seed number occur in various other inflorescence types, such as terminal racemes (center).

reproduction, as in the decoupling of male and female gametogenesis in water-stressed maize.

The net effect of the phenological response to stress depends on the balance between these trade-offs and on plant life history. In general, perennials have more options for delaying their reproductive timing because flowering time is more variable within the growing season and can, depending on the species, either precede or follow leaf expansion. Also, perennials can defer reproduction for an entire year. For annuals, delaying reproduction entails the risk of dying before the stress is relieved. In addition, several mechanisms tend to reduce reproductive yield in annuals that flower late. Seed yield may be reduced merely because of a shortened reproductive period or increasing competition among plants through time.

Phenological response may also be affected by family or genus membership, which are major determinants of a species' flowering time. For example, members of the Asteraceae tend to be longer growing and later maturing than members of the Brassicaceae (Kochmer and Handel, 1985). Constraints imposed by growth form, phenology, and taxonomic identity may explain why similar stresses have different effects on reproductive partitioning in different species. Although these constraints are evident in studies of wild plants, their effects are best seen in comparisons among crop cultivars.

C. Plastic Responses in Crop Species

High yields by modern cultivars of grain crops necessitate significant input of resources and high planting densities. Comparisons between grain crops and their wild progenitors have identified evolutionary shifts in physiological and morphological traits associated with maximizing yield under these conditions. More than any other single plant variable, changes in biomass partitioning, especially harvest index, have been the greatest source of improvements of seed yield. Therefore, the sensitivity of reproductive partitioning is important in determining the effect of stress on yield in grain crops. For comparison with ecological studies of wild plants, we focus on plastic changes in reproductive partitioning in grain crops under single- and multiple-resource limitation.

For a number of crop species, cultivars differing in height, growth form, or maturity have been compared in experimental settings that reveal the importance of these factors in shaping partitioning responses to stress. The response of harvest index to varying nitrogen differs among grain crop species and among cultivars of a given grain crop. Studies on rice cultivars ranging from tall to semidwarf varieties suggest that the partitioning response to nitrogen is related to plant stature and canopy type. Jennings and de Jesus (1968) observed that rice cultivars receiving no nitrogen amendment were shorter and produced less aboveground biomass than treatments receiving 50 kg/ha, but partitioning responses varied with plant height. The taller the cultivar, the more it increased harvest index under low nitrogen (Fig. 7). Across all cultivars, there was a negative relationship between harvest index and height.

A somewhat different pattern was observed in studies of 12 winter wheat cultivars ranging from tall varieties introduced before 1910 to more recently developed dwarf varieties (Austin *et al.*, 1980). Under both high and low nitrogen, dwarf varieties outyielded tall varieties because of increased partitioning to grain versus stems, rather than higher dry matter production. Grain yield of dwarf varieties was also more sensitive to nitrogen. Across all cultivars, harvest index was again negatively related to plant height (Fig. 8), as in the rice cultivars in Figure 7. Within individual cultivars, however, the response of harvest index to nitrogen was not related to plant height. In four cultivars that together spanned the height range of the entire group, harvest index decreased with low nitrogen, despite decreases in plant height.

Height differences may interact with other factors that affect partitioning. Among sunflower cultivars, height is correlated with maturity. Taller cultivars flower later and are longer lived. In studies of five sunflower cultivars spanning a twofold height range, Rawson and Turner (1982) found that the shortest cultivar had a significantly higher harvest index than taller cultivars, except when all were grown without any irrigation (Fig. 9). For treatments allowed to recover from the dry treatment by switching

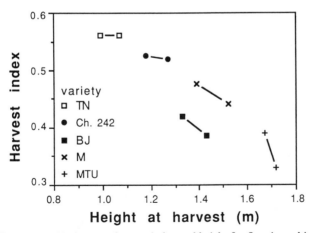

Figure 7 Relationship between harvest index and height for five rice cultivars grown in hill plots and receiving 0 or 50 kg/ha nitrogen. Each symbol corresponds to one variety, identified at lower left. The lines join the 0 and 50 kg/ha nitrogen additions; in all cases, height increased at the higher nitrogen. [Calculated from Table 3 in Jennings and de Jesus (1968).]

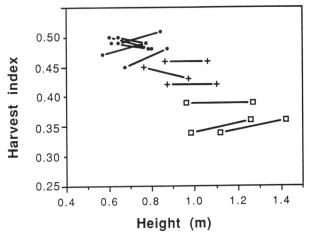

Figure 8 Relationship between harvest index and height (to the base of the ear) for twelve cultivars of winter wheat receiving high or low nitrogen and grouped into three height classes. Each pair of symbols joined by a line represents one cultivar; in all cases, height increased under high nitrogen. The three symbols designate three height classes, which do not overlap within each nitrogen treatment. Overall, harvest index decreased with height increases caused by nitrogen, but individual cultivars varied in their response. Under both high and low nitrogen, the highest grain yields were from dwarf cultivars. [Data from Table 1 in Austin *et al.* (1980).]

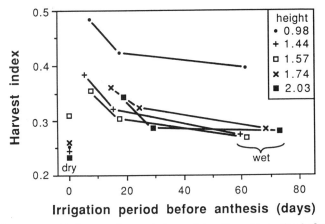

Figure 9 Relationship between harvest index and duration of irrigation before anthesis for five sunflower (*Helianthus annuus* L.) cultivars grown under four watering regimes in field plots. Mean height under all treatments is given at the upper right for each cultivar. The five symbols correspond to the five cultivars; each point represents the mean for a cultivar under a given watering treatment. The "dry" treatment (0 days irrigation) received no irrigation after seedling establishment. The "wet" treatment was irrigated frequently throughout the growing season. Intermediate points reflect "recovery" treatments switched from dry to wet during the vegetative state. Harvest index decreased with irrigation period (earliness of "recovery") in watered treatments but was generally lowest in the dry treatment. [Data from Tables 3 and 5 in Rawson and Turner (1982).]

to frequent irrigation, harvest index increased the later the recovery in the vegetative phase. In the irrigated treatments, seed yield was positively correlated with time to anthesis and with cultivar height.

These three examples illustrate three different patterns of yield response to resource limitation across cultivars differing in height. Increased harvest index under stress occurred within cultivars in two examples (but not one) and across cultivars in two examples (but not one). The examples also differ in the relationship of harvest index to absolute yield. In wheat, cultivars with the highest harvest index have the highest yields, both at low and high nitrogen. In rice, this was true only at low nitrogen, and in irrigated sunflower cultivars, the opposite was true. Differences such as these may reflect a variety of species-specific controls on partitioning.

VI. Other Forms of Whole-Plant Integration

At a finer level than biomass partitioning, a variety of interactions between vegetative and reproductive processes determine the response of reproductive yield to stress. Vegetative characters may affect the sensitivity of reproductive structures, such as the effect of rootstock scion on the survival of *Prunus* flower buds after freezing (Brown and Cummins, 1988). Also, the stress response of reproduction may depend on developmental changes in stress effects on leaves. In soybeans, water stress of -23 bars causes abortion of flowers but not pods, yet soybean yields are more sensitive to water stress during pod filling because soybean leaves are less able to recover from water stress durng this phase (Sionit and Kramer, 1977). Flowering may induce developmental changes in leaf characters that affect seed filling. In pearl millet (*Pennisetum americanum* [L.] Leeke), stomatal conductance is more sensitive to bulk leaf water potential in preflowering plants than in flowering plants, even controlling for factors such as leaf age and osmotic adjustment (Henson *et al.*, 1984). In wheat, progressive osmotic adjustment decreases the sensitivity of stomata to water potential through time (Teare *et al.*, 1982). Developmental changes such as these conserve water during the vegetative state and tend to keep stomata open during grain filling. These examples suggest that many aspects of whole-plant integration affect reproductive windows of sensitivity.

VII. Yield Components

As we have discussed in Section II, reproductive development is a linear progression of stages, each of which is sensitive to stress. As a result, stress effects can ramify through development and sometimes cause complete reproductive failure. In most cases, however, reproductive yield is linearly related to vegetative biomass, even when biomass ranges over several orders of magnitude along a resource gradient. The robustness of these relationships, despite the inherent sensitivity of many developmental events, is due to several factors: excess reproductive capacity, plasticity of yield components, reproductive abortion, and yield component compensation.

Plants typically initiate more reproductive meristems than eventually develop and mature. This is true even for inflorescences that are morphologically determinate, except for a small number of single-flowered species with single-flowered ramets, such as *Podophyllum peltatum*. In most higher plants, a significant fraction of flowers, ovules, seeds, and/or fruits aborts.

A. Abortion

Abortion has been viewed as a mechanism for tailoring reproductive output to the level of resources available for reproductive growth (Lloyd, 1980; Stephenson, 1981). If ovules, seeds, or fruits are aborted early in their development, the energetic cost to the plant may be small relative to the potential benefit of having additional reproductive sinks when resources permit, when seed predators reduce the seed crop, or both. Consistent with this idea, a number of studies have found that seed abortion is related to plant resource status (Nakamura, 1988). A variety of other factors, however, may also determine the pattern and degree of abortion. Large inflorescences, in which few flowers are destined to set any seed, may be necessary for pollinator attraction or may confer high fitness through pollen production (Willson and Price, 1977; Willson and Rathcke, 1974). Breeding system and the level of deleterious allelic combinations may also account for a significant fraction of abortion, as evidenced by the typically higher rate of ovule abortion in outbreeding (perennial) species than inbreeding (annual) species (Wiens, 1984). Recent studies suggest that multiple factors operate at once in determining abortion rates. For the multiply sired seed crop of *Raphanus raphanistrum*, stress during the reproductive phase induces selective abortion that alters seed paternity relative to unstressed controls (Marshall and Ellstrand, 1988).

Although resource stress plays a role in many cases of abortion, not all abortion in resource-limited plants is due to resource stress. When plants experience continuous resource limitation, they often respond by producing fewer flowers per inflorescence, rather than aborting a larger fraction of them. Thus, *Lolium perenne* plants grown without fertilizer have the same rate of ovary and seed (i.e., caryopsis) abortion (ca. 50%) as plants receiving 150 kg N h^{-1} (Marshall and Ludlam, 1989). Similarly, Stephenson (1984) observed that unfertilized *Lotus corniculatus* yielded less than fertilized plants, but not because of higher seed abortion. Rather, their ramets initiated fewer inflorescences, each with slightly fewer flowers. These studies suggest that plants initiate reproductive meristems commensurate with resource availability, and that a baseline level of abortion may be due to seed genotype or other factors. Stress-induced abortion is probably most important when the stress is abrupt and occurs after reproductive development has begun.

B. Plasticity of Yield Components

In most plant species, all or nearly all of the components of yield are phenotypically plastic. The relative plasticity of yield components may depend on a variety of factors, including plant phenology, inflorescence type, and the axis of stress to which the plant responds. When resources are abundant, each component of yield tends toward its maximum value. Therefore, growing wild plants under enhanced resource conditions often produces increases in every yield component (Fig. 10). Although seed weight tends to be the most conserved yield component (Harper, 1977), it is quite plastic in some species, such as *Prunella vulgaris* (Fig. 10; Winn and Werner, 1987). In sorghum, the plasticity of seed weight depends on position within the panicle (Hamilton *et al.,* 1982).

The developmental state of a plant determines the target of short-term stress, as demonstrated by studies of water stress and shoot chilling in soybeans (Sionit and Kramer, 1977). Sionit and Kramer (1977) observed that plants that were water stressed to -23 bars for about one week had lower yields than unstressed controls, but the cause of yield reductions depended on the timing of stress. Water stress during flower formation and flowering caused abortion of some flowers and decreased the flowering period but did not affect seed weight. The same degree of stress during pod filling did not affect flower or pod number but reduced seed weight. In the same variety of soybean, Musser *et al.* (1986) found that during photoinduction of flowering, one week of shoot chilling at $10°$ C reduced the production of floral primordia. The following week, the same treatment increased the number of fused or malformed pods. The third week, the treatment caused abscission of three-fourths of early flowers and pods. As these studies illustrate, the impact of short-term stress is strongest on the yield component being formed at the time of stress.

C. Yield Compensation

Because of the plasticity of yield components, the effects of stress on one yield component can be compensated by subsequent yield components if stress is relieved. Yield compensation is often observed in crop species and may result following a variety of stress effects. It can be due to plasticity in virtually every yield component. For example, in the shoot-chilling experi-

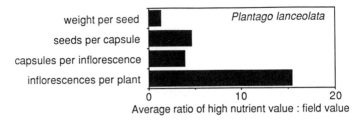

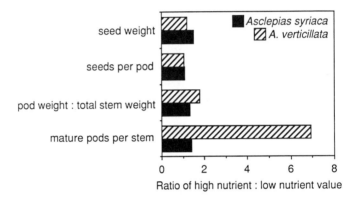

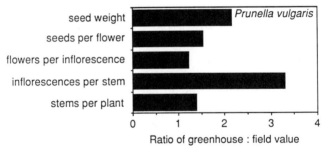

Figure 10   Examples of the plasticity of yield components measured as the ratio of the value under greenhouse or high nutrient conditions relative to the value from field-collected plant or plants grown under low nutrients. [Data for *Plantago* are from Primack and Antonovics (1981); data for *Asclepias* are from Willson and Price (1980); data for *Prunella* are from Winn and Werner (1987).]

ments by Musser *et al.* (1986), increases in seed weight compensated for reductions in other yield components and restored the total seed yield to the control value. In pearl millet, early flowering results in more panicles, with fewer but larger grains. Midseason water stress often has no effect on final yield because of these interactions (Bidinger *et al.,* 1987). Removal of spikelets from sorghum panicles has a similar effect. Removal of up to 20%

of spikelets from a panicle can be fully compensated through increased seed weight and number (Hamilton *et al.*, 1982). Increased grain weight may also compensate for reduced numbers of wheat grains or ears under SO_2 fumigation (Murray and Wilson, 1988).

The ability of later yield components to compensate for reduction in early yield components results in negative correlations among yield components. Such negative correlations may also occur when separate yield components compete for resources, as in plants that undergo prolonged stress (Adams, 1967). When barley is grown under competitive conditions in row plots, yield components are negatively correlated, but under hill plot cultivation, they are not (Rasmusson and Cannell, 1970). Row cultivation of field beans *(Phaseolus vulgaris)* shows a similar pattern with increased planting density (Adams, 1967). Under conditions of high resource availability, each component of yield tends toward its maximum value, and so negative correlations disappear. Although negative correlations have been reported mainly for morphological components of yield, yield itself may be negatively correlated with nutritional characteristics, such as grain protein concentration (Kibite and Evans, 1984) or nitrogen content (Harder *et al.*, 1982).

Compensation can also occur via reproductive cohorts initiated after a stress. In cotton, herbivory on early flower cohorts can be compensated for by increased retention of surviving structures from either the same cohort or later cohorts (Stewart and Sterling, 1988). This is most likely to occur in indeterminate plants exposed to pests of limited activity period. In pearl millet, delayed flowering by tillers provides late cohorts that can compensate for reductions in main shoot yields caused by water stress or grain removal (Mahalakshmi and Bidinger, 1985, 1986).

Although wild plants are typically resource limited, negative correlations among yield components are generally rare or very weak in wild plants. Marshall *et al.* (1985) found that yield component correlations were negative in roughly a third of cases for three legumes native to Texas floodplains, but all were weak ($r < 0.25$); positive correlations were comparable in number and sometimes higher (e.g., $r = 0.34$, $r = 0.55$). When the two annual legumes were subjected to nine stress treatments, each showed only one significant yield component correlation, opposite in sign (Marshall *et al.*, 1986). In wild highbush blueberries, Pritts and Hancock (1985) found one significant (positive) yield component correlation.

This difference between cultivated and wild plants appears to be due to several factors. One is that plant size in natural populations varies far more (e.g., up to several orders of magnitude for synchronous cohorts) than in crops. Because reproductive yield is closely correlated with vegetative biomass, studies of wild plants have consistently shown that variation in vegetative biomass determines yield to a far greater extent than variation in

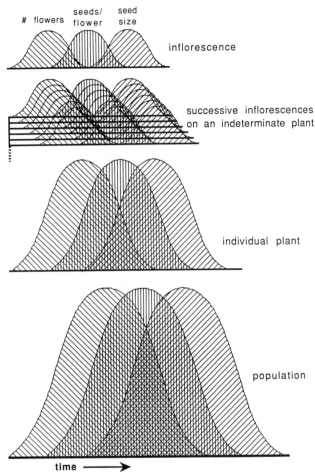

Figure 11 Temporal distribution of yield components in wild plants at levels ranging from the inflorescence to the individual plant to the population. Within a single inflorescence, successive stages of yield component determination overlap relatively little. Continued production of inflorescences produces overlap between early stages in one inflorescence and later stages in another on the same plant. Similarly, variability in flowering time among plants in a population produces more overlap at this level.

yield components. This may drive positive correlations among yield components positively correlated with vegetative biomass. Second, wild plants are not as synchronous as crops in their reproductive development. Synchronous fruit and seed maturation occurs in crops through a variety of means, such as basipetal development following acropetal differentiation (Evans and Wardlaw, 1976). Although the development of yield components within the individual inflorescence of a wild plant may follow a pattern similar to a crop, the reproductive phase may be more protracted in wild plants, with staggered initiation of new inflorescences and staggered seed maturation. At the level of the plant, this tends to increase the temporal overlap between successive yield component stages (on separate inflorescences, and still more, on separate plants) and exposes more stages to short-term stresses (Fig. 11). Prolonged flowering also makes compensation possible through successive cohorts, rather than successive developmental stages, which makes yield components more variable and less likely to be correlated. Finally, negative correlations are also eroded by parallel declines through a season in most yield components, such as seed weight (Cavers and Steel, 1984).

VIII. Conclusions

Two themes we have examined are that reproductive development entails windows of sensitivity to stress and that reproductive yield and vegetative growth of plants are strongly coupled. The coupling is implicit when reproduction is viewed as a form of resource partitioning. Stresses arising from unpredictable variation in the environment elicit partitioning changes that are plastic, but the degree of response is under genetic control and may vary among individuals, populations, and species. Therefore reproductive partitioning may be an important component in analyzing a species' response to stress, but general patterns across species, including differences between monocarps and polycarps, are blurred by a variety of species-specific constraints.

   Both experimental and modeling studies suggest that relative measures of reproduction provide limited insight unless coupled to absolute measures. Simple optimization models suggest that partitioning should remain constant in response to stresses that reduce resource acquisition, but this constancy is rarely observed. Among crop cultivars, height and maturity interact with partitioning so that partitioning responses to belowground resources vary among species. The importance of interspecific differences in stress responses is also evident in studies of aboveground stresses, such as air pollutants. Although atmospheric pollutants tend to have consistent effects on partitioning between root and vegetative shoot, the effects on partitioning to reproductive structures are quite variable (Miller, 1988). The role of constraints on partitioning responses related to growth form, phenology, and taxonomy needs further study.

   Major buffering mechanisms against stress include overinitiation of reproductive structures, prolonged flowering, and plasticity of yield components. Together they enable a plant to rebound from short-term stress by providing a sufficient number of reproductive sinks of sufficient strength. These mechanisms interact to determine the period of sensitivity to particular stresses and can have enormous agronomic importance. Prolonged flowering may be undesirable in many crops because of harvesting procedures, yet it may increase the likelihood of flowering without stress. In plants that vastly overinitiate reproductive structures, even a small extension of the flowering period may provide enough flowers for a full crop (as in peaches, which need as little as 10% flower survival; Brown and Cummins, 1988). The plasticity of individual yield components interacts with other buffering mechanisms in similar ways. Differences between wild and cultivated plants in habitat variability and developmental synchrony within populations result in distinct patterns of yield compensation in response to stress.

A consequence of the overinitiation of reproductive structures is a high rate of abortion of flowers, ovules, fruits, and seeds. The potential for selective abortion has received increasing examination, mainly with regard to the genotype and position of embryos. The significance of this potential with regard to stress responses needs further study. The physiological impact of different types of resource limitation may, through their impact on vegetative structures, have different consequences for selective abortion and the plasticity of yield components.

References

Abrahamson, W. G. (1975). Reproductive strategies in dewberries. *Ecology* 56: 721–726.

Adams, M. W. (1967). Basis of yield component compensation in crop plants with special reference to the field bean, *Phaseolus vulgaris. Crop Sci.* 7: 505–510.

Andel, J. van, and Vera, F. (1977). Reproductive allocation in *Senecio sylvaticus* and *Chamaenerion angustifolium* in relation to mineral nutrition. *J. Ecol.* 65: 747–758.

Andrews, P. K., Proebsting, E. L., and Gross, D. C. (1983). Differential thermal analysis and freezing injury of deacclimating peach and sweet cherry reproductive organs. *J. Am. Soc. Hortic. Soc.* 108: 755–759.

Aspinall, D., and Hussain, I. (1970). The inhibition of flowering by water stress. *Aust. J. Biol. Sci.* 23: 925–936.

Aspinall, D., Nichols, P. B., and May, L. H. (1964). The effects of soil moisture stress on the growth of barley. I. Vegetative development and grain yield. *Aust. J. Agric. Res.* 15: 729–745.

Austin, R. B., Bingham, J., Blackwell, R. D., Evans, L. T., Ford, M. A., Morgan, C. L., and Taylor, M. (1980). Genetic improvements in winter wheat yields since 1900 and associated physiological changes. *J. Agric. Sci. Camb.* 94: 675–689.

Beatley, J. C. (1970). Perennation in *Astragalus lentiginosus* and *Tridens pulchellus* in relation to rainfall. *Madroño* 20: 326–332.

Begg, J. E., and Turner, N. C. (1976). Crop water deficits. *Adv. Agron.* 28: 161–217.

Bidinger, F. R., Mahalakshmi, V., and Rao, G. D. P. (1987). Assessment of drought resistance in pearl millet [*Pennisetum americanum* (L.) Leeke]. I. Factors affecting yields under stress. *Aust. J. Agric. Res.* 38: 27–48.

Bonte, J. (1982). Effects of air pollutants on flowering and fruiting. *In* "Effects of Gaseous Air Pollution in Agriculture and Horticulture" (M. H. Unsworth and D. P. Ormrod, eds.), pp. 207–223. Butterworth, London.

Boyer, J. S., and McPerson, H. G. (1975). Physiology of water deficits in cereal crops. *Adv. Agron.* 27: 1–23.

Brown, S. K., and Cummins, J. N. (1988). Rootstock influenced peach flower bud survival after a natural freeze. *HortScience* 23: 846–847.

Burke, M. J., and Stushnoff, C. (1979). Frost hardiness: A discussion of possible molecular causes of injury with particular reference to deep supercooling of water. *In* "Stress Physiology in Crop Plants" (H. Mussell and R. C. Staples, eds.), pp. 197–225. Wiley, New York.

Cartica, R. J., and Quinn, J. A. (1982). Resource allocation and fecundity of populations of *Solidago sempervirens* along a coastal dune gradient. *Bull. Torrey Bot. Club* 109: 299–305.

Cavers, P. B., and Steel, M. G. (1984). Patterns of change in seed weight over time on individual plants. *Am. Nat.* 124: 324–335.

DeJong, T. J., Klinkham, P. G., and Prins, A. H. (1986). Flowering behavior of the monocarpic perennial *Cynoglossum officinale* L. *New Phytol.* 103: 219–229.

Douglas, D. A. (1981). The balance between vegetative and sexual reproduction of *Mimulus primuloides* (Scrophulariaceae) at different altitudes in California. *J. Ecol.* 69: 295–310.

Downey, L. A., and Miller, J. W. (1971). Rapid measurements of relative turgidity in maize. *New Phytol.* 70: 555–560.

Du Plessis, D. F., and Dijkhuis, F. J. (1967). The influence of the time lag between pollen shedding and silking on the yield of maize. *S. Afr. J. Agric. Sci.* 10: 667–674.

Durner, E. F., and Gianfagna, T. J. (1988). Fall ethephon application increases peach flower bud resistance to low-temperature stress. *J. Am. Soc. Hortic. Sci.* 113: 404–406.

Ernst, W. H. O., Tonneijck, A. E. C., and Pasman, F. J. M. (1985). Ecotypic response of *Silene cucubalus* to air pollutants (SO_2, O_3). *J. Plant Physiol.* 118: 439–450.

Evans, L. T., and Wardlaw, I. F. (1976). Aspects of the comparative physiology of grain yield in cereals. *Adv. Agron.* 28: 301–359.

Fenner, M. (1986). The allocation of minerals to seeds in *Senecio vulgaris* plants subjected to nutrient shortage. *J. Ecol.* 74: 385–392.

Foulds, W. (1978). Response to soil moisture supply in three leguminous species. I. Growth, reproduction and mortality. *New Phytol.* 80: 535–545.

Gay, G., Kerhoas, C., and Dumas, C. (1987). Quality of a stress-sensitive *Cucurbita pepo* L. pollen. *Planta* 171: 82–87.

Godzik, S., and Krupa, S. V. (1982). Effects of sulfur dioxide on the growth and yield of agricultural and horticultural crops. *In* "Effects of Gaseous Air Pollution in Agriculture and Horticulture" (M. H. Unsworth and D. P. Ormrod, eds.), pp. 247–265. Butterworth Scientific, London.

Grandtner, M. M., and Gervais, C. (1985). Extreme precocity and thermal conditions of apical and floral development in *Claytonia caroliniana* var. *caroliniana. Can. J. Bot.* 63: 1516–1520.

Grime, J. P. (1979). "Plant Strategies and Vegetation Processes." Wiley, New York.

Haase, P. (1986). An ecological study of the subalpine shrub *Senecio bennettii* (Compositae) at Arthur's Pass, South Island, New Zealand. *N. Z. J. Bot.* 24: 247–262.

Hamilton, R. I., Subramanian, B., Reddy, M. N., and Rao, C. H. (1982). Compensation in grain yield components in a panicle of rainfed sorghum. *Ann. Appl. Biol.* 101: 119–125.

Harder, H. J., Carlson, R. E., and Shaw, R. H. (1982). Yield, yield components, and nutrient content of corn grain as influenced by post-silking moisture stress. *Agron. J.* 74: 275–278.

Harper, J. L. (1977). "Population Biology of Plants." Academic Press, London.

Harris, W. (1970). Yield and habit of New Zealand populations of *Rumex acetosella* at three altitudes in Canterbury. *N. Z. J. Bot.* 8: 114–131.

Henson, I. E., Mahalakshmi, N., Alagarswamy, G., and Bidinger, F. R. (1984). The effect of flowering on stomatal response to water stress in pearl millet (*Pennisetum americanum* [L.] Leeke). *J. Exp. Bot.* 35: 219–226.

Hermesh, R., and Acharya, S. N. (1987). Reproductive response to three temperature regimes of four *Poa alpina* populations from the Rocky Mountains of Alberta, Canada. *Arct. Alp. Res.* 19: 321–326.

Herrero, M. P., and Johnson, R. R. (1981). Drought stress and its effects on maize reproductive systems. *Crop Sci.* 21: 105–110.

Hickman, J. C. (1975). Environmental unpredictability and plastic energy allocation strategies in the annual *Polygonum cascadense* (Polygonaceae). *J. Ecol.* 63: 689–701.

Houston, D. B., and Dochinger, L. S. (1977). Effects of ambient air pollution on cone, seed, and pollen characteristics in eastern white and red pines. *Environ. Pollut.* 12: 1–5.

Jennings, P. R., and de Jesus, J. J. (1968). Studies on competition in rice. I. Competition in mixtures of varieties. *Evolution* 22: 119–124.

Karnosky, D. F., and Stairs, G. R. (1974). The effects of SO_2 on in vitro forest tree pollen germination and tube elongation. *J. Environ. Qual.* 3: 406–409.

Kibite, S., and Evans, L. E. (1984). Causes of negative correlations between grain yield and grain protein concentration in common wheat. *Euphytica* 33: 801–810.

Kochmer, J. P., and Handel, S. N. (1985). Constraints and competition in the evolution of flowering phenology. *Ecol. Monogr.* 56: 303–325.

Kowithayakorn, L., and Humphreys, L. R. (1987). Effects of severity and repetition of water stress on seed production of *Macroptilium atropurpureum* cv. Siratro. *Aust. J. Agric. Res.* 38: 529–536.

Lee, T. D., and Bazzaz, F. A. (1980). Effects of defoliation and competition on growth and reproduction in the annual plant *Abutilon theophrasti*. *J. Ecol.* 68: 813–821.

Lee, T. D., and Bazzaz, F. A. (1982). Regulation of fruit and seed production in an annual legume, *Cassia fasciculata*. *Ecology* 63: 1363–1373.

Lloyd, D. G. (1980). Sexual strategies in plants. I. An hypothesis of serial adjustment of maternal investment during one reproductive session. *New Phytol.* 86: 69–79.

Ma, T.-H., and Khan, S. H. (1976). Pollen mitosis and pollen tube growth inhibition by SO_2 in cultured pollen tubes of *Tradescantia*. *Environ. Res.* 12: 144–149.

Ma, T.-H., Isbandi, D., Khan, S. H., and Tseng, X.-S. (1973). Low levels of SO_2-enhanced chromatid aberrations in *Tradescantia* pollen tubes and seasonal variations of the aberration rates. *Mutat. Res.* 21: 93–100.

Mahalakshmi, V., and Bidinger, F. R. (1985). Flowering response of pearl millet to water stress during panicle development. *Ann. Appl. Biol.* 106: 571–578.

Mahalakshmi, V., and Bidinger, F. R. (1986). Water deficit during panicle development in pearl millet: Yield compensation by tillers. *J. Agric. Sci. Camb.* 106: 113–119.

Marc, J., and Palmer, J. H. (1976). Relationship between water potential and leaf and inflorescence initiation in *Helianthus annuus*. *Physiol. Plant.* 36: 101–104.

Marshall, C., and Ludlam, D. (1989). The pattern of abortion of developing seeds in *Lolium perenne* L. *Ann. Bot.* 63: 19–27.

Marshall, D. L., and Ellstrand, N. C. (1988). Effective mate choice in wild radish: Evidence for selective seed abortion and its mechanism. *Am. Nat.* 131: 739–756.

Marshall, D. L., Fowler, N. L., and Levin, D. A. (1985). Plasticity in yield components in natural populations of three species of *Sesbania*. *Ecology* 66: 753–761.

Marshall, D. L., Levin, D. A., and Fowler, N. L. (1986). Plasticity of yield components in response to stress in *Sesbania macrocarpa* and *Sesbania vesicaria* (Leguminosae). *Am. Nat.* 127: 508–521.

Masaru, N., Syozo, F., and Saburo, K. (1976). Effects of exposure to various injurious gases on germination of lily pollen. *Environ. Pollut.* 11: 181–187.

McNaughton, S. J. (1966). Ecotype function in the *Typha* community-type. *Ecol. Monogr.* 36: 297–325.

Miller, G. A. (1986). Pubescence, floral temperature, and fecundity in species of *Puya* (Bromeliaceae) in the Ecuadorian Andes. *Oecologia* 70: 155–160.

Miller, J. E. (1988). Effects on photosynthesis, carbon allocation, and plant growth associated with air pollutant stress. *In* "Assessment of Crop Loss from Air Pollutants" (W. W. Heck, O. C. Taylor, and D. T. Tingey, eds.), pp. 287–324. Elsevier, London.

Mooney, H. A., and Billings, W. D. (1961). Comparative physiological ecology of arctic and alpine populations of *Oxyria digyna*. *Ecol. Monogr.* 31: 1–29.

Moss, G. I., and Downey, L. A. (1971). Influence of drought stress on female gametophyte development in corn (*Zea mays* L.) and subsequent grain yield. *Crop Sci.* 11: 368–372.

Murray, F., and Wilson, S. (1988). The joint action of sulphur dioxide and hydrogen fluoride on the yield and quality of wheat and barley. *Environ. Pollut.* 55: 239–249.

Musser, R. L., Kramer, P. J., and Thomas, J. F. (1986). Periods of shoot chilling sensitivity in soybean flower development, and compensation in yield after chilling. *Ann. Bot.* 57: 317–329.

Nakamura, R. P. (1988). Seed abortion and seed size variation within fruits of *Phaseolus vulgaris:* Pollen donor and resource limitation effects. *Am. J. Bot.* 75: 1003–1010.

Neuffer, B., and Hurka, H. (1986). Variation in development time until flowering in natural poulations of *Capsella bursa-pastoris* (Cruciferae). *Plant Syst. Evol.* 152: 277–296.

Ohlson, M. (1986). Reproductive differentiation in a *Saxifraga hirculus* population along an environmental gradient on a central Swedish mire. *Holarct. Ecol.* 9: 205–213.

Patterson, B. D., Mutton, L., Paull, R. E., and Nguyen, V. Q. (1987). Tomato pollen development stages sensitive to chilling and a natural environment for the selection of resistant genotypes. *Plant, Cell Environ.* 10: 363–368.

Primack, R. B., and Antonovics, J. (1981). Experimental ecological genetics in *Plantago*. V. Components of seed yield in the ribwort plantain *Plantago lanceolata* L. *Evolution* 35: 1069–1079.

Pritts, M. P., and Hancock, J. F. (1985). Lifetime biomass partitioning and yield component relationships in the highbush blueberry, *Vaccinium corymbosum* L. (Ericaceae). *Am. J. Bot.* 72: 446–452.

Quamme, H. A. (1978). Mechanism of supercooling in overwintering peach flower buds. *J. Am. Soc. Hortic. Sci.* 103: 57–61.

Quamme, H. A., and Gusta, L. V. (1987). Relationship of ice nucleation and water status to freezing patterns in dormant peach flower buds. *HortScience* 22: 465–467.

Rasmusson, D. C., and Cannell, R. Q. (1970). Selection for grain yield and components of yield in barley. *Crop Sci.* 10: 51–54.

Rawson, H. M., and Turner, N. C. (1982). Recovery from water stress in five sunflower (*Helianthus annuus* L.) cultivars. I. Effects of the timing of water application on leaf area and seed production. *Aust. J. Plant Physiol.* 9: 437–448.

Reader, R. J. (1984). Comparison of the annual flowering schedules for Scottish heathland and Mediterranean-type shrublands. *Oikos* 43: 1–8.

Reinartz, J. A. (1984). Life-history variation of common mullein (*Verbascum thapsus*) I. Latitudinal differences in population dynamics and timing of reproduction. *J. Ecol.* 72: 897–912.

Roos, F. H., and Quinn, J. A. (1977). Phenology and reproductive allocation in *Andropogon scoparius* (Gramineae) populations in communities of different successional stages. *Am. J. Bot.* 64: 535–540.

Royce, C. L., and Cunningham, G. L. (1982). The ecology of *Abronia angustifolia* Greene (Nyctaginaceae) I. Phenology and perennation. *Southwest Nat.* 27: 413–423.

Saliba, M. R., Schrader, L. E., Hirano, S. S., and Upper, C. D. (1982). Effects of freezing field-grown soybean plants at various stages of podfill on yield and seed quality. *Crop Sci.* 22: 73–78.

Samson, D. A., and Werk, K. S. (1986). Size-dependent effects in the analysis of reproductive effort in plants. *Am. Nat.* 127: 667–680.

Simpson, R. L., Leck, M. A., and Parker, V. T. (1985). The comparative ecology of *Impatiens capensis* Meerb. (Balsaminaceae) in central New Jersey. *Bull. Torrey Bot. Club* 112: 295–311.

Sionit, N., and Kramer, P. J. (1977). Effect of water stress during different stages of growth of soybean. *Agron. J.* 69: 274–278.

Sparks, D. (1989). Drought stress induces fruit abortion in pecan. *HortScience* 24: 78–79.

Stephenson, A. G. (1981). Flower and fruit abortion: Proximate causes and ultimate functions. *Annu. Rev. Ecol. Syst.* 12: 253–279.

Stephenson, A. G. (1984). The regulation of maternal investment in an indeterminate flowering plant *(Lotus corniculatus). Ecology* 65: 113–121.

Stewart, S. D., and Sterling, W. L. (1988). Dynamics and impact of cotton fruit abscission and survival. *Environ. Entomol.* 17: 629–635.

Teare, I. D., Sionit, N., and Kramer, P. J. (1982). Changes in water status during water stress at different stages of development in wheat. *Physiol. Plant.* 55: 296–300.

Varshney, S. R. K., and Varshney, C. K. (1981). Effect of sulphur dioxide on pollen germination and pollen tube growth. *Environ. Pollut. Ser. A Ecol. Biol.* 24: 87–92.

Wertheim, F. S., and Craker, L. E. (1988). Effects of acid rain on corn silks and pollen germination. *J. Environ. Qual.* 17: 135–138.

Wiens, D. (1984). Ovule survivorship, brood size, life history, breeding systems, and reproductive success in plants. *Oecologia* 64: 47–53.

Willson, M. F., and Price, P. W. (1977). The evolution of inflorescence size in *Asclepias* (Asclepiadaceae). *Evolution* 31: 495–511.

Willson, M. F., and Price, P. W. (1980). Resource limitation and seed production in some *Asclepias* species. *Can. J. Bot.* 58: 2229–2233.

Willson, M. F., and Rathcke, B. J. (1974). Adaptive design of the floral display in *Asclepias syriaca* L. *Am. Midl. Nat.* 92: 47–57.

Winn, A. A., and Werner, P. A. (1987). Regulation of seed yield within and among populations of *Prunella vulgaris. Ecology* 68: 1224–1233.

Wyatt, R. (1981). Components of reproductive output in five tropical legumes. *Bull. Torrey Bot. Club* 108: 67–75.

Yegappan, T. M., Paton, D. M., Gates, C. T., and Müller, W. J. (1980). Water stress in sunflower *(Helianthus annuus* L.): I. Effect on plant development. *Ann. Bot.* 46: 61–70.

9

Multiple Stress-Induced Foliar Senescence and Implications for Whole-Plant Longevity

Eva J. Pell   Michael S. Dann

I. Introduction

The process of maturing, aging, and senescence is a carefully timed sequence of events in the life of a plant. Many stresses have been associated with the premature senescence of plants, plant parts, or both, but the mechanisms are still obscure. Stress can arise at any time during plant growth, but the effect on the lifespan of the plant may differ depending on the plant's developmental stage at the time the stress occurs. If a plant is in a linear growth phase, with synthetic processes operating fully, stress

Response of Plants to Multiple Stresses. Copyright © 1991 by Academic Press, Inc. All rights of reproduction in any form reserved.

effects are likely to be minimal, repair probable, and longevity of the plant left unaffected. If, on the other hand, senescence is already under way when the stress arises, enhancement might result in a shortened lifespan of the organ or organism. A third scenario might involve stress imposition late in the maturation process. If plant synthetic potential and repair processes are minimal, senescence might be initiated prematurely and lifespan reduced as a result.

The goal of this chapter is to explore how multiple stresses induce plant senescence. First, we will consider the molecular changes that occur within foliage when plants are challenged by different stresses. We will use this molecular framework to explain and predict how multiple stresses interact to influence foliar senescence. In the second portion of the chapter, we will consider the influence of foliar longevity on the senescence of the whole plant.

II. Foliar Senescence: A Molecular Sequence of Events

To explore how stresses affect senescence, we need to think about a sequence of molecular changes leading to cell death. Figure 1 shows one view of the possible interrelationships among molecular events in a senescing leaf. Cytokinins have long been associated with the nonsenescent state, and a decrease in cytokinins may be an early event in leaf aging (Leopold, 1980). Since cytokinins are known to stimulate synthesis of polyamines, a withdrawal of the hormones may lead to a reduction in concentration of these secondary metabolites. Senescence is associated with a drop in polyamines; the implications are multifold. Reduction in polyamines, e.g., spermine and spermidine, is associated with senescence, perhaps because of the role these compounds play in scavenging free radicals and superoxide ions and in stabilizing membranes, tRNA, and mRNA (Galston and K-Sawhney, 1987; Drolet et al., 1986).

Polyamines and ethylene are related by a shared synthetic pathway. Ethylene content is directly correlated with advancing senescence and foliar chlorosis, whereas polyamines are associated with the nonsenescent condition. The synthesis of ethylene depends on the same precursor, S-adenosylmethionine (SAM), as do the polyamines (Fig. 2). The conversion from 1-aminocyclopropane-1-carboxylic acid (ACC) to ethylene is a superoxide-mediated reaction (Drolet et al., 1986). Polyamines inhibit conversion of ACC to ethylene by reducing synthesis of ACC synthase and scavenging oxygen free radicals involved in catalytic conversion of ACC to ethylene. As a plant ages, polyamines are no longer synthesized, making more SAM available as a precursor for ACC. More superoxide is available because free-radical scavenging is suppressed in the absence of polyamines. Consequently, ACC conversion to ethylene can progress.

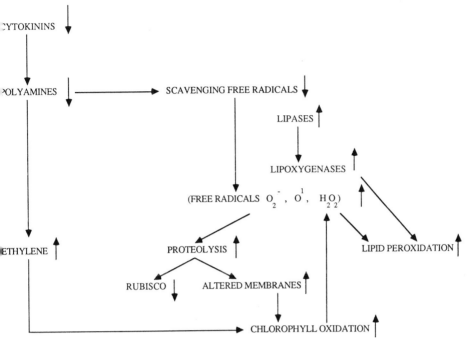

Figure 1   Senescence pathway reflecting cellular changes leading to foliar senescence.

Lipids and proteins are the targets of degradation during plant aging. Senescing membranes are subject to increased action of three membrane-associated lipases (Thompson *et al.*, 1987). Two of these, phosphatidic acid phosphatase and phospholipase D, are stimulated by free Ca^{2+} mediated by calmodulin. Through deesterification of membranes, linoleic and linolenic acid are released, resulting in increased activity of membranous and cytosolic lipoxygenase, which use these fatty acids as substrates. Lipoxygenase is thought to initiate lipid peroxidation, which leads to formation of activated oxygen in the form of superoxide and singlet oxygen (Thompson *et al.*, 1987; Lynch and Thompson, 1984). Once these activated forms of oxygen have been generated, lipid peroxidation can occur nonenzymatically as well, by direct reaction between fatty acids and superoxide or singlet oxygen.

The generation of free radicals and other activated oxygen species may play a role in the enhanced proteolysis that occurs during senescence. Many proteins are hydrolyzed during senescence, including ribulose 1,5-bisphosphate carboxylase/oxygenase (rubisco) (Thayer *et al.*, 1987). This enzyme makes up 50% or more of total soluble leaf protein. In addition to its role in carbon dioxide fixation, rubisco acts as a major nitrogen source for developing plant parts. Rubisco is synthesized during leaf expansion and de-

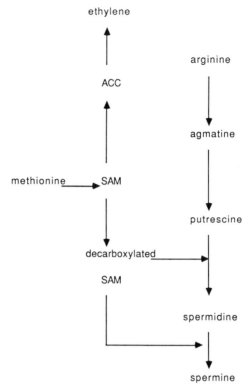

Figure 2 Pathways of ethylene and polyamine (spermine and spermidine) biosynthesis. ACC, 1-aminocyclopropane-1-carboxylic acid; SAM, S-adenosylmethionine. [Modified from Roberts *et al.* (1984).]

graded during aging and resulting senescence (Lauriere, 1983). Dalling (1987) has suggested that free radicals react with rubisco to produce structural changes in the protein that make it more physically acceptable to proteases. The reaction between free radicals, which are generated within the cell, and rubisco would offer a mechanism for altering the enzyme (Dalling, 1987).

Vulnerability of proteins in general to degradation during senescence and the lipid peroxidation discussed above may lead to altered membrane configuration. Such structural changes could expose compounds that are protected from the external environment earlier in the life of the leaf. One candidate for such exposure could be the chlorophyll–protein complex, which is normally embedded in the thylakoid (McRae and Thompson, 1983). Once chlorophyll is exposed, it becomes vulnerable to oxygen attack in the light (Harbour and Bolton, 1978). Chlorophyll oxidation will lead to pigment destruction associated with senescence and the emission of super-oxide. Superoxide can then react further as described earlier.

A. Mechanisms Inducing Senescence

The phenomena described above reflect changing cellular constituents during senescence. There are probably several ways in which senescence is initiated. Below we will examine the role of chemical oxidation and hormonal regulation in inducing senescence. These mechanisms should not be viewed as mutually exclusive but rather as potentially interactive.

1. Oxidative Mechanism The evolution of organisms into an aerobic biosphere brought new risks along with the gains. Oxygen has great importance as an electron acceptor in many metabolic processes. Complete reduction of oxygen to water requires four electrons and is a sequential, univalent process. Several reactive intermediates can be produced, including superoxide anions, hydrogen peroxide, and hydroxyl radicals (Fridovich, 1978). These species are produced biotically by such processes as lipoxygenase-catalyzed lipid peroxidation (Thompson *et al.*, 1987) and oxygen reduction by electron acceptors associated with photosystem I (Asada, 1984). Free radicals may also be xenobiotic in origin. When ozone enters the cell, it degrades into a variety of oxygen species including those mentioned above. Many important environmental toxins, in turn, exert observed effects through radical-mediated reactions. Other stresses, such as desiccation, freezing, and anoxia, may perturb metabolism in such a way as to disrupt electron transport reactions and promote formation of free radicals, such as superoxide and the hydroxyl radical (McKersie *et al.*, 1988).

The role of reactive oxygen species and free radicals in the life of aerobes has been discussed for many types of life-forms. When plants are young, antioxidant systems, including superoxide dismutase, catalase, α-tocopherol, carotenoids, and so on, provide detoxifying mechanisms. As plants age, many of these compounds decline in prominence, leading to elevated levels of free radicals. The free radicals generated through oxidative processes have been associated with aging and death for many years (Fantone and Ward, 1985). For example, it has been shown in an oat-leaf model system that when oxygen concentrations in the atmosphere are artificially increased, senescence accelerates (Trippi and De Luca d'Oro, 1985). Since many stresses increase oxidizing potential, either directly or indirectly, it seems appropriate to examine the role in senescence of oxidizing free radicals and other reactive oxygen species.

One of the early events in senescence appears to be the increased in vivo activity of lipoxygenases (Thompson *et al.*, 1987; Lynch and Thompson, 1984). The activated oxygen species, superoxide, and singlet oxygen are products of lipoxygenase catalysis. Once these free radicals are formed, a number of possible reactions may result. Certainly, additional toxic radical and nonradical compounds may be generated. For example, superoxide and hydrogen peroxide can react to yield the very toxic hydroxyl radicals through the Haber-Weiss reaction (1934). Reactions between free radicals and lipids or proteins are likely due to (1) potential reactivity of the biochemicals, (2) ubiquity of the biochemicals within the cell, and (3) proximity of lipids and proteins to sites of lipoxygenase activity.

2. Hormonal Mechanisms   The regulation of the generation and unchecked action of free radicals may be mediated by hormones and related secondary metabolites, whose presence or absence may thus play a pivotal role in the progress of senescence. The association between hormone function and senescence has been known for many years. Cytokinins were first characterized as antisenescents in 1957 (Richmond and Lang), although the mechanism by which cytokinins perform this function is not well understood. Leshem *et al.* (1979) proposed an antioxidant function for cytokinins, demonstrating that cytokinin behaved analogously to α-tocopherol, a well-defined antioxidant. The antioxidant response characteristic of cytokinins was shown by treating detached senescing pea leaves with the hormone; lipoxygenase activity decreased and chlorophyll retention was enhanced. Since antioxidant activity is independent of endogenous cytokinins, Leshem *et al.* (1979) proposed that cytokinins may act at the plasma membrane, scavenging superoxide produced by lipoxygenase activity. This hypothesis is supported by the observation that cytokinin analogues can convert amines to amides while detoxifying superoxide, thereby preventing further oxidation. Alternatively, cytokinin may inactivate lipoxygenases by altering an active site on the enzyme. This idea is supported by the observation that lipoxygenases purified on an affinity column can subsequently be inactivated by cytokinins. By either mechanism, senescence-associated superoxide anion action, production, or both is suppressed.

Ethylene has been associated with senescence, as discussed at the beginning of Section II, but the mechanism by which this hormone stimulates aging is not known. It has been proposed that ethylene action depends on binding to membranes, a phenomenon that increases with leaf age and development (Goren *et al.*, 1984). When detached tobacco leaves are incubated in silver nitrate, ethylene binding is prevented, and chlorophyll breakdown associated with senescence is delayed. Ethylene levels increase with enhanced senescence, but the gas must be bound to have an effect. In the beginning of Section II, we discussed the role of chlorophyll oxidation in senescence. Only when thylakoid membranes are oriented in a way that permits chlorophyll to be exposed would senescence related to oxidation occur. Perhaps these membranes represent an important binding site for ethylene.

B. How Do Stresses Trigger Senescence?

Many natural and anthropogenic stresses have been associated with accelerated senescence. Our objective is to examine the kinds of changes that plants experience in response to stresses and how these changes contribute to senescence. Many stresses induce protein degradation, an important feature of senescence. Nitrogen supply correlates directly with degradation of rubisco. When Makino *et al.* (1984) treated rice plants with three levels of nitrogen (sufficient, control, and deficient), they observed that leaves with the highest level of nitrogen had the most rubisco at full leaf expansion. Although degradation began at the same time for all three treatments, it peaked early for "control" and "sufficient" nitrogen treatments and continued in a curvilinear fashion for 80 days. Degradation of rubisco in the "deficient" treatment was relatively constant until complete degradation was achieved by about 60 days. Rubisco declined more than total nitrogen did, perhaps reflecting synthesis of some new proteins during senescence. Since rubisco acts as an easily mobilized source of nitrogen, a leaf with a low nitrogen supply will senesce more rapidly.

Water stress also affects protein degradation. When barley is water stressed by floating leaf segments on polyethylene glycol, total protein synthesis is depressed, and, to a greater degree, protein degradation is enhanced when compared with nonstressed leaf segments (Dungey and Davies, 1982). Some proteins are degraded more rapidly by water stress, and others are unaffected. Cooke *et al.* (1980) proposed the vacuole as the source of proteinases; as long as the tonoplast is intact, selective protein degradation is ensured. When the tonoplast is breached, as in the case of water stress, protein degradation begins in a less controlled manner. Water stress also stimulates stomatal closure and increased concentration of abscisic acid. Increases in abscisic acid have been correlated with increases in ethylene (Sexton *et al.*, 1985), which has been associated with senescence.

Activity and quantity of rubisco decrease faster in the foliage of potato and radish when plants are stressed with the oxidizing air pollutant ozone (Fig. 3) (Dann and Pell, 1987, 1989; Pell *et al.*, unpublished data). We have speculated that ozone, a free-radical producer (Grimes *et al.*, 1983), may create a more oxidizing environment in the cell. This phenomenon could result from a variety of nonenzymatic reactions that lead to increased concentrations of superoxide, hydroxyl radicals, and so on. Small changes could lead to increased proteolysis of rubisco, and early senescence might result. Dann (1988) has shown that when purified rubisco is exposed to ozone in vitro, the enzyme is more susceptible to subsequent proteolysis than when exposed to oxygen. This is consistent with the work of Wolff and Dean (1986), who have shown that hydroxyl radicals and hydrogen peroxide increased the proteolytic susceptibility of in vitro concentrations of a number of animal proteins.

Above we discussed mechanisms by which stresses could induce proteolysis through structural modifications of proteins. Available evidence also indicates that proteinase activity may be enhanced. Watanabe and Kondo (1984a,b) injected plants with sodium sulfite, a surrogate for sulfur dioxide

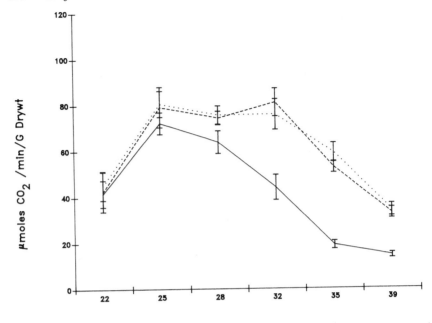

Figure 3 Total activity of ribulose 1,5-bisphosphate carboxylase/oxygenase (rubisco) extracted from radish foliage grown in charcoal-filtered air with or without supplemental ozone. Ozone treatments were delivered from 1000 to 1930 hours each day from plant emergence (day 7) through day 39. ---, charcoal-filtered air; —, average ozone exposure of 38 ppb; ..., average ozone exposure of 66 ppb.

and inducer of superoxide and hydrogen peroxide. When cell extracts were studied, proteinase activity increased and proteinase inhibitor decreased in the treated plants. Watanabe and Kondo (1984a,b) proposed that active oxygen may have deactivated the inhibitor.

Many stresses, including mechanical wounding, viruses, cadmium, ozone, chilling, flooding, and drought, induce ethylene and associated premature senescence (Yang and Hoffman, 1984). Ozone-induced foliar injury and depressed net photosynthesis can be prevented if ethylene synthesis is inhibited (Melhorn and Wellburn, 1987; Taylor *et al.*, 1988). Stress ethylene emission is a transitory response typically declining within 24 hours of stress induction. How then could it lead to premature senescence? Melhorn and Wellburn (1987) suggest that ozone may induce ethylene emission and then react with ethylene to generate reactive species responsible for most of the foliar injury that results. The possibility that free radicals are produced as a result of reactions between ozone and ethylene and that these free radicals initiate a chain reaction leading to accelerated senescence seems plausible. Alternatively, ethylene production might exert its effects on the level of antioxidants present in the cell. It has been demonstrated that exogenously supplied ethylene can inhibit polyamine synthesis (Ickeson *et al.,* 1986). During the few hours of stress ethylene synthesis, a suppression of polyamine synthesis could allow for free-radical reactions that might otherwise be suppressed. Whether a rise in free radicals for a few hours would be sufficient to trigger accelerated senescence is open to speculation.

The mechanism by which stress affects lipids is unclear, but there are many examples of alterations of this key membrane component, including membrane changes in fluidity in response to cold, elevated temperature, and salt stress (Raison *et al.,* 1980; Levitt, 1980). It is fairly clear that free radicals, e.g., hydroxyl radicals, can peroxidize lipids (Fantone and Ward, 1985), and many stresses, including ozone, generate these free radicals (Grimes *et al.,* 1983). Although there have been some reports on effects of ozone on sterols, fatty acids, and phospholipids (Travathan *et al.,* 1979; Fong and Heath, 1981; Spotts *et al.,* 1975), the results have little bearing on the hypothesized events leading to increased lipoxygenase activity. Changes in fatty acids driving the lipoxygenase system are transitory events that may not be reflected in easily discernible changes in lipid content.

C. How Do Multiple Stresses Affect Aging and Senescence?

Little information bears directly on the impact of multiple stresses on senescence. Several stresses could all target the same process, or a number of different cellular events could be influenced by various stresses. In either case, senescence would be accelerated, but the mechanism(s) for arriving there would be different.

Certain stresses, like nitrogen deficiency, would influence both the size of the protein pool and protein degradation (Makino *et al.,* 1984). Addition of an external oxidant might stimulate proteinase activity. A nitrogen-deficient plant has uniform proteolysis, and ozone-stimulated proteolysis could enhance the process. Since nitrogen-deficient plants complete the process of proteolysis more rapidly (Makino *et al.,* 1984), properly timed ozone-induced proteolysis could further reduce the photosynthetic life of a leaf. When senescence begins, leaves with higher levels of nitrogen reflect higher initial rates of proteolysis, followed by a slower relative rate of proteolysis—in other words, a curvilinear response. Oxidant stress that occurs at the onset of proteolysis might not influence the already-high rate of protein degradation or related senescence. If oxidant pressure occurs later during aging, enhanced proteolysis might have a greater effect in accelerating senescence.

If ozone is the initial stress, it might enhance lipid peroxidation and subsequent degradation of proteins, particularly rubisco in the chloroplast. If water stress is imposed later, resulting cytoplasmic proteolysis could occur because enzymes would be released from the vacuole. In this case, the resultant accelerated senescence would reflect interaction between two stresses acting at different cellular sites.

III. Whole-Plant Senescence

Up to this point we have discussed induction of foliar senescence with an emphasis on the molecular mechanisms by which this may occur. The influence of stress on foliage must now be related to senescence of the whole plant. The mechanisms of induction of monocarpic and polycarpic senescence are not well understood, although multiple theories have been proposed (Kelly and Davies, 1988; Nooden, 1988). Regardless of the triggering mechanisms, it seems apparent that whole-plant senescence is influenced by the assimilatory capacity of the foliage, also referred to as the source, and the size and demand of the sink for carbon within the plant (Kelly and Davies, 1988).

Leaves senesce as nitrogen is translocated from the source to a sink, be it a younger leaf or a storage organ. Many signals can lead to the removal of nitrogen from a leaf. There are genetically programmed limits to longevity; therefore, movement of nitrogen within the plant is predetermined (Mooney *et al.*, 1981). When monocarpic plants shift into the reproductive phase, nitrogen demand by the seed is met both by root uptake of the nutrient and redistribution of amino acids from vegetative parts of the plant (Sinclair and de Wit, 1975). Once supply from the source is depleted, the plant is fully mature, and senescence is complete.

A. Influence of Environmental Stress on Senescence

Many environmental stresses influence the onset of reproduction (see Chapter 8). Environment can also play a direct role in influencing the onset of foliar senescence. Nitrogen supply can have an important effect on the longevity of many monocarpic species. For example, when potato plants are grown with a low nitrogen supply, tuberization is initiated earlier than when plants are grown with more nitrogen (Marschner, 1986). The shift to sink formation results in movement of carbon and nitrogen from leaves to tubers. The leaves of plants grown with low nitrogen supply are relatively small, and depletion of source metabolites will occur more rapidly than in plants grown with more available nitrogen. The result is accelerated whole-plant senescence. Similar examples can be drawn from many other species. Other nutrients also have the potential for controlling plant longevity, but nitrogen is most often the limiting factor.

Water stress will induce similar responses as low nitrogen supply, perhaps directly by causing less of the nutrient to be imported from the soil (Wolfe *et al.*, 1988). Alternatively, water stress may trigger premature leaf drop by stimulating shifts in hormones (increase in the abscisic acid : cytokinin ratio) (Simpson, 1981). Regardless of the mechanism, less nitrogen and carbon are available for translocation to sinks, and the duration of the process is abbreviated, again leading to accelerated senescence.

Light also has a profound influence on the nitrogen export of aging leaves. Mooney *et al.* (1981) demonstrated that as leaves of annual plants in Illinois aged and became shaded, both leaf specific weight and the percentage of nitrogen in leaves diminished. In contrast, desert annuals, which did not experience shading of foliage, exhibited increases or neglible shifts in leaf specific weight. Percentages of leaf nitrogen declined in aging desert annuals but not as dramatically as was observed in the more temperate-adapted species. Even though plants grown at the two environments were in the early reproductive phase of development, leaf loss in temperate annuals was significantly greater than in desert annuals. Thus, shading reduces the size of the source, and, in the case of determinant annuals, leads to senescence of the plant.

B. Timing of Stress Affects Its Impact on Senescence

The discussion above focused on the examples of how plant species respond to chronic stress. In these cases, the longevity of the source is shortened, and, as sink strength increases, the plant senesces. In many cases, the scenario is more complex. The stress may be acute, only affecting leaves that are vulnerable at the time of environmental aberration. Under these conditions, effects on whole-plant senescence may be more difficult to define. The plant species may be a determinant annual, as in many of the examples above, or an indeterminant annual or perennial with greater inherent potential for repair or compensation.

Two kinds of stresses influence foliar senescence: (1) those that are transient and may influence only a small percentage of all foliage or all of the foliage within a given year (e.g., an ozone episode or an annual drought in a temperate climate) and (2) long-term stresses, such as low nutrient status or chronic water stress, as occurs in a desert environment. An acute stress like an ozone episode may induce early abscission of the leaves that were vulnerable during the exposure. If the stress is isolated, compensation mechanisms may produce new foliage, and assimilatory capacity may be restored (Walmsley *et al.*, 1980). With source strength recovered, accelerated senescence of the whole plant would not be predicted. Alternatively, if ozone stress is chronic, and accelerated senescence occurs on foliage throughout the plant, any compensatory repair process would eventually be compromised as the source yielded to the strength of the sink. This phenomenon has been observed on determinant potato plants, which, when stressed by chronic ozone exposures, exhibited significant acceleration of senescence, in contrast to plants grown in charcoal-filtered air (Pell *et al.*, 1988).

The timing of stress in relation to the onset of senescence in polycarpic species becomes particularly complex. The notion that ozone may induce decline in several forest species has been proposed, but the mechanism remains undefined (Reich and Amundson, 1985). For ozone to induce senescence in polycarpic species, it would have to diminish source strength repetitively over many years, ultimately resulting in a major reduction in the ability of foliage to sustain nonphotosynthetic sinks. Alternatively, ozone could induce senescence of a polycarpic species if the stressed individual(s) were at a stage of development chronologically close to the genetically determined lifespan of the species. In such a case, the source strength would already be reduced at the expense of the sink. Ozone-induced foliar abscission would contribute to reduced source strength, leading to accelerated senescence.

C. Multiple-Stress Effects on Whole-Plant Senescence: Hypothetical Pathways

The potential effects of single stresses on whole-plant senescence are summarized in Figure 4. Stresses may be acute or chronic. If source strength is diminished, sink strength will eventually drain the species and hasten senescence of the plant (IIa). Alternatively, reduction in source strength because of stress could lead to adjustment of the sink so that the source : sink ratio is maintained, and senescence of the plant will not be affected (IIb). Rather than diminish source strength, some stress regimes may induce compensation, so that source strength is maintained (Ia and b). Depending on the cost of compensation, sink strength will be either diminished or left unaffected. Regardless of the effect on sink strength, senescence of the plant would be unaffected in this case.

If we now superimpose multiple stresses on the schematic describing plant response to single stresses, we can develop some hypotheses to predict multiple-stress effects on plant senescence. Plants that follow path Ia (Fig. 4) in response to stress, decreasing root : shoot ratio, will be more vulnerable to subsequent stress requiring greater sink strength; thus a plant capable of compensating for a stress like ozone could become more vulnerable to drought. Although plant senescence would not be affected by ozone alone, a subsequent drought could trigger proteolysis in the foliage and lead to more rapid aging of the plant. Alternatively, a determinant plant challenged by a foliar stress like ozone at a developmental time when compensation was no longer an option (Fig. 4; path IIa) may be less

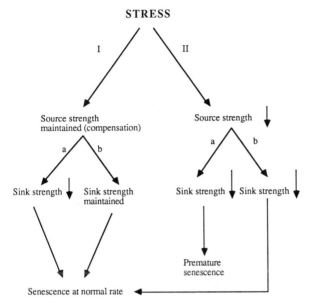

Figure 4 Schematic proposal to describe stress-induced shifts in size of source and sink and their impact on whole-plant senescence.

vulnerable to interacting stresses demanding greater sink strength, such as drought. In this case, the source would already be translocating nitrogen to the sink, and water stress would be less likely to influence the rate of foliar senescence and associated whole-plant aging.

Information on interactions between stresses and plant senescence is scant. The observation of forest decline throughout the world is followed by speculation that multiple-stress interactions could explain the early dying of several tree species. To understand and ultimately predict the effects of multiple stresses on whole-plant senescence, we need to consider plants with different life-history strategies: (1) determinant annuals, (2) indeterminant annuals, and (3) perennials. The nature of the stress, whether chronic or acute, will be critical to the magnitude of plant response. Finally, the timing of stress in relation to physiological phase in the development of the plant will determine the magnitude of any potential interaction.

Acknowledgments

Some of the research reported herein was supported by EPRI contract RR 2799-4 and USDA Grant No. 84-CSRS-22394.

References

Asada, K. (1984). Chloroplasts: Formation of active oxygen and its scavenging. *Methods Enzymol.* 105:422–429.

Cooke, R. J., Roberts, K., and Davies, D. D. (1980). Model for stress-induced protein degradation in *Lemna minor. Plant Physiol.* 66: 1119–1122.

Dalling, M. J. (1987). Proteolytic enzymes and leaf senescence. *In* "Plant Senescence: Its Biochemistry and Physiology" (W. W. Thomson, E. A. Nothnagel, and R. C. Huffaker, eds.), pp. 54–70. American Society of Plant Physiologists, Rockville, Maryland.

Dann, M. S. (1988). Impact of ozone on the activity and quantity of ribulose bisphosphate carboxylase/oxygenase in potato foliage and its relation to premature senescence. Ph.D. dissertation. Pennsylvania State University, University Park.

Dann, M. S., and Pell, E. J. (1987). Decline of rubisco activity and net photosynthesis in ozone-treated potato foliage. *Plant Physiol.* 83: 111.

Dann, M. S., and Pell, E. J. (1989). Impact of ozone on ribulose bisphosphate carboxylase/oxygenase in an ozone-sensitive potato cultivar. *Plant Physiol.* 91: 427–432.

Drolet, G., Dumbroff, E. B., Legge, R. L., and Thompson, J. E. (1986). Radical scavenging properties of polyamines. *Phytochemistry* 25: 367–371.

Dungey, N. O., and Davies, D. D. (1982). Protein turnover in isolated barley leaf segments and the effects of stress. *J. Exp. Bot.* 33: 12–20.

Fantone, J. C., and Ward, P. A. (1985). Oxygen-derived radicals and their metabolites: Relationship to tissue injury. Upjohn Company, Kalamazoo, Michigan.

Fong, F., and Heath, R. L. (1981). Lipid content in the primary leaf of bean *(Phaseolus vulgaris)* after ozone fumigation. *Z. Pflanzenphysiol.* 104: 109–115.

Fridovich, I. (1978). The biology of oxygen radicals. *Science* 201: 875–880.

Galston, A. W., and K-Sawhney, R. (1987). Polyamines and senescence in plants. *In* "Plant Senescence: Its Biochemistry and Physiology" (W. W. Thomson, E. A. Nothnagel, and R. C. Huffaker, eds.), pp. 167–181. American Society of Plant Physiologists, Rockville, Maryland.

Goren, R., Mattoo, A. K., and Anderson, J. D. (1984). Ethylene binding during leaf development and senescence and its inhibition by silver nitrate. *J. Plant Physiol.* 117: 243–248.

Grimes, H. D., Perkins, K. K., and Boss, W. F. (1983). Ozone degrades into hydroxyl radical under physiological conditions. *Plant Physiol.* 72: 1016–1020.

Haber, F., and Weiss, J. (1934). The catalytic decomposition of hydrogen peroxide by iron salts. *Proc. R. Soc. Lond.* A 147: 332–351.

Harbour, J. R., and Bolton, J. R. (1978). The involvement of the hydroxyl radical in the destructive photooxidation of chlorophylls in vivo and in vitro. *Photochem. Photobiol.* 28: 231–234.

Ickekson, I., Bakhnashvili, M., and Apelbaum, A. (1986). Inhibition by ethylene of polyamine biosynthetic enzymes enhanced lysine decarboxylase activity and cadaverine accumulation in pea seedlings. *Plant Physiol.* 82: 607–609.

Kelly, M. O., and Davies, E. P. J. (1988). The control of whole-plant senescence. *In CRC Crit. Rev. Plant Sci.* 7: 139–173.

Laurière, C. (1983). Enzymes and leaf senescence. *Physiol. Veg.* 21: 1159–1177.

Leopold, A. C. (1980). Aging and senescence in plant development. *In* "Senescence in Plants" (K. V. Thimann, R. C. Adelman, and G. S. Roth, eds.), pp. 1–12. CRC Press, Boca Raton, Florida.

Leshem, Y. Y., Grossman, S., Frimer, A., and Ziv, J. (1979). Endogenous lipoxygenase control and lipid-associated free-radical scavenging as modes of cytokinin action in plant senescence retardation. *In* "Advances in the Biochemistry and Physiology of Plant Lipids" (L. A. Appelqvist and C. Liljenberg, eds.), pp. 193–198. Elsevier, New York.

Levitt, J. (1980). "Responses of Plants to Environmental Stresses," Vol. I and II. Academic Press, New York.

Lynch, D. V., and Thompson, J. E. (1984). Lipoxygenase-mediated production of superoxide anion in senescing plant tissue. *FEBS Lett.* 173: 251–254.

Makino, A., Mae, T., and Ohira, K. (1984). Relation between nitrogen and ribulose 1,5-bis-phosphate carboxylase in rice leaves from emergence through senescence. *Plant Cell Physiol.* 25: 429–437.

Marschner, H. (1986). "Mineral Nutrition of Higher Plants," p. 674. Academic Press, New York.

McKersie, B. D., Senaratna, T., Walker, M. A., Kendall, E. S., and Hetherington, P. R. (1988). Deterioration of membranes during aging in plants: Evidence for free-radical mediation. *In* "Senescence and Aging in Plants" (L. D. Nooden and A. C. Leopold, eds.), pp. 441–464. Academic Press, New York.

McRae, D. G., and Thompson, J. E. (1983). Senescence-dependent changes in superoxide anion production by illuminated chloroplasts from bean leaves. *Planta* 158: 185–193.

Melhorn, H., and Wellburn, A. R. (1987). Stress ethylene formation determines plant sensitivity to ozone. *Nature* 327: 417–418.

Mooney, H. A., Field, C., Gulmon, S. L., and Bazazz, F. (1981). Photosynthetic capacity in relation to leaf position in desert versus old-field annuals. *Oceologia* 50: 109–112.

Nooden, L. D. (1988). Whole-plant senescence. *In* "Senescence and Aging in Plants" (L. D. Nooden and A. C. Leopold, eds.), pp. 391–439. Academic Press, New York.

Pell, E. J., Pearson, N. S., and Vinten-Johansen, C. (1988). Qualitative and quantitative effects of ozone and/or sulfur dioxide on field-grown potato plants. *Environ. Pollut.* 53: 171–186.

Raison, J. K., Berry, J. A., Armond, P. A., and Pike, C. S. (1980). Membrane properties in relation to the adaptation of plants to temperature stress. *In* "Adaptation of Plants to Water and High-Temperature Stress" (N. C. Turner and P. J. Kramer, eds.), pp. 261–273. Wiley, New York.

Reich, P. B., and Amundson, R. G. (1985). Ambient levels of ozone reduce net photosynthesis in tree and crop species. *Science* 230: 566–570.

Richmond, A. E., and Lang, A. (1957). Effect of kinetin on protein content and survival of detached *Xanthium* leaves. *Science* 125: 650–651.

Sexton, R., Lewis, L. N., Trewavas, A. J., and Kelly, P. (1985). Ethylene and abscission. *In* "Ethylene and Plant Development" (J. A. Roberts and G. A. Tucker, eds.), pp. 173–196. Butterworth, London.

Simpson, G. M. (1981). "Water Stress on Plants," p. 324. Praeger, New York.

Sinclair, J. R., and de Wit, C. T. (1975). Photosynthate and nitrogen requirements for seed production. *Science* 189: 565–567.

Spotts, R. A., Lukezic, F. L., and LaCasse, N. L. (1975). The effect of benzimidazole, cholesterol, and a steroid inhibitor on leaf sterols and ozone resistance of bean. *Phytopathology* 65: 45–49.

Taylor, G. E., Ross-Todd, B. M., and Gunderson, C. A. (1988). Action of ozone on foliar gas exchange in *Glycine max* L. Merr.: Potential role for endogenous stress ethylene. *New Phytol.* 110: 301–307.

Thayer, S. S., Choe, H. T., Tang, A., and Huffaker, R. C. (1987). *In* "Plant Senescence: Its Biochemistry and Physiology" (W. W. Thomson, E. A. Nothnagel, and R. C. Huffaker, eds.), pp. 71–80. American Society of Plant Physiologists, Rockville, Maryland.

Thompson, J. E., Paliyath, G., Brown, J. H., and Duxbury, C. L. (1987). The involvement of activated oxygen in membrane deterioration during senescence. *In* "Plant Senescence: Its Biochemistry and Physiology" (W. W. Thomson, E. A. Nothnagel, and R. C. Huffaker, eds.), pp. 146–155. American Society of Plant Physiologists, Rockville, Maryland.

Trevathan, L. E., Moore, L. D., and Orcutt, D. M. (1979). Symptom expression and free sterol and fatty acid composition of flue-cured tobacco plants exposed to ozone. *Phytopathology* 69: 582–585.

Trippi, V. S., and De Luca d'Oro, G. M. (1985). The senescence process in oat leaves and its regulation by oxygen concentration and light irradiance. *Plant Cell Physiol.* 26: 1303– 1311.

Walmsley, L., Ashmore, M. R., and Bell, J. N. B. (1980). Adaptation of radish, *Raphanus sativus* L., in response to continuous exposure to ozone. *Environ. Pollut.* 23: 165–177.

Watanabe, T., and Kondo, N. (1984a). The change in leaf proteinase and proteinase inhibitor activities by air pollutant. I. Participation of proteinases in cellular and molecular damages of plant leaves by SO_3^{2-} and H_2O_2. *Res. Rep. Natl. Environ. Stud. Jpn.* 65: 241–250.

Watanabe, T., and Kondo, N. (1984b). The change in leaf proteinase and proteinase inhibitor activities by air pollutant. II. Purification and some properties of proteinase and its inhibitor in the leaf of *Ricinus communis*. *Res. Rep. Natl. Environ. Stud. Jpn.* 65: 253–262.

Wolfe, D. W., Henderson, D. W., Hsiao, T. C., and Alvino, A. (1988). Interactive water and nitrogen effects on senescence of maize. I. Leaf area duration, nitrogen distribution, and yield. *Agron. J.* 80: 859–864.

Wolff, S. P., and Dean, R. T. (1986). Fragmentation of proteins by free radicals and its effect on their susceptibility to enzymic hydrolysis. *Biochem. J.* 234: 399–403.

Yang, S. F., and Hoffman, E. E. (1984). Ethylene biosynthesis and its regulation in higher plants. *Annu. Rev. Plant Physiol.* 35: 155–189.

10

Modelling Integrated Response of Plants to Multiple Stresses

Peter J. H. Sharpe   **Edward J. Rykiel, Jr.**

Modelling, like scientific observation and experimentation, is a method for increasing our understanding of cause-and-effect relationships. Of the modelling approaches available, the approach selected for a given problem depends on implied objectives, which can be expressed as a combination of precision, generality, and reality (Levins, 1966). Sharpe (1990) defined *precision* as the degree of exactness in a measurement or prediction, *general-*

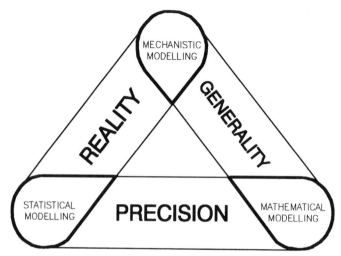

Figure 1 Three general modelling approaches—statistical, mathematical, and mechanistic—represented as the intersection of reality, generality, and precision. The side opposite the modelling method indicates the likely compromise for that approach.

ity as the applicability of a concept to a whole range of instances, and *reality* as the collection of true causes and effects that underlies superficial appearances. Levins (1966) suggests that it is not possible to maximize more than two of these objectives at the same time. Given this limitation, we propose three general approaches (Fig. 1). The intersection of precision and reality can be identified with statistical modelling, generality and precision with mathematical modelling, and generality and reality with mechanistic modelling. Not understanding the limitations of each of these approaches may lead to unrealistic expectations of what a model can do. Considering these limits is important for choosing a modelling approach to integrate the effects of multiple stresses on plant growth. In this chapter, we emphasize mathematical and mechanistic models.

I. Modelling Compromises

A. Statistical Modelling

If the objective of the model is to provide a precise numerical or graphical representation of data without regard for underlying mechanisms, then statistical modelling procedures are the most appropriate. The compromise is the loss of generality (see Fig. 1). Prerequisites for using this approach include adequate data sets that have been collected with a well-designed sampling procedure. Many of the well-known growth and yield regression models used in agriculture fall into this category. The drawbacks to this approach are: (1) because it is data specific, it precludes generalization beyond the data sets and the specific conditions for which the model parameters were estimated, and (2) as more stress and environmental factors are introduced into the model, the uncertainty associated with each ultimately outweighs the extra precision gained by adding new variables. The dimensionality of the data-based models required for representing and integrating stress interactions eventually limits their utility. Because of these limitations, statistical modelling, when used by itself, has a restricted appeal as a tool for integrating multifactor responses of plants.

B. Mathematical Modelling

Mathematical modelling involves abstraction, which results in the creation of a simplified reality. The compromise in mathematical modelling is the idealization of a phenomenon, which may or may not reflect true cause-and-effect relationships (i.e., reality; Fig. 1). One of the most elegant examples is the theoretical model of Cowan (1977, 1982). In this model, Cowan uses calculus of variations to determine optimal patterns of stomatal opening. The goal is to give the best balance between water loss and carbon assimilation. The value of this mathematical model is that it establishes a theoretical standard for comparison with real plant behavior. Gross (1989) has provided a short review of topics where theory has contributed new perspectives on plant growth processes.

Mathematical models have also been proposed to optimize allocation of biomass increment to leaves, stems, roots, and fruits. Cost–benefit analysis that takes biomass or a limited nutrient as a currency to predict the ideal allocation pattern is an example of this type of model. Bloom *et al.* (1985) have proposed the optimum resource allocation hypothesis, which states that plants respond homeostatically to imbalances in resource availability by allocating new biomass to organs that enhance the acquisition of those resources most strongly limiting growth. Again, this theoretical method provides an ideal with which to compare actual plant growth.

Levin (1981) argues that theoreticians play a larger role than that of mere participants in the induction–deduction–verification process. The theorist aims at a general understanding, not at precise prediction or representation of processes. Thus, fundamental properties of a system may be revealed that otherwise might be obscured by more detailed models.

C. Mechanistic Modelling

Documenting and testing true cause-and-effect relationships is the ratio-nale for developing mechanistic models that explain observed behavior. The compromise in this context is that, even though the mechanistic model structure and its predictions are qualitatively correct, the results will not be quantitatively exact (i.e., no precision; Fig. 1). For a model to have explanatory power, its hypotheses must be formulated at a mechanistic level. But the constituents of mechanism vary according to the level at which the system is observed. Consider a simple ecological hierarchy; mechanism at one level is empiricism at another. Therefore, it is not possible to define an absolute level that can be called mechanistic; mechanistic conjectures underlying one conceptual framework may be empirical observations in another. The important distinction is the relationships among hierarchical levels (Rykiel *et al.*, 1988) for a given problem.

One of the challenges in designing models with explanatory power is to avoid a mechanistic regress. Physiologists, in particular, look for a one-to-one correspondence between the physiology in the plant and in the model. Ideally, they desire each mechanism to be described in terms of more primitive mechanisms. Sometimes this can be legitimately achieved, and a more robust model results. More often, mechanisms are understood in isolation from the rest of the plant's control system. This is not surprising because reductionistic techniques have been used to understand the particular mechanism in the first place. The modeller now wants to construct a holistic system in which to view the contribution of the different mechanisms. Because linkages between mechanisms are largely unknown, it is essential to keep the number of these linkages as low as possible; otherwise, it is difficult to recognize inadequate or inappropriate linkages. Mechanistic models need to maintain a balance between simplicity and complexity (Fig. 2). The goal is to recreate realistic behavior without unnecessary detail.

D. Criteria for Testing Models

A common misconception is that all models can be tested using the same criteria. Each of the three modelling approaches — statistical, mathemati-

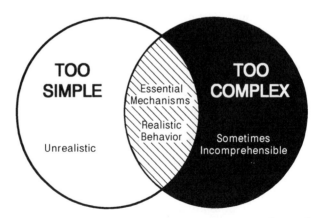

Figure 2 Explanatory model design must be balanced between simplicity and complexity.

cal, and mechanistic—has different test criteria. Statistical models are appropriately tested against particular data sets and should not be expected to provide generality beyond these data (see Fig. 1). In turn, mechanistic models should not be expected to provide precise quantitative results, and mathematical models should not be judged on the basis of their biological detail. In an actual modelling project, all of these approaches are likely to be used, but one approach will generally predominate, based on the goals of the project.

Validation of mechanistic models is a two-step task. The first step is to consider the validity of basic mechanisms included in the model (test the pieces). The validity of these reductionistic components should be tested outside the model. Most explanatory models have many basic mechanisms, and it is totally inappropriate to attempt to test these within the context of the overall model. The second step is to evaluate how the mechanisms are linked together (test the integrated whole). The biochemical photosynthesis model of Farquhar *et al.* (1980) is a good example. The rate constants for rubisco included in the model have been determined by in vitro studies and thus cannot be changed when setting parameters of the model. The structure of the model represents a new hypothesis. In contrast to the typical hypotheses investigated in most experimental studies, this hypothesis takes the form of a systems design describing the structure and function of a complex suite of interacting biochemical and biophysical mechanisms (the integrated whole).

Mathematical and, especially, mechanistic models cannot always be represented in a form such that experimenters can design empirical tests of the model in toto. In most cases, it is only possible to test particular model predictions and to gain confidence in the model as these results are supported. In general, mathematical and mechanistic models can produce far more behavior than can be tested experimentally. We therefore have to be content with limited testing and must rely on the validity of the component mechanisms for justifying confidence in the model's behavior over a wide range of conditions, many of which may not have been experimentally observed or tested. General circulation models (GCMs) of the atmosphere (Henderson-Sellers and McGuffie, 1987) may be the quintessential example of this difficulty.

II. Adaptive Growth Processes

We examine three examples of adaptive biochemical and biophysical processes in this section. A survey of basic metabolic features of natural adaptive responses to multiple stresses can provide a template for model design by increasing its realism. The mechanisms chosen are cyclical pathways with dynamic pools, carbon flow homeostasis, and compensatory allocation of photosynthate.

A. Cycles with Dynamic Pools

For modelling purposes, biochemical processes can be viewed either as linear pathways or as regenerative cycles. In a linear pathway, biochemical substrates enter in one form and leave in another. A linear system has little capability to adjust to changes in the availabilities of input substrates. A closer look at actual biochemical systems in nature reveals the widespread occurrence of a hierarchy of regenerative cycles with dynamic pools. This hierarchy of cycles is expressed in terms of particles, catalysts, substrates, and mass flow systems. Although these cycles are described in most biochemistry texts, both biochemists and ecologists often fail to recognize the adaptive nature of the cyclic design in meeting the demands imposed by multiple stresses. Rather than referring readers to the detail of biochemistry texts, we provide a simple visual summary of the hierarchy of representative growth cycles.

Particle cycles include both electron (e^-) and proton (H^+) transport systems. These two systems are coupled to form cyclic photophosphorylation processes in the grana of chloroplasts (Fig. 3). In this cycle, electrons released by splitting water in photosystem II are recycled between high and low energy states in photosystem I. Energy in the form of incident photons raises the free electron to a higher energy state. This energy is used to transport protons into the inner regions of the thylakoid. The gradient of protons across the thylakoid membrane creates an electric potential for ATP synthesis. The linkage of these two particle cycles synthesizes ATP from light energy.

Molecular reactions involve a cyclic process of enzyme transition from a free state to enzyme–substrate complex with the arrival of substrate at the

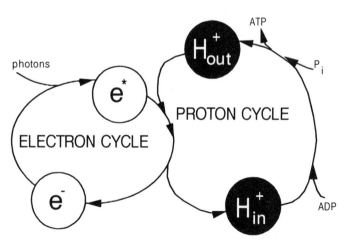

Figure 3   Electron and proton transport cycles. Together they form the cyclic photophosphorylation system.

active site. Release of product returns the enzyme to the free state again. Mechanistic modelling of the temperature dependence of enzyme-based biochemical processes has been described by Sharpe and DeMichele (1977). In biochemical cycles, an enzyme product from one reaction becomes the substrate for the next. In the Calvin cycle, enzyme reactions regenerate C_5 carbon skeletons, which combine with carbon dioxide to form two C_3 sugars. The Calvin cycle is driven by chemical energy input in the form of NADPH and ATP. Six revolutions of the cycle result in the synthesis of one C_6 sugar (Fig. 4).

Whole-plant cycles are also evident. The translocation cycle (Fig. 5) circulates water and inorganic ions (such as potassium) to transport carbon from leaf synthesis sites to growth sinks. This cycle is driven by active loading of sugars and inorganic ions in the leaf. Osmotic pressure causes the sugar–water solution to move by mass flow along the phloem sieve-tube elements. The difference in water potential between the xylem and phloem causes additional water to enter the phloem, resulting in increased velocities and decreasing sugar concentrations (Goeschl *et al.*, 1976; Magnuson *et al.*, 1979, 1986; Goeschl and Magnuson, 1986). Potassium and other nutrient ions are recycled in the translocation cycle or, if leached from the leaves, by nutrient cycling in the ecosystem.

Figure 6 shows a synthesis of biochemical and biophysical cycles at the whole-plant level. Additional cycles included are the Tolbert, glycine–serine–ammonia, Krebs, and amino-acid synthesis. The dependence of these coupled cycles on the input of water, light, nutrients, and carbon dioxide is shown. Integration of resource inputs occurs by coupling cyclical processes, regulated by dynamic metabolic pools. From a modelling viewpoint, it is not the identity of these innumerable intermediate compounds that is important, but the manner by which the overall system is structured to be regenerative and self-regulating in response to changes in resource

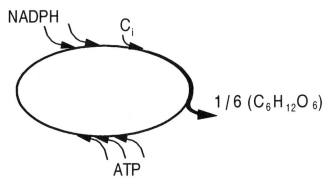

Figure 4   The Calvin cycle, showing cyclic flow of carbohydrate intermediates. Inputs include carbon dioxide, 2NADPH, and 3ATP. Outputs include 1/6 C_6 sugar, 2NADP, and 3ADP.

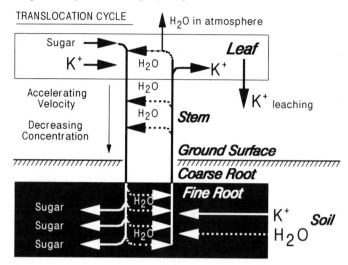

Figure 5 Translocation cycle showing transport of sugars and cyclical movement of water and monovalent ions such as potassium.

inputs. Later in this chapter, we will re-examine the integrated structure of cyclic processes as a guide to adaptive model design.

B. Carbon Flow Homeostasis

Physiological knowledge regarding regulation of carbon flow is limited. What fragmentary information we have focuses on stomatal regulation. In many respects, we know more about the behavior of stomata than the mechanisms controlling this behavior. For example, the physiology of the guard cell is well described in terms of stomatal mechanics, carbon metabolism, ion fluxes, bioenergetics, light, and hormones and age (Zeiger *et al.*, 1987). Yet this detailed physiology does not explain observed homeostasis. There is no alternative at present to developing phenomenological models. Mechanistic models are in the future.

An ideal control strategy for regulating conductance has been defined by Cowan (1977, 1982). It is a trajectory that minimizes water loss, subject to the constraint that carbon assimilation satisfies the plant's carbon needs for growth and reproduction. To test the concept that stomata behave in a manner to optimize carbon supply to meet plant growth demands, Ball *et al.* (1987) measured stomatal conductance in relation to four variables. They found that stomatal conductance g varies directly with assimilation rate A scaled by the ratio of relative humidity at the leaf surface h_s to the mole fraction of CO_2 at the leaf surface C_s:

$$g = kAh_s/C_s \tag{1}$$

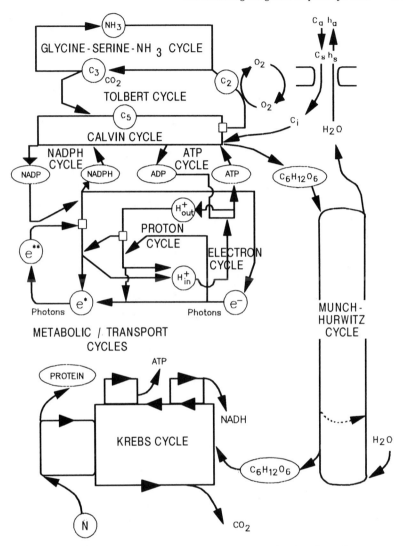

Figure 6 Coupling metabolic and transport cycles for the whole plant.

where k is a proportionality constant. Eliminating A from Equation (1) by substituting the diffusion gradient equation (Nobel, 1983)

$$A = g(C_s - C_i) \qquad (2)$$

where C_i is the internal leaf concentration of CO_2 yields the relationship

$$(1 - C_i/C_s) = 1/(h_s k) \qquad (3)$$

Mechanistically, it is difficult to explain the regulation of internal CO_2 by surface relative humidity as Equation (3) suggests. Rearrangement of Equation (1) also yields

$$A = (gC_s)/(h_s k) \tag{4}$$

This relationship indicates that carbon assimilation rate is proportional to stomatal conductance. Experience with the model of Farquhar *et al.* (1980) suggests that this proportionality is only valid in the CO_2 range from 200 to 600 ppm.

The stomata are an example of a key physiological component for which we have insufficient information to develop an explanatory model. Without a mechanistic understanding, we are forced to use empirical relationships such as Equations (1), (3), and (4). The restricted range over which empirical relationships are valid is a significant limitation to model design.

C. Compensatory Allocation

The optimum resource allocation hypothesis states that plants respond homeostatically to imbalances in resource availability by allocating newly acquired carbon to organs that enhance the acquisition of resources most strongly limiting growth (see review by Bloom *et al.*, 1985). The ideal is that plants adjust their allocation so that their limitation for growth is more nearly equal for all resources. Experimental data reviewed by Bloom *et al.* indicate that plants from resource-rich environments have highly flexible allocation patterns in response to environmental stress. In contrast, plants from resource-poor environments have more rigid allocation patterns. When resources temporarily become plentiful, plants from poor environments typically respond by increased storage.

A flexible pattern of compensatory allocation enables a plant to adapt to resource limitations caused by neighbors and environmental stresses. The review of Bloom *et al.* (1985) documents situations where light and carbon limitations result in proportionally more shoot than root biomass. Conversely, nutrient stress leads to increasing proportional allocation to root growth. Water stress can either increase or decrease root : shoot allocation, depending on the species and the severity of the stress. The physiological mechanisms controlling the allocation system probably involve an interaction between the phloem transport system and hormonal messengers carried in the xylem-phloem water flow cycle. The mechanistic principles controlling these crucial physiological processes are unknown (see also Chapters 1, 5, 6).

III. Model Design Using Cyclic Processes

The problems in building plant growth models have been reviewed by Acock and Reynolds (1990). They point out that building a process model is a complex and difficult task. A large systems model, comprising numerous subsystems, can daunt those interested in modifying or adapting it for other purposes. Acock and Reynolds suggest that growth-process modelling is still at the master-craftsman stage. They propose that the components of process models should be standardized into interchangeable units. This is an excellent idea but one that is difficult to implement given the dynamic nature of computer hardware, modelling software, biological understanding, and funding sources.

Gross (1989) points to a "lack of expertise" in conceptual methods to deal with interactions involving integrated biochemical control systems. Most systems modellers are acutely aware of the difficulties of integrating submodels for water, carbon, nitrogen, and energy. These separate submodels must be linked in a manner that reflects the natural functional integration of the plant's biochemical and biophysical processes. Ford and Kiester (1990) used the phloem translocation model of Smith *et al.* (1980) to integrate water and carbon submodels to describe branch growth in trees. They also point out that transport has not been studied adequately, either empirically or mathematically, except in short plants. What is required is a synergistic approach that uses holism as a template for arranging reductionistic components. The history of incremental progress in designing photosynthesis models may serve as an example.

The first attempt to develop a photosynthesis model based on a simplified version of the Calvin cycle was that of Chartier (1966, 1970). This model made no provision for photorespiration because biochemical mechanisms controlling photorespiration were at that time unknown. This was rectified by plant biochemists in the late 1960s. A simplified coupled three-cycle biochemical schema was subsequently proposed by Hall (1971). A flood of simplified coupled biochemical cycle models appeared over the next few years (Lommen *et al.*, 1971; Charles-Edwards and Ludwig, 1974; Hall and Björkmkan, 1975; Peisker, 1974; Chartier and Prioul, 1976; Laisk, 1977; Tenhunen *et al.*, 1977, 1980). Modelling of temperature responses of photosynthesis also used coupled biochemical cycles as a template (Tenhunen *et al.*, 1979; Sharpe, 1983). Cyclic models became widely accepted with Farquhar *et al.* (1980) and Farquhar and von Caemmerer (1982). Farquhar and colleagues coupled photosynthetic carbon reduction and photorespiratory carbon oxidation with the cycle for regeneration of NADPH and ATP linked to light-driven electron transport. Thus, a holistic cyclic template was used to organize a few key reductionistic mechanisms into a model that gave realistic responses.

The growth cycle can be used as a model template for integrating multiple stresses. Growth occurs as the result of coupling of the interdependent cycles shown in Figure 6. The cyclic nature of the crucial biochemical processes facilitates self-regulation of rates of production and consumption by adjusting the concentrations of various intermediates in response to changes in inputs. The growth-cycle concept has been used to describe natural and pollutant stress effects on plant susceptibility to insect attack (Sharpe and Wu, 1985; Sharpe *et al.*, 1985b; Sharpe and Scheld, 1986; Sharpe *et al.*, 1986b).

In its simplest form, growth can be viewed as a three-compartment cycle, as shown in Figure 7. Compartment 0 is the "regenerative" intermediate (analogous to C_5 sugars in the Calvin cycle). Compartment 1 is the first aggregate biochemical product of the growth cycle (analogous to C_3 products in the Calvin cycle). Compartment 2 is the second aggregate biochemical product of the growth cycle (analogous to C_6 sugars in the Calvin cycle). Flows between compartments represent the biochemical processes of the plant. Flow from 0 to 1 represents resource acquisition; flow from 1 to 2 represents temperature-dependent assimilation processes; and flow from 2 to 0 represents allocation and growth. Total biomass synthesized is assumed to be proportional to the concentration of the second product [2]. A branch in the biochemical flow occurs at the point where products are allocated to either root or shoot biomass.

The compartments represent the fractions of the total amount of intermediates in the system and add up to one. That is, a normalizing equation (Eq. 5) is required to provide integrated control of the system. Therefore, when the concentration in any compartment changes, the other compartments must adjust accordingly to preserve the relationship

$$[0] + [1] + [2] = 1 \tag{5}$$

The flow rate f_{01} is assumed to be a first-order reaction proportional to the product of resource availability R and intermediate concentration [0]:

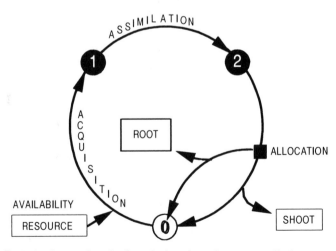

Figure 7   A simple growth cycle. Growth viewed as a three-step cyclical process: resource acquisition, assimilation, and allocation.

$$f_{01} = k_{01} [0] R \qquad (6)$$

where k_{01} is the biochemical pathway rate constant, and f_{01} is the flow from node 0 to node 1. The remaining flows are assumed to be proportional to the fraction of total intermediate in each compartment

$$f_{12} = k_{12} [1] \qquad (7)$$

$$f_{20} = k_{20} [2] \qquad (8)$$

where k_{12} and k_{20} are the rate constants for flows 12 and 20. The four equations, Equations (5)–(8), are sufficient to obtain a solution (the concentrations are eliminated in the solution and consequently do not need to be measured). Solving the compartment model yields a growth rate (Sharpe *et al.*, 1985a) G_s for the shoot

$$G_s = \sigma k_{01} k_{12} k_{20} R / (k_{01} R + k_{12} + k_{20}) \qquad (9)$$

where σ is the allocation fraction for shoots. Equation (9) possesses saturation-rate kinetics similar to Michaelis–Menten enzyme kinetics.

The three-compartment model, while useful for generating ideas, is too simple to deal with multiple stress effects, as it includes only one resource. Nevertheless, it provides a framework for development. The next logical step is to extend the model to include multiple resources explicitly (light, carbon, water, and a limiting nutrient such as nitrogen) (Fig. 8) as factors controlling the flows between compartments. The flows between compart-

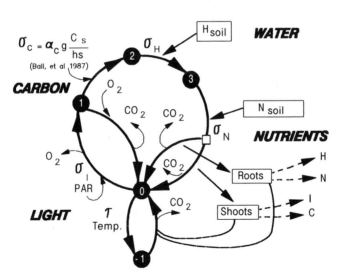

Figure 8 Coupling growth and maintenance respiration cycles. Cyclic flow between compartments regulated by light (PAR), oxygen (O_2), carbon dioxide (C_s), relative humidity (hs), soil water (H), soil nutrients (N), and temperature (τ). Predictions include shoot and root growth rates, maintenance respiration, and carbon assimilation rate.

ments are shown as heavy arrows; thin arrows show resource inputs and outputs. The flow 0 to 1 is light absorption; flow 1 to 0 is photorespiration. Flow 1 to 2 is carbon dioxide uptake; flow 2 to 3 is water uptake; and flow 3 to 0 is nutrient uptake, allocation, and root–shoot biomass synthesis. The compartment -1 has been added to represent the respiration cycle. The flow to compartment -1 represents maintenance respiration demand, and the flow from compartment -1 represents respiratory carbon dioxide release. The coupling of day–night respiration with a growth cycle was proposed by Sharpe *et al.* (1985a). This coupling scheme implies that maintenance respiration is a function of light. Recently, Brooks and Farquhar (1985) have been able to verify that maintenance respiration is a function of light. They have also been able to measure "day" and "night" respiration as separate processes. Night respiration is at least 100% greater than day respiration.

The holistic growth cycle shown in Figure 8 results in a complex saturation-rate-kinetics equation with a form similar to Equation (9) (Sharpe *et al.*, 1985a, 1986a, 1987). To capture the complexity of rubisco regulation of stress interactions, the next step is to couple the model of Farquhar *et al.* (1980) to a growth-cycle model.

IV. A Simplified Model of Partitioning under Multiple Stresses

A greatly simplified model can be derived from the cyclic-process models by restricting consideration to the branch along the flow from compartment 3 to compartment 0 in Figure 8. This model will yield qualitative predictions of changes in root and shoot fractions resulting from stress interactions. Determining plant growth responses to anthropogenic stresses such as air pollutants depends to a large extent on being able to predict growth responses to combinations of naturally occurring stresses (Weinstein and Beloin, 1990; Ford and Kiester, 1990).

The simplified model is a straightforward statement of the optimum resource allocation hypothesis in terms of the ratio of resource availabilities. For convenience, we consider only roots and shoots and four resources: a limiting nutrient, N; carbon, C; water, H; and light, L. Deficits in these resources are considered to be natural environmental stresses. We also consider ozone and sulfur dioxide as anthropogenic environmental stresses. The model computes root and shoot fractions, ρ and σ:

$$\rho = 2\,\rho_0\,(L + C)/(L + C + H + N) \tag{10}$$

where ρ_0 is the root fraction under nonlimiting resource conditions. Natural resource availabilities N, C, H, and L are scaled from 0 to 1. Zero availability for any resource means that total growth goes to zero and the plant dies [mathematically, the root fraction is undefined when the denominator of Equation (10) becomes zero]. Therefore, values for resources have a lower bound, which is defined to be 0.1 for this example. A resource availability of 1 represents the point on the growth curve at which a particular resource first becomes saturating (i.e., further addition of the resource does not produce more growth). The shoot fraction σ is calculated as

$$\sigma = 1 - \rho \tag{11}$$

Changing the parameter ρ_0 allows the model predictions to be compared with actual species. As a starting point, let $\rho_0 = 0.5$ when there are no limitations on growth. Nonlimiting carbon availability corresponds to 1000 ppm because the carbon dioxide response curve saturates at that level. This condition would only apply in a growth chamber or in a future world of global carbon dioxide elevation. The model predictions are therefore calculated for carbon availabilities of 1 (elevated CO_2) and 0.36 (ambient CO_2) in Table I. In the examples, where $C = 0.36$ and other conditions are nonlimiting, shoot biomass shows a greater dominance. The general behavior of the model is consistent with the optimum-allocation hypothesis. Light and carbon limitations lead to proportionally more shoot, whereas water and nitrogen limitations lead to more roots.

How much of the observed responses of plants to combinations of natural and anthropogenic stresses can be explained by Equations (10) and (11)? We first assume that ozone causes premature senescence of leaf proteins, particularly that of rubisco. Reducing the half-life of rubisco releases nitrogen and is equivalent to a combination of a nutrient subsidy

Table I Model Predictions for Root–Shoot Response to Resource Imbalances[a]

Light	Carbon	H (water)	Nitrogen	ρ	σ
1	1	1	1	0.5	0.5
0.1	1	1	1	0.28	0.72
1	1	1	0.1	0.65	0.35
1	1	0.1	1	0.65	0.35
1	1	0.1	0.1	0.91	0.09
1	0.36	1	1	0.40	0.60
0.1	0.36	1	1	0.19	0.81
1	0.36	1	0.1	0.55	0.45
1	0.36	0.1	1	0.55	0.45
1	0.36	0.1	0.1	0.87	0.13

[a] Results calculated from Equations (10) and (11) with $\rho_0 = 0.5$ (see text for explanation).

Table II   Effect of Ozone on Model Predictions of
Root–Shoot Response to Resource Imbalances[a]

Light	Carbon	H (water)	Nitrogen	ρ	σ
1	0.1	1	1	0.35	0.65
0.1	0.1	1	1	0.09	0.91
1	0.1	1	0.1	0.5	0.5
1	0.1	0.1	1	0.5	0.5
1	0.1	0.1	0.1	0.84	0.16

[a] Results calculated from Equations (10) and (11) with $\rho_0 = 0.5$ (see text for explanation).

($N = 1$) and a reduction in carbon supply ($C = 0.1$). Partitioning patterns are now calculated using $C = 0.1$ (Table II). The root fraction shows a consistent decrease below that for similar conditions with greater carbon availability (Table I). Ozone-exposed plants are more susceptible to drought stress than plants not exposed to ozone. They are less susceptible to low light because shoot growth is already high and can be further increased.

A special case exists where pollutants cause premature senescence of leaves. Under these circumstances, internal nitrogen is remobilized and becomes an additional source of nutrient supply. This internal nutrient source bypasses the normal controls, establishing a supersaturation of nutrients and leading to distortions in resource balance in the mesophyll cells. This supersaturation can be relieved by a further reduction in root biomass.

Sulfur dioxide may have both a positive and a negative impact on partitioning. Sulfur is a nutrient and stimulates growth processes, but at higher concentrations it has an inhibitory effect on biochemical processes. Its qualitative impact is therefore similar to ozone, and, when evaluated by Equation (10), its effect on partitioning is expected to be similar. Experimental data shows ozone to cause a greater reduction in growth than sulfur dioxide.

V. Holism versus Reductionism in Plant Growth Modelling

Our knowledge at present does not permit us to predict the behavior of whole plants from component mechanisms. A successful mechanistic model must be constructed from reductionist knowledge of low-level processes (plant parts or biochemistry) and also from observations of the behavior of whole organisms (whole plants). Thus, achieving a useful mechanistic model requires an integration of holistic, top-down modelling with reductionistic, bottom-up modelling.

Plant growth, from a holistic viewpoint, can be defined as the accumulation of biomass by means of five interdependent processes: resource acquisition, assimilation and allocation, maintenance, and partitioning. These processes are tightly linked and demonstrate adaptive feedback control. Ideally, these processes should be integrated within a single conceptual framework to represent the dynamics of plant function. Realistically, however, the conceptual framework inevitably proliferates into a linking of subsystems as the number and complexity of mechanistic components increase.

The first requirement of a plant growth model is that it has a holistic framework. The second requirement is that the processes of the model use mechanisms whenever they are known in sufficient detail to contribute to the reality of the model. The third requirement is that the holistic and reductionistic features of the model be integrated.

VI. Concluding Remarks

Mechanistic modelling has now established itself as an important and useful tool for handling interactions among multiple environmental stresses. Forest-process models are among the best examples. Dixon *et al.* (1990) review in depth a wide variety of forest growth models that integrate environmental stresses. Weinstein and Beloin (1990) model the imbalances in plant carbon, water, and nutrient acquisition and storage systems resulting from interactions with atmospheric pollutants. Chen and Gomez (1990) set up a series of hypotheses and integrate their predictions with a model of tree responses to interacting stresses.

We will never have enough information to model plant stress response using only a single approach. A combination of statistical, mathematical, and mechanistic models will continue to be needed to improve our understanding of how multiple stresses interact. The plant itself demonstrates an ability to compensate for the effects of various simultaneous stresses. We do not yet understand the mechanistic basis for this ability. We are therefore forced to build behavioral (phenomenological) models rather than mechanistic ones. The cyclic model is relatively simple and possesses many of the characteristics of the compensatory behavior exhibited by a plant. Compensatory behavior is intrinsic to the model structure; thus, this type of model may warrant further study.

Acknowledgments

The ideas presented in this chapter were developed as part of ongoing growth-modelling studies supported by NSF grant BSR-86-14911 and

EPRI project PR5880. The authors wish to thank Hsin-i Wu and Doug Spence for their discussion of earlier drafts, Katy McKinney for review of the present draft, and Hal Mooney, Bill Winner, and Eva Pell for their consistent encouragement and support.

References

Acock, B., and Reynolds, J. R. (1990). Model structure and data-base development. *In* "Process Modelling of Forest Growth Responses to Environmental Stress" (R. K. Dixon, R. S. Meldahl, G. A. Ruark, and W. G. Warren, eds.), pp. 169–179. Timber Press, Portland, Oregon.

Ball, J. T., Woodrow, I. E., and Berry, J. A. (1987). A model predicting stomatal conductance and its contribution to the control of photosynthesis under different environmental conditions. *Prog. Photosynth. Res.* 5: 221–224.

Bloom, A. J., Chapin, F. S., III, and Mooney, H. A. (1985). Resource limitation in plants—an economic analogy. *Annu. Rev. Ecol. Syst.* 16: 363–392.

Brooks, A., and Farquhar, G. D. (1985). Effect of temperature on the CO_2/O_2 specificity of ribulose 1,5-bisphosphate carboxylase/oxygenase and the rate of respiration in the light: Estimates from gas-exchange measurements on spinach. *Planta* 165: 397–406.

Charles-Edwards, D. A., and Ludwig, L. J. (1974). A model of leaf carbon metabolism. *Ann. Bot.* 38: 921–930.

Chartier, P. (1966). Etude théorique de l'assimilation brute de la feuille. *Ann. Physiol. Veg.* 8: 167–196.

Chartier, P. (1970). A model of CO_2 assimilation in the leaf. *In* "Prediction and Measurement of Photosynthetic Productivity" (I. Setlik, ed.), pp. 307–315. Pudoc, Wageningen.

Chartier, P., and Prioul, J. L. (1976). Effects of irradiance, carbon dioxide, and oxygen on the net photosynthetic rate of the leaf: A mechanistic model. *Photosynthetica* 10: 20–24.

Chen, C. W., and Gomez, L. E. (1990). Modeling tree responses to interacting stresses. *In* "Process Modelling of Forest Growth Responses to Environmental Stress" (R. K. Dixon, R. S. Meldahl, G. A. Ruark, and W. G. Warren, eds.), pp. 180–190. Timber Press, Portland, Oregon.

Cowan, I. R. (1977). Stomatal behavior and environment. *Adv. Bot. Res.* 4: 1176–1227.

Cowan, I. R. (1982). Regulation of water use in relation to carbon gain in higher plants. *In* "Encyclopedia of Plant Physiology, New Series" (O. L. Lange, P. S. Nobel, C. B. Osmond, and H. Ziegler, eds.), Vol. 12B, pp. 589–613. Springer-Verlag, Berlin.

Dixon, R. K., Meldahl, R. S., Ruark, G. A., and Warren, W. G. (1990). "Process Modelling of Forest Growth Responses to Environmental Stress." Timber Press, Portland, Oregon.

Farquhar, G. D., and von Caemmerer, S. (1982). Modelling of photosynthetic response to environmental conditions. *In* "Encyclopedia of Plant Physiology, New Series" (O. L. Lange, P. S. Nobel, C. B. Osmond, and H. Ziegler, eds.), Vol. 12B, pp. 549–587. Springer-Verlag, Berlin.

Farquhar, G. D., von Caemmerer, S., and Berry, J. A. (1980). A biochemical model of photosynthetic CO_2 assimilation in leaves of C_3 species. *Planta* 149: 78–90.

Ford, E. D., and Kiester, R. (1990). Modeling the effects of pollutants on the processes of tree growth. *In* "Process Modelling of Forest Growth Responses to Environmental Stress" (R. K. Dixon, R. S. Meldahl, G. A. Ruark, and W. G. Warren, eds.), pp. 180–190. Timber Press, Portland, Oregon.

Goeschl, J. D., and Magnuson, C. E. (1986). Physiological implications of the Munch–Horwitz theory of phloem transport: Effects of loading rates. *Plant, Cell Environ.* 9: 95–102.

Goeschl, J. D., Magnuson, C. E., DeMichele, D. W., and Sharpe, P. J. H. (1976). Concentration-dependent unloading as a necessary assumption for a closed-form mathematical model of osmotically driven pressure flow in phloem. *Plant Physiol.* 58: 556–562.

Gross, L. J. (1989). Plant physiological ecology: A theoretician's perspective. *In* "Perspectives in Ecological Theory" (J. Roughgarden, R. M. May, and S. A. Levin, eds.), pp. 11–24. Princeton Univ. Press, Princeton, New Jersey.

Hall, A. (1971). A model of leaf photosynthesis and respiration. *Carnegie Inst. Washington Year Book* 70: 530–540.

Hall, A., and Björkman, O. (1975). A model of leaf photosynthesis and respiration. *In* "Perspectives of Biophysical Ecology" (D. M. Gates and R. B. Schmerl, eds.), pp. 55–72. Springer-Verlag, New York.

Henderson-Sellers, A., and McGuffie, K. (1987). "A Climate-Modelling Primer." Wiley, New York.

Laisk, A. (1977). Modelling of the closed Calvin cycle. *In* "Biophysikalische Analyse pflanzlicher Systeme" (K. Unger, ed.), pp. 175–182. Gustav Fischer, Jena, Germany.

Levin, S. (1981). Mathematics, ecology, ornithology. *Auk* 97: 422–425.

Levins, R. (1966). The strategy of model building in population biology. *Am. Sci.* 54: 421–431.

Lommen, P. W., Schwintzer, C. R., Yocum, C. S., and Gates, D. M. (1971). A model describing photosynthesis in terms of gas diffusion and enzyme kinetics. *Planta* 98: 195–220.

Magnuson, C. E., Goeschl, J. D., Sharpe, P. J. H., and DeMichele, D. W. (1979). Consequences of insufficient equations in models of the Munch hypothesis of phloem transport. *Plant, Cell Environ.* 2: 181–188.

Magnuson, C. E., Goeschl, J. D., and Fares, Y. (1986). Experimental tests of the Munch-Horwitz theory of phloem transport: Effects of loading rates. *Plant, Cell Environ.* 9: 103–109.

Nobel, P. S. (1983). "Biophysical Plant Physiology and Ecology." Freeman, New York.

Peisker, M. (1974). A model describing the influence of oxygen on photosynthetic carboxylation. *Photosynthetica* 8: 47–50.

Rykiel, E. J., Jr., Coulson, R. N., Sharpe, P. J. H., Allen, T. F. H., and Flamm, R. O. (1988). Disturbance propagation by bark beetles as an episodic landscape phenomenon. *Landscape Ecol.* 1: 129–139.

Sharpe, P. J. H. (1983). Responses of photosynthesis and dark respiration to temperature. *Ann. Bot.* 52: 325–343.

Sharpe, P. J. H. (1990). Forest modelling approaches: Compromises between generality and precision. *In* "Process Modelling of Forest Growth Responses to Environmental Stress" (R. K. Dixon, R. S. Meldahl, G. A. Ruark, and W. G. Warren, eds.), pp. 180–190. Timber Press, Portland, Oregon.

Sharpe, P. J. H., and DeMichele, D. W. (1977). Reaction kinetics of poikilotherm development. *J. Theor. Biol.* 64: 649–670.

Sharpe, P. J. H., and Scheld, H. W. (1986). Role of mechanistic modelling in estimating long-term pollution effects upon natural and man-influenced forest ecosystems. *In* "Proc. of a Workshop on Controlled Exposure Techniques and Evaluation of Tree Responses to Airborne Chemicals," pp. 72–82. NCASI Technical Bulletin No. 500. National Council for Air and Stream Improvement, New York.

Sharpe, P. J. H., and Wu, H. (1985). A preliminary model of host susceptibility to bark beetle attack. *In* "Proceedings of the International Union of Forest Research Organizations: Host Insect Work Group" (L. Safranyik and A. A. Berryman, eds.), pp. 108–127. Canadian Forestry Service, Victoria, British Columbia.

Sharpe, P. J. H., Walker, J., Penridge, L. K., and Wu, H. (1985a). A physiologically based continuous-time Markov approach to plant growth modelling in semiarid woodlands. *Ecol. Modell.* 29: 189–214.

Sharpe, P. J. H., Wu, H., Cates, R. G., and Goeschl, J. D. (1985b). Energetics of pine defense systems to bark beetle attack. *U.S. For. Serv. Gen. Tech. Rep. SO* 56: 206–223.

Sharpe, P. J. H., Walker, J., Penridge, L. K., Wu, H., and Rykiel, E. J., Jr. (1986a). Spatial considerations in physiological models of tree growth. *Tree Physiol.* 2: 403–421.

Sharpe, P. J. H., Newton, R. J., and Spence, R. D. (1986b). Forest pests: The role of phloem osmotic adjustment in the defensive response of conifers to bark beetle attack. *In* "Stress Physiology and Forest Productivity" (T. C. Hennessey, P. M. Dougherty, S. V. Kossuth, and J. D. Johnson, eds.), pp. 113–131. Martin Nijhoff, Dordrecht.

Sharpe, P. J. H., Rykiel, E. J., Jr., Olson, R. L., and Wu, H. (1987). A continuous-time Markov model of plant growth. *In* "Models in Plant Physiology and Biochemistry" (D. W. Newman and K. G. Wilson, eds.), Vol. II, pp. 81–85. CRC Press, Boca Raton, Florida.

Smith, K. C., Magnuson, C. E., Goeschl, J. D., and DeMichele, D. W. (1980). A time-dependent mathematical expression of the Munch hypothesis of phloem transport. *J. Theor. Biol.* 86: 493–505.

Tenhunen, J. D., Weber, J. A., Filipek, L. H., and Gates, D. M. (1977). Development of a photosynthesis model with an emphasis on ecological applications. III. Carbon dioxide and oxygen dependencies. *Oecologia* 30: 189–307.

Tenhunen, J. D., Weber, J. A., Yocum, C. S., and Gates, D. M. (1979). Solubility of gases and the temperature dependency of whole-leaf affinities for carbon dioxide and oxygen. *Plant Physiol.* 63: 916–923.

Tenhunen, J. D., Hesketh, J. D., and Harley, P. C. (1980). Modeling C_3 leaf respiration in the light. *In* "Predicting Photosynthate Production and Use for Ecosystem Models" (J. D. Hesketh and J. W. Jones, eds.), Vol. II, pp. 17–47. CRC Press, Boca Raton, Florida.

Weinstein, D. A., and Beloin, R. (1990). Evaluating effects of pollutants on integrated tree processes: A model of carbon, water, and nutrient balances. *In* "Process Modelling of Forest Growth Responses to Environmental Stress" (R. K. Dixon, R. S. Meldahl, G. A. Ruark, and W. G. Warren, eds.), pp. 180–190. Timber Press, Portland, Oregon.

Zeiger, E., Farquhar, G. D., Cowan, I. R. (eds.) (1987). "Stomatal Function." Stanford Univ. Press, Stanford, California.

II

Biotic Interactions

11

Growth Responses Induced by Pathogens and Other Stresses

P. G. Ayres

I. Introduction

The interactions among plants, their pathogens, and the abiotic environment can be illustrated as a disease triangle, or as a pyramid if the influence of ontogeny or another temporal aspect is included (Fig. 1). Each of the relationships is two-way, although a plant's ability to modify its abiotic environment is limited, for example, to local effects on the pool of soil water or nutrients. This chapter focuses on one corner of the triangle and examines growth responses to pathogenic microorganisms. Are these re-

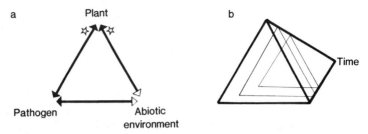

Figure 1 A disease triangle (a) can represent interactions between plant, pathogen, and the abiotic environment. Starred relationships are the subject of this chapter; open arrowheads show weak interactions. The disease pyramid (b) emphasizes that interactions in the triangle are modified by time.

sponses similar to those induced by abiotic stresses, particularly drought, nutrient deficiency, and atmospheric pollution? Do these abiotic stresses interact with pathogen stress through the induction of common responses in the plant? Do common responses at the cellular level play a role in stress interactions?

Although this chapter does not cover other aspects of the triangle, one should remember that abiotic stresses can affect pathogens directly, or they may influence the host's resistance to the pathogen, which may in turn affect the occurrence or extent of infection. For example, sulfur dioxide is known to inhibit the growth of a range of leaf-surface fungi in vitro (Magan and McCleod, 1988) and to reduce the extent of infection by several fungal pathogens on winter cereals (McCleod, 1988). Growth in dry soil enhances the development of adult plant resistance to powdery mildew disease in barley (Ayres and Woolacott, 1980).

From the array of phytopathogenic microorganisms, I have chosen fungi for special attention here. They cause the most frequently encountered diseases, which have the greatest biological and social impact, and their physiological effects are best documented. The general conclusions reached, however, may apply to a greater or lesser extent to pathogenic infections caused by other microorganisms. Even among the fungi, patterns of infection differ widely. Each species has its own environmental requirements for infection, and each produces to greater or lesser extent its own characteristic array of extracellular enzymes, toxins, and plant growth regulators, which together help to determine its host range. Closely related fungal species may show strikingly contrasting behavior. Thus *Botrytis cinerea* is a weak opportunistic pathogen, mainly attacking senescing tissue of a wide range of plants, whereas *Botrytis fabae* can attack younger tissues but is restricted to *Vicia faba* and closely related species (Harrison, 1988). Nevertheless, the immediate results of the stress presented by different fungal pathogens share two important characteristics: they involve structural injury, i.e., irreversible damage, and they are localized in particular tissues or organs. Localized damage typically also arises from other biotic stresses such as insect herbivory, but not from abiotic stresses such as drought, nutrient deficiency, or pollution. The latter involve no wounding and have their initial effect on a large portion, if not the whole, of the plant.

II. Cellular Interactions and Responses to Injury

Fungal pathogens cause physical damage as they invade their host. They trigger active defense responses in surrounding tissues either by generating an action potential and increasing cytosolic Ca^{2+} concentrations, or by liberating plant cell-wall fragments (endogenous elicitors of defense reactions; de Wit, 1987), or both. Similar events follow abiotic wounding (Davies, 1987), which, like any wounding, may be accompanied by increased ethylene production by the plant. Pathogens also release "disease determinants," which may include fragments of microbial cell walls or membranes (exogenous elicitors), polysaccharidases specific for host cell walls, toxins, and plant growth regulators (de Wit, 1987; Keon *et al.*, 1987). Although such enzymes and toxins help extend infection sites, the pathogen's growth is restricted by a range of localized host responses that elicitors may induce, such as modification of neighboring cell walls and the synthesis of antimicrobial chemicals, including the de novo synthesis of phytoalexins (Halverson and Stacey, 1986). Both wounding and infection are associated with increased dark respiration in the host, which provides the carbon skeletons and energy required for defense and repair.

Damage to host cells can occur in only a limited number of ways. Injury and death typically result from disrupted membrane function, which disturbs vital processes such as energy transduction and osmoregulation (Ayres, 1984). Pathogens bring about such changes relatively rapidly, but progressive deterioration of membranes can also result from normal senescence or senescence promoted by severe or prolonged abiotic stresses. Disruption of membranes is produced by a self-perpetuating wave of free radicals emanating from peroxidation within the lipid bilayer (Thompson *et al.*, 1987) and ultimately leads to cell death (see Chapter 9 for more detail). Common cellular responses exist, therefore, through which pathogens and other stresses may interact within the plant, typically in an additive manner.

To an extent yet to be fully determined, membranes may be protected by common mechanisms. Recent studies of mildewed and rusted plants, for example, show an accumulation of solutes such as proline (Murray and Ayres, 1986) and polyamines (Walters *et al.*, 1985), which also accumulate under abiotic stresses, such as drought or nutrient deficiency, and are widely regarded as membrane-stabilizing agents (Smith, 1985). The role of polyamines is intriguing, since they and ethylene are alternative metabolic products of S-adenosylmethionine. Ethylene is produced by plants subjected to many abiotic stresses (Chapter 9) and by some infected plants, e.g., leaves of *Poa pratensis* infected with *Bipolaris sorokiniana* (Sacc) Shoem (*Helminthosporium sativum* P. K. and B.), where it enhances pathogen-induced chlorosis (Hodges and Coleman, 1984). Different stresses might interact through regulation of this pathway.

III. Integrated Growth Responses

Long-term growth responses that enable a plant to minimize the reduction in dry matter accumulation following a sublethal stress are not well documented for plants infected by pathogens, and such responses have rarely been studied in plants subjected to combinations of pathogen and abiotic stresses. In this section I examine experimental investigations involving two host–pathogen combinations—groundsel (*Senecio vulgaris* L.) infected by rust (*Puccinia lagenophorae* Cooke), and faba bean (*Vicia faba* L.) infected by rust (*Uromyces viciae-fabae* [Pers] Shroet.)—as well as interactions with drought or nutrient deficiency. The conclusions probably apply to other foliar diseases as well. I argue that the responses affecting photosynthesis and partitioning of dry matter between roots and shoots are likely to occur in other infected plants as well as in those stressed by atmospheric pollutants.

A. Photosynthesis

Foliar pathogens typically reduce a plant's accumulation of dry matter. In well-watered groundsel infected by rust this reduction was small; it was much greater in plants from which water had been withheld (Fig. 2a; Paul and Ayres, 1984). Pathogens reduce the rate of net photosynthesis Pn at the sites of infection, but often, as in rust-infected groundsel, Pn may be temporarily stimulated in younger leaves that emerge from the shoot apex after inoculation. As a result, in the groundsel experiment, average Pn per unit leaf area in the whole plant was similar for rust-infected and healthy plants (Fig. 2b). Growth of dry matter in rust-infected plants was less than that of healthy controls because leaf area growth was slower (Fig. 2c).

Since differences between the growth of leaf area and dry matter in rust-infected control plants were even greater when plants were water stressed (water potentials at 15 days after inoculation, or dai, were −1.6 and −1.1 MPa in rust-infected and control plants, respectively), the authors concluded that this slower growth was in turn related to the lower water potential and probably lower turgor in rust-infected plants (water

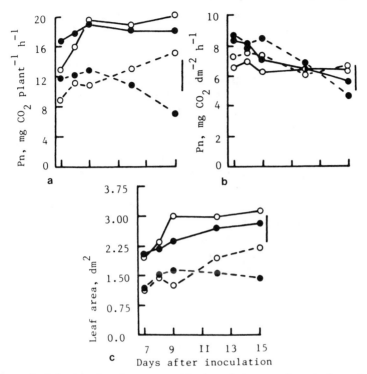

Figure 2 Infection, drought, or a combination of stresses reduces leaf growth and a plant's capacity for photosynthesis. Net photosynthesis *Pn* is reduced per plant (a), per unit leaf area (b), and per leaf area (c) in healthy groundsel (O) or groundsel infected by rust (●). Plants were well watered daily (——) or deprived of water from 2 to 10 days after inoculation and then watered only to replace transpirational losses (------). Bars indicate LSD at *P* = 0.05. [Modified from Paul and Ayres (1984).]

potential at 15 dai was −0.8 MPa in rust-infected plants and −0.2 MPa in controls). Clearly, photosynthetic compensation between infected and un-infected leaves was less effective during drought than otherwise, largely because drought prevented the growth of new, uninfected leaves.

Stimulation of net photosynthesis has also been observed in uninfected leaves of pea (Ayres, 1981a) and in barley (Williams and Ayres, 1981) infected by powdery mildew fungi (*Erysiphe pisi* and *E. graminis* f. sp. *hordei*, respectively). Stimulation may also occur in healthy plants after herbivores or humans have removed older leaves; in such cases the increased *Pn* has been attributed to an altered source : sink ratio (Kramer, 1981). A pathogen is an effective sink for fixed carbon, but in barley infected by powdery mildew, ${}^{14}C$ tracer studies have shown no evidence of increased carbon export from the youngest, uninfected leaf (Ayres, 1982). The increased photosynthesis has instead been attributed to the increased synthesis of ribulose 1,5-bisphosphate carboxylase and the increased activity of this and another carbon-fixing enzyme, phosphoenolpyruvate carboxylase (Walters and Ayres, 1983). These researchers found more protein-nitrogen and less nitrate-nitrogen in infected plants than in comparable leaves on healthy plants, suggesting that nitrogen metabolism, rather than carbon transport, regulated photosynthetic capacity during the maturation of leaves on infected plants.

Relative growth rate (dry matter increment per unit of plant mass) is the product of net assimilation rate (dry matter increment per unit of leaf area) and leaf area ratio (leaf area divided by plant mass). Compensation for the inhibition of net photosynthesis in infected leaves can be achieved by an increase in leaf area ratio. In rust-infected groundsel growing in sand culture, an increase in leaf area ratio occurred both in plants supplied with adequate nutrition (50% of full nutrient solution) and in nutrient-stressed plants (5% of full nutrient solution; Fig. 3). The increase was sufficient to maintain relative growth rates similar to those of healthy controls except for short periods between weeks two and four after inoculation. In groundsel lightly infected with powdery mildew *(E. fischeri)* (Ben Kalio,

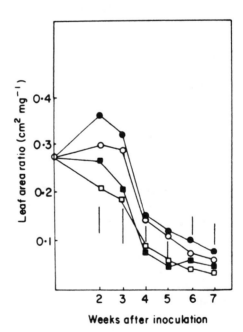

Figure 3   Leaf area ratio increases in response to infection. Healthy (open symbols) and rust-infected (closed symbols) groundsel were grown in sand culture and fed 50% (O, ●) or 5% (□, ■) of full nutrient solution. Bars indicate LSD at $P = 0.05$. [Modified from Paul and Ayres (1986a).]

1976) and in King Edward potato infected with the vascular wilt fungi *Verticillium albo-atrum* and *V. dahliae* (Harrison and Isaac, 1968), leaf area ratio was also higher than in healthy controls. An increase in leaf area ratio is not an invariable response to infection, however; for example, in *V. faba* infected either by the biotroph *U. viciae-fabae* or by the necrotroph *Botrytis fabae*, leaf area decreased (Williams, 1975, 1978).

The contribution that healthy leaves can make to the relative growth rate of a plant suffering localized stress depends on their size and age or position. In groundsel, as described above, the localized nature of the infection allowed a rapid increase in net assimilation rate in uninfected leaves and a slow increase in leaf area ratio. The nonlocalized nature of abiotic stresses probably precludes such compensation through net assimilation rate, but it still allows slower compensation through an increase in leaf area ratio, which is achieved through a change in the partitioning of newly fixed dry matter. Again, a stressed plant may be able to reach a relative growth rate similar to that of larger, unstressed controls. For example, leaf area ratio increased in radish *(Raphanus sativus)* treated with 170 ppb ozone (Walmsley *et al.*, 1980), allowing the relative growth rate to recover to control rates (Fig. 4). The same phenomenon occurred in three pasture grasses, *Poa pratensis, Dactylis glomerata,* and *Phleum pratense,* fumi-

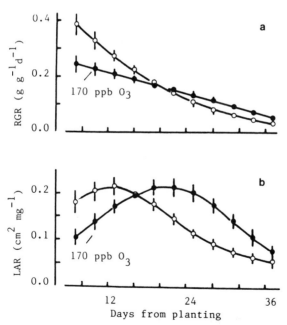

Figure 4 Leaf area ratio increases in response to ozone pollution, as illustrated by relative growth rate (a) and leaf area ratio (b) of radishes grown in clean air (O) or in 170 ppb ozone (●). Bars indicate 95% confidence limits. [Modified from Walmsley *et al.* (1980).]

gated in the glasshouse over long periods with low concentrations of sulfur dioxide (40–100 ppb; Whitmore, 1982) and in winter barley fumigated with 40–200 ppb in the field (Baker *et al.*, 1987).

B. Dry Matter Partitioning

Increased leaf area ratio in plants exposed to atmospheric pollution or infected by fungi is essentially a consequence of the reduced partitioning of dry matter to roots. Such reduced partitioning occurs in rust-infected groundsel under both nutrient-rich (low root:shoot ratios) and nutrient-poor conditions (high root:shoot ratios; Fig. 5) and is a frequent consequence of foliar infections, although not universal. Lupton and Sutherland (1973), for example, observed an increased root:shoot ratio in wheat infected by *E. graminis hordei*. Whereas decreased root:shoot ratios are noted in plants exposed to pollution (Kasana and Mansfield, 1986), most notably those fumigated with a mixture of sulfur dioxide and nitrogen dioxide (Fig. 6), increased root:shoot ratios are frequently found in plants growing under dry or nutrient-deficient conditions (e.g., see Fig. 5). Grime (1979) proposed that species characteristic of high-nutrient habitats show the greatest plasticity in partitioning in response to nutrient stress; it is not known whether these species show the greatest or least response to stresses such as infection or atmospheric pollution.

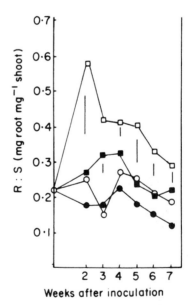

Figure 5 Infection inhibits the increase in root:shoot dry matter ratio that is a response of healthy plants to nutrient stress. Healthy and rust-infected groundsel were grown as described in the legend to Fig. 2. [Modified from Paul and Ayres (1986a).]

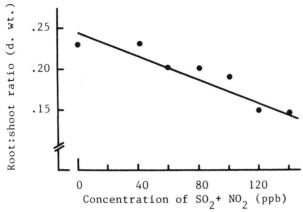

Figure 6 Atmospheric pollutants decrease the root:shoot dry matter ratio of spring barley *(Hordeum vulgare)*. Seedlings were fumigated for two weeks, beginning 3–4 days after germination. Concentrations of SO_2 and NO_2 were equal. [Modified from Pande and Mansfield (1985).]

C. Root Function

The uptake of nutrients or water by roots is more closely related to the length or surface area of the roots than to their dry mass. Measurements of the dry mass of root systems fail to account for adaptive changes in length per unit mass (specific root length) and therefore provide only a poor indication of potential root function in stressed plants. In rust-infected groundsel (Paul and Ayres, 1986b), faba bean (Tissera and Ayres, 1988), and wheat (Martin and Hendrix, 1974), lengthening of roots is inhibited much less than growth of dry weight, resulting in an increase in specific root length. Specific root length also increases in some healthy grasses in response to low soil fertility (*Poa annua, Phleum pratense,* and barley), and in other grasses (*Agrostis capillaris* but not *P. pratensis*) in response to drought (Fitter, 1985). Although specific root length doubled in well-fertilized groundsel after rust infection, reaching the high values found in nutrient-stressed uninfected plants (Fig. 7), specific root length was unaltered in rust-infected, nutrient-stressed plants, suggesting that their adaptability was exhausted.

In well-fertilized rust-infected groundsel in which each gram of dry matter produced a greater length of root than in healthy groundsel, infection increased the absorption rates of potassium (Fig. 8), phosphorus, and nitrate-nitrogen per unit of root dry mass but had no effect on absorption rates per unit of root length (Paul and Ayres, 1988b). The increase per unit mass could be explained largely by an increased shoot demand (i.e., a decrease in the root:shoot ratio), whereas the absence of increased uptake per unit length resulted because the higher root

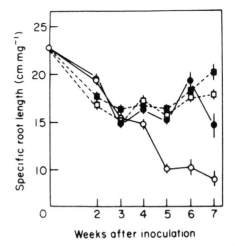

Figure 7 Specific root length increases in response to nutrient deficiency and infection, but responses to the two stresses are not additive. Healthy (open symbols) and rust-infected (closed symbols) groundsel were grown in sand culture and fed 50% (O——●) or 5% (□- - - -■) of the full nutrient solution. Bars indicate SE of means. [Modified from Paul and Ayres (1986b).]

length : root mass ratio of infected plants led to stable values of shoot mass per unit root length (i.e., shoot demand was not increased on this basis).

In soil, although nutrient uptake per unit length of root increases with root radius, the volume of soil exploited by a given volume (or mass) of roots is greater for finer roots. Changes in specific root length and the efficiency with which roots explore the soil are probably even more important for immobile ions, such as phosphate, than for mobile ions such as nitrate.

It is clear that the internal anatomy of roots may be modified by environmental stress, but the functional significance of such changes is less clear. Radin and Eidenbock (1986) found that the vascular anatomy of roots responded to nutrient stress. Roots of young cotton plants normally had tetrarch steles (Fig. 9). When growing under nitrogen- or phosphorus-deficient conditions, however, they produced triarch or pentarch steles. It was not reported whether these changes were related to a change in specific root length, but observations by other authors (Oosterhuis and Wullschleger, 1987) indicated improved functioning. These researchers found that under constant pressure, water flux through open-ended segments of roots of cotton with tetrarch steles and the largest xylem vessels was greater than through those with pentarch steles, with narrower but more numerous vessels and the greatest total vessel area. (Xylem and root cross-sectional areas are closely correlated both within and between species; Fitter, 1987.)

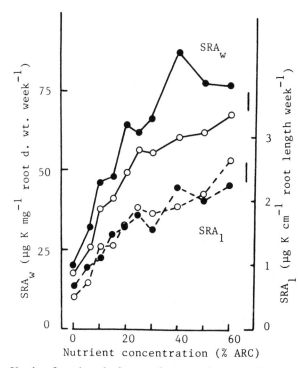

Figure 8 Uptake of nutrients both per unit mass and per root length decreases as the external concentration of nutrients decreases, as shown by specific K absorption rates by roots of healthy (open symbols) or rust-infected (closed symbols) groundsel plants, grown at a range of nutrient concentrations. An increase in the ratio of root length to dry matter in response to infection increases the uptake of nutrients per unit of root mass but not length. Rates are expressed per unit dry weight (SRA_w) (——) and per unit root length (SRA_l)(- - - -). Bars indicate LSD, $P = 0.05$. [Modified from Paul and Ayres (1988b).]

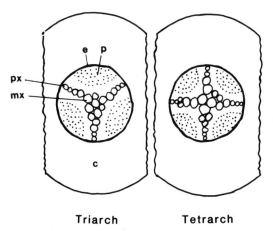

Figure 9 Cross-sections of roots showing different patterns of development of the primary xylem. c, Cortex; e, endodermis; px, protoxylem; mx, metaxylem; p, phloem.

Because the radial resistance to water flow across intact (closed) roots is often higher than axial resistance (see the example below), an increase in the number of protoxylem poles close to the endodermis may shorten the main pathway of water movement between endodermis and xylem and so significantly reduce both the radial and total root resistance. The number of protoxylem poles is also probably important because it is related to the number of lateral roots initiated and to root system architecture. Such potentially important relationships among specific root length, vascular anatomy, and root function in stressed plants deserve more attention. We do know something about these relationships in infected plants.

When water was forced under pressure through excised root systems of faba bean, infection slightly reduced the volume of water transported to the shoot stump. On the other hand, infection markedly increased the hydraulic conductance of the system; that is, it increased the amount of water transported per unit area of root or endodermal surface (Tissera and Ayres, 1988).

To determine whether these changes occurred in tissues that had matured before infection or only in those that developed after infection, the volume of water transported and hydraulic conductance were compared in the two regions of lateral roots. A partial vacuum was applied to one end of five-centimeter-long excised root segments (the cut end of each segment was sealed with wax) so as to draw water into and through each segment. Apical segments formed after inoculation transported more water and had a greater hydraulic conductance than subapical segments formed immediately before inoculation (Table I). The difference between the segments of different age may lie in the fact that root systems from rust-infected bean plants had a greater specific root length than controls, and the diameters of individual roots and their endodermal cylinders were reduced. Similar

Table I Exudation Rate and Hydraulic Conductance of Lateral Roots from Healthy or Rust-Infected *Vicia faba*[a]

	Healthy	Rust-infected
Exudation rate ($m^3 s^{-1} \times 10^{-9}$)		
Subapical segment	0.58 ± 0.08	0.58 ± 0.08[b]
Apical segment	0.70 ± 0.14	1.02 ± 0.02[c]
Hydraulic conductance ($m^3 m^{-2} s^{-1} MPA^{-1} \times 10^{-6}$)		
Subapical segment	8.3 ± 1.2	8.3 ± 1.3[b]
Apical segment	9.6 ± 1.4	14.4 ± 2.7[d]

[a] Values for 5-cm root segments under partial vacuum (40 kPA). [Modified from Tissera and Ayres (1988).]
[b] No statistical significance.
[c] $P = 0.05$.
[d] $P = 0.01$.

changes had been noted previously in barley infected by powdery mildew (Walters and Ayres, 1982), whose roots were narrower than controls. In bean we found a small but statistically significant reduction in the diameter of the largest xylem vessel and a small but statistically insignificant reduction in the number of smaller xylem vessels, which together would have increased the axial resistance to water movement across the root. Since the major resistance is located in the radial pathway between the external solution and the xylem vessels (Reid and Hutchison, 1986), the narrowing of root and endodermal cylinder may have shortened this pathway and been the overriding factor determining the change in hydraulic conductance.

Alterations in specific root length are important because they influence the maintenance of functional efficiency in the presence of pathogen stress. Changes in specific root length alleviate, but do not totally offset, the disturbance caused by infection. One reason for this failure, as noted above for rust-infected groundsel under nutrient stress, is that root growth in infected plants seems less able to cope with a second environmental stress. This inability has also been shown in the roots of mildewed pea, which could not maintain turgor and extension growth at low external water potentials and so develop the high root:shoot ratios characteristic of healthy water-stressed plants (Ayres, 1981b). The root tips of mildewed pea plants could not generate solute potentials as low as those of controls (-0.93 and -1.23 MPa, respectively). Within the root system, lateral roots in the rapidly drying upper regions of the soil showed the greatest inhibition of growth. Similar responses have been seen in rust-infected faba bean (Tissera and Ayres, 1986).

A second reason for the failure of increased specific root length to neutralize the effects of infection is that root systems of infected plants remain smaller in total mass and length than those of healthy plants. In short-term laboratory experiments, taproot growth in rust-infected faba bean (like that in mildewed pea) was slightly reduced by infection, and fewer lateral roots penetrated the deepest layers of the soil, below 50 cm. Such shallower rooting may be significant if, as Hamblin and Tennant (1987) have argued, water uptake by water-stressed crops (e.g., pea) is more closely related to maximum rooting depth than to root length per unit area of ground. However, the inhibition of growth in upper regions of the soil profile may actually be more significant. In the laboratory, rust inhibited root growth of faba bean plants in precisely the zone (18–36 cm deep) from which bean crops in the field extract the most water over a growing season. Mildew had similar effects on peas and, in drying soil, infection reduced the shoot water potential of both plant species.

Relationships among root growth, water extraction, and soil drying are complex. To understand the effects of infection alone, as well as its possible interactions with other stresses, more studies are needed of the pattern of water extraction by different root system components. Two important aspects of this subject are (a) the relationship between root architecture and water uptake and (b) the extent of interference among root systems of neighboring plants.

Fitter *et al.* (1988) reported that the root system of red clover *(Trifolium pratense)* under water stress was of a simple herringbone pattern, consisting of main axis plus primary laterals, a relatively efficient design for exploring soil. Significantly more branching occurred in well-watered plants. The same species showed a comparatively small response to variation in the supply of phosphorus or nitrogen. Little is known about the effect of pathogen stress on root system architecture, but one investigation of rust *(P. recondita)* on wheat showed reduced branching (Martin and Hendrix, 1974), suggesting that the plant was maximizing the volume of soil explored, possibly to obtain enough water to meet the increased evapotranspirational demands of the shoot. In barley infected by powdery mildew, where water consumption and, consequently, the demands on roots were reduced, roots were branched more than in controls (Walters and Ayres, 1981). The limited evidence available suggests that infection and drought, and possibly other stresses, may interact through their effects on root branching.

IV. Pathogens and Plant Populations

A. Effects on Competition

If population density is such that there is competitive interference for resources and the growth of individuals is affected, each plant represents a stress factor to its neighbors. Infection magnifies differences in both inter- and intraspecific competitive fitness.

In a mixed culture of lettuce and rust-infected groundsel, for example, the relative mixture response for groundsel (R_x = yield in monoculture minus yield in mixture, all divided by yield in monoculture; Joliffe *et al.*, 1984) increased over a range of planting densities (Paul and Ayres, 1987a). At the highest density, the R_x for lettuce was lower when it was grown with rust-infected groundsel than when grown with healthy groundsel, whatever the proportions of the two species (Fig. 10a), indicating that lettuce was inhibited less by rust-infected groundsel than by healthy groundsel. The R_x for healthy groundsel increased as the ratio of lettuce to groundsel increased, indicating that groundsel was growing progressively less well; R_x was only significantly increased by infection when the ratio of lettuce to groundsel was 1 : 3 (Fig. 10b). Thus, rust affected interspecific competition, but its main effect was seen in the enhanced growth of the uninfected

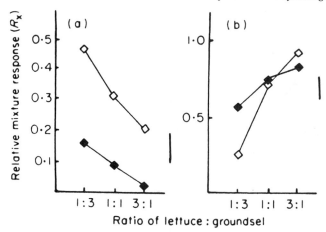

Figure 10 (a) Relative mixture response (R_x) of lettuce grown with healthy (◊) and rust-infected (♦) groundsel; (b) R_x of healthy (◊) or rust-infected (♦) groundsel grown with lettuce. Infection alters the responses of plants to interspecific competition when species are grown in mixtures of different proportions but of constant density; in 1:1 mixtures the stronger competitor has an $R_x < 0.5$; the weaker competitor has an $R_x > 0.5$. Bars indicate LSD, $P = 0.05$. [Modified from Paul and Ayres (1987a).]

competitor. Growing a resistant and a healthy cultivar of perennial ryegrass *(Lolium perenne)* in monoculture and mixture with or without crown rust *(P. coronata)* gave similar results (Potter, 1987).

The damaging effects of infection on growth in intraspecific competition are seen in Figure 11, which shows the production of flowers by healthy and rust-infected groundsel. Similar diagrams could be presented for most dry-weight parameters. Rust-infected plants grew less well in monoculture than did healthy plants, as earlier sections of this chapter would predict. Rust-infected plants yielded less in mixed culture, and healthy plants much more, than was predicted from their monoculture yields. The little evidence we have (Paul and Ayres, 1986c) suggests that harvest index (seed yield divided by vegetative yield) does not increase in response to infection. Crop plants can adjust the parameters of harvest index; such adjustment can alleviate the effects of water stress on yield (Austin *et al.*, 1980). Clarke (1986) has suggested that similar adjustment might be effective in wild plants, which typically have a lower harvest index than crop plants, and thus more scope for acclimation, but supporting evidence is lacking.

Members of a population, even of one species, vary naturally in size. Infection increases this diversity in both monoculture and mixture, leading to populations with a large proportion of exceptionally small individuals that are associated, especially in mixtures, with a few plants that grow exceptionally large (Fig. 12).

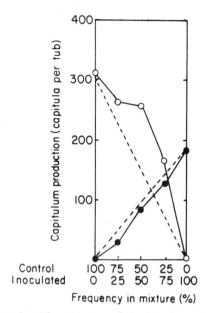

Figure 11 Infection alters the responses of plants to intraspecific competition when healthy and infected individuals are grown in mixtures of different proportions but of constant density. When groundsel is grown in such mixtures, capitulum production is promoted in healthy plants (O) and inhibited in rust-infected plants (●). The broken line indicates the yield expected in mixed culture on the basis of the yield in monoculture. [Modified from Paul and Ayres (1986c).]

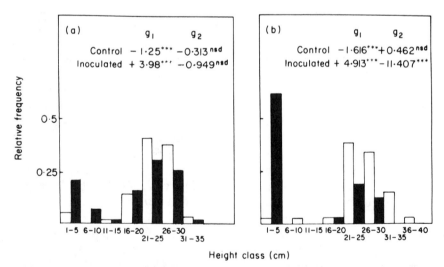

Figure 12 Infection increases the diversity of sizes within the host population, particularly in mixtures of healthy and infected plants, as illustrated by the height of healthy (open bars) and rust-infected (solid bars) groundsel grown in (a) monoculture and (b) 1 : 1 mixtures. In mixtures, the distribution of healthy plants is skewed to the right; that of infected plants is skewed to the left; coefficients of skewness differ significantly at $P = 0.001$. [Modified from Paul and Ayres (1986c).]

It is interesting to note that when healthy and rust-infected monocultures were grown in winter, infected cultures had lower population densities at winter's end than did healthy cultures because of the high mortality induced by rust (30%, compared with only 5% in controls, 3 weeks after sowing; and 60%, compared with 35% in controls, 28 weeks after sowing; Paul and Ayres, 1986d). Yet since severe winter weather virtually eliminated rust infection sites and disease was no longer a constraint upon growth, the stunted but surviving rust-infected plants were able to take advantage of their lower population densities. The overall outcome was that reproduction in the groundsel population was less affected by rust in winter, despite high mortality, than in summer, despite low mortality. In the winter, floret production per unit area of land in the infected population (approximately equal to seed production) was 55% of that in the healthy population (Paul and Ayres, 1986e), but in the summer it was only 40% of that in the healthy population (Paul and Ayres, 1987b).

B. Competition, Infection, and Other Stresses

The possibility that abiotic stresses could induce or exacerbate competitive interactions has rarely been studied directly, although there is evidence that such effects do occur. For example, lack of potassium gave the tropical grass *Setaria* a competitive advantage over the legume *Desmodium intortum* that was absent when plants were well fertilized (Hall, 1974). Drought enhanced the small competitive advantage that perennial ryegrass had over white clover *(Trifolium repens)* grown in mixture (Table II), an advantage probably associated with its greater rooting depth (Thomas, 1984).

It is not surprising, then, that drought and rust infection had additive effects on intraspecific competition in groundsel. In monoculture, drought exacerbated the tendency of rust to make the size of individuals in a population more diverse (Paul and Ayres, 1987c); rust also lowered plant water potentials. Rust had its most marked effects on infected plants grown

Table II   Competitive Ability of *Trifolium repens* Grown in Mixture with *Lolium perenne*[a,b]

	Before drought (0–28 days)	During drought (28–56 days)	Recovery (56–92 days)
Wet			
Shoot	0.04	−0.53	−0.34
Root	−0.02	−0.25	—
Dry			
Shoot	—	−1.37	−1.05
Root	—	0.70	—

[a] Competitive ability = $\dfrac{\ln\,(2 \times \text{growth in mixture})}{(\text{growth in monoculture})}$

[b] [Modified from Thomas (1984)].

in unwatered mixtures, however, producing a preponderance of individuals with low shoot dry weights and exceptionally low water potentials (Fig. 13). Although the size of the plants' root systems could not be measured in this study, rust infection may have stunted root growth and increased diversity among root systems. Rust-infected plants, particularly those in the smallest size classes, would thus have competed poorly belowground for the limiting resource, water. The outcome of competition was no doubt influenced by other factors, such as competition for light among shoots of different height. Here again, infected plants would have been at a disad-

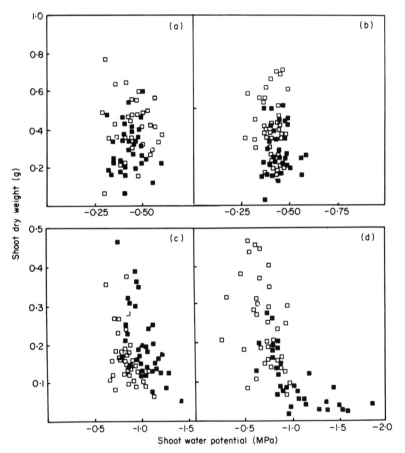

Figure 13 Shoot dry weight and shoot water potential of healthy (□) and rust-infected (■) groundsel in (a) well-watered monocultures, (b) well-watered mixtures, (c) water-stressed monocultures, and (d) water-stressed mixtures. Infection increases size diversity in the host population, making plants susceptible to other stresses, such as drought. Rust infection lowers host water potential, particularly in drought conditions, with the most severe effects on the smallest plants in the population. [Modified from Paul and Ayres (1987c).]

vantage because they were short; on the other hand, vapor-pressure deficits are relatively low in the bottom layers of the canopy so the forces driving transpiration would have been less than for taller plants. Nevertheless, damaging effects of rust on the size and functional efficiency of the root system would seem to have been of overriding importance.

V. General Conclusions

Among the plant characteristics responsible for high competitive ability, Grime and Hodgson (1987) cite high morphological plasticity during leaf and root differentiation. The plants examined in this chapter are not, in the main, those naturally having what Grime (1979) terms a competitive strategy. Nevertheless, some evidence suggests that morphological plasticity appears after some infections; this chapter highlighted increased leaf area ratio in shoots and increased specific root length, but other mechanisms undoubtedly exist as well. It is not yet known how much plasticity is affected by the age of the plant at infection and the degree of infection. In the laboratory, changes in leaf area ratio and specific root length can partially counteract the damaging effects of infection on functional efficiency. Similar responses may help the plant cope with other stresses, including atmospheric pollution, that limit the amount of dry matter the plant has available to produce structures for capturing environmental resources, or with stresses such as drought, in which fewer resources are available for capture. It is likely that simultaneous infection and abiotic stress can exhaust a plant's responsiveness. In general, pathogens and other stresses have additive deleterious effects because each reduces the plant's capacity for plastic responses, although there are some exceptions (e.g., rust infection of groundsel grown under extreme nutrient deficiency promotes increased phosphorus concentrations in tissues; Paul and Ayres, 1988a).

The effects of stress on populations of plants may differ significantly from the effects on individual plants if resources are reallocated within a population to ensure its success at the expense of the individual. Differences between the population and the individual may be especially marked if success is measured in terms of seed production or the size of the next population, since both can be maintained at relatively normal levels as long as the frequency of stress-induced mortality is relatively low, or the maturation of only a small fraction of the population is suppressed. Although much work has shown that the grain yield of disease-susceptible cultivar mixtures is superior to that of susceptible cultivar monocultures (Wolfe, 1984), those studies have not explored the physiology of such mixtures. Thus the possibility remains that leaf area or root system volume of lightly infected individuals may extend to exploit resources left untapped by more heavily infected neighbors.

The challenge for the future is to assess the importance of morphological plasticity under realistic conditions, where plants grow together and pathogens and other stresses increase natural diversity within and among populations.

References

Austin, R. B., Morgan, C. L., Ford, M. A., and Blackwell, R. D. (1980). Contributions to grain yield from preanthesis assimilation in tall and dwarf barley phenotypes in two contrasting seasons. *Ann. Bot.* 45: 309–319.

Ayres, P. G. (1981a). Powdery mildew stimulates photosynthesis in uninfected leaves of pea plants. *Phytopathol. Z.* 100: 312–318.

Ayres, P. G. (1981b). Root growth and solute accumulation in pea in response to water stress and powdery mildew. *Physiol. Plant Pathol.* 19: 169–180.

Ayres, P. G. (1982). Water stress modifies the influence of powdery mildew on root growth and assimilate import in barley. *Physiol. Plant Pathol.* 21: 283–293.

Ayres, P. G. (1984). Interaction between environmental stress injury and biotic disease physiology. *Annu. Rev. Phytopathol.* 22: 53–75.

Ayres, P. G., and Woolacott, B. (1980). Effect of soil water level on the development of adult plant resistance to powdery mildew in barley. *Ann. Appl. Biol.* 94: 255–263.

Baker, C. K., Fullwood, A. E., and Colls, J. J. (1987). Tillering and leaf area of winter barley exposed to sulphur dioxide in the field. *New Phytol.* 107: 373–385.

Ben Kalio, V. D. (1976). Effects of powdery mildew *Erysiphe cichoracearum* D.C. ex Merat on the growth and development of groundsel (*Senecio vulgaris* L.). Ph. D. dissertation, University of Glasgow.

Clarke, D. D. (1986). Tolerance of parasites and disease in plants and its significance in host–parasite interactions. *In* "Advances in Plant Pathology" (D. S. Ingram and P. H. Williams, eds.), Vol. 5, pp. 161–197. Academic Press, London.

Davies, E. (1987). Action potentials as multifunctional signals in plants: A unifying hypothesis to explain apparently disparate wound responses. *Plant, Cell Environ.* 10: 623–631.

de Wit, P. G. M. (1987). Specificity of active resistance mechanisms in plant–fungus interactions. *In* "Fungal Infection of Plants" (G. F. Pegg and P. G. Ayres, eds.), pp. 1–24. Cambridge Univ. Press, Cambridge.

Fitter, A. H. (1985). Functional significance of root morphology and root system architecture. *In* "Ecological Interaction in Soil: Plants, Microbes and Animals" (A. H. Fitter, D. Atkinson, D. J. Read, and H. B. Usher, eds.), pp. 87–106. Blackwell, Oxford.

Fitter, A. H. (1987). An architectural approach to the comparative ecology of plant root systems. *New Phytol.* 106 (Suppl.): 61–78.

Fitter, A. H., Nichols, R., and Harvey, M. L. (1988). Root system architecture in relation to life history and nutrient supply. *Funct. Ecol.* 2: 345–351.

Grime, J. P. (1979). "Plant Strategies and Vegetation Processes." Wiley, New York.

Grime, J. P., and Hodgson, J. G. (1987). Botanical contributions to contemporary ecological theory. *New Phytol.* 106 (Suppl.): 283–295.

Hall, R. L. (1974). Analysis of the nature of interference between plants of different species. II. Nutrient relations in a Nandi *Setaria* and green leaf *Desmodium* association with particular reference to potassium. *Aust. J. Agric. Res.* 25: 749–756.

Halverson, L. J., and Stacey, G. (1986). Signal exchange in plant–microbe interactions. *Microbiol. Rev.* 50: 193–225.

Hamblin, A., and Tennant, D. (1987). Root length density and water uptake in cereals and grain legumes: How well are they correlated? *Aust. J. Agric. Res.* 38: 513–527.

Harrison, J. G. (1988). The biology of *Botrytis* spp. on *Vicia* bean and chocolate spot disease. *Plant Pathol.* 37: 168–201.

Harrison, J. A. C., and Isaac, I. (1968). Leaf area development in King Edward potato plants inoculated with *Verticillium albo-atrum* and *V. dahliae. Ann. Appl. Biol.* 61: 217–230.

Hodges, C. F., and Coleman, L. W. (1984). Ethylene-induced chlorosis in the pathogenesis of *Bipolaris sorokiniana* leaf spot of *Poa pratensis. Plant Physiol.* 75: 462–465.

Joliffe, P. A., Minjas, A. N., and Runeckles, V. C. (1984). A reinterpretation of yield relationships in replacement-series experiments. *J. Appl. Ecol.* 13: 513–521.

Kasana, M. S., and Mansfield, T. A. (1986). Effects of air pollutants on the growth and functioning of roots. *Proc. Indian Acad. Sci. Plant Sci.* 96: 429–441.

Keon, J. P. R., Byrde, R. J. W., and Cooper, R. M. (1987). Some aspects of fungal enzymes that degrade cell walls. *In* "Fungal Infection of Plants" (G. F. Pegg and P. G. Ayres, eds.), pp. 133–157. Cambridge Univ. Press, Cambridge.

Kramer, P. J. (1981). Carbon dioxide concentration, photosynthesis, and dry matter production. *Bioscience* 31: 29–33.

Lupton, F. G. H., and Sutherland, J. (1973). The influence of powdery mildew *(Erysiphe graminis)* infection on the development of four spring wheats. *Ann. Appl. Biol.* 74: 35–39.

McCleod, A. R. (1988). Effects of open-air fumigation with sulphur dioxide on the occurrence of fungal pathogens in winter cereals. *Phytopathology* 78: 88–94.

Magan, N., and McCleod, A. R. (1988). In vitro growth and germination of phylloplane fungi in atmospheric sulphur dioxide. *Trans. Br. Mycol. Soc.* 90: 571–575.

Martin, N. E., and Hendrix, J. W. (1974). Anatomical and physiological responses of Baart wheat roots affected by stripe rust. *Wash. Agric. Exp. Stn. Tech. Bull.* 77: 1–17.

Murray, A. J., and Ayres, P. G. (1986). Infection with powdery mildew can enhance the accumulation of proline and glycinebetaine by salt-stressed barley seedlings. *Physiol. Mol. Plant Pathol.* 29: 271–277.

Oosterhuis, D. M., and Wullschleger, S. D. (1987). Water flow through cotton roots in relation to xylem anatomy. *J. Exp. Bot.* 38: 1866–1874.

Pande, P. C., and Mansfield, T. A. (1985). Responses of spring barley to SO_2 and NO_2 pollution. *Environ. Pollut.* A38: 87–97.

Paul, N. D., and Ayres, P. G. (1984). Effects of rust and postinfection drought on photosynthesis, growth, and water relations in groundsel. *Plant Pathol.* 33: 561–570.

Paul, N. D., and Ayres, P. G. (1986a). The effects of infection by rust *(Puccinia lagenophorae* Cooke) on the growth of groundsel *(Senecio vulgaris* L.) cultivated under a range of nutrient conditions. *Ann. Bot.* 58: 321–331.

Paul, N. D., and Ayres, P. G. (1986b). The effects of nutrient deficiency and rust infection on the relationship between root dry weight and length in groundsel *(Senecio vulgaris* L.). *Ann. Bot.* 57: 353–360.

Paul, N. D., and Ayres, P. G. (1986c). Interference between healthy and rusted groundsel *(Senecio vulgaris* L.) within mixed populations of different densities and proportions. *New Phytol.* 104: 257–269.

Paul, N. D., and Ayres, P. G. (1986d). The impact of a pathogen *(Puccinia lagenophorae)* on populations of groundsel *(Senecio vulgaris* L.) overwintering in the field. I. Mortality, vegetative growth, and development of size hierarchies. *J. Ecol.* 74: 1069–1084.

Paul, N. D., and Ayres, P. G. (1986e). The impact of a pathogen *(Puccinia lagenophorae)* on populations of groundsel *(Senecio vulgaris* L.) overwintering in the field. II. Reproduction. *J. Ecol.* 74: 1085–1094.

Paul, N. D., and Ayres, P. G. (1987a). Effect of rust infection of *Senecio vulgaris* on competition with lettuce. *Weed Res.* 27: 431–441.

Paul, N. D., and Ayres, P. G. (1987b). Survival, growth, and reproduction of groundsel *(Senecio vulgaris* L.) infected by rust *(Puccinia lagenophorae)* in the field during summer. *J. Ecol.* 75: 61–71.

Paul, N. D., and Ayres, P. G. (1987c). Water stress modifies intraspecific interference between rust *(Puccinia lagenophorae)*–infected and healthy groundsel *(Senecio vulgaris). New Phytol.* 106: 555–566.

Paul, N. D., and Ayres, P. G. (1988a). Nutrient relations of groundsel *(Senecio vulgaris* L.) infected by rust *(Puccinia lagenophorae)* at a range of nutrient concentrations. I. Concentrations, content, and distribution of N, P, and K. *Ann. Bot.* 61: 489–498.

Paul, N. D., and Ayres, P. G. (1988b). Nutrient relations of groundsel *(Senecio vulgaris* L.) infected by rust *(Puccinia lagenophorae* Cooke) at a range of nutrient concentrations. II. Uptake of N, P, and K and root–shoot interactions. *Ann. Bot.* 61: 499–506.

Potter, L. R. (1987). Effect of crown rust on regrowth, competitive ability, and nutritional quality of perennial and Italian ryegrasses. *Plant Pathol.* 36: 455–461.

Radin, J. W., and Eidenbock, M. P. (1986). Vascular patterns in roots of phosphorus- and nitrogen-deficient cotton seedlings. *Proceedings Cotton Physiology Research Conference.* pp. 85–88, Beltsville.

Reid, J. B., and Hutchinson, B. (1986). Soil and plant resistances to water uptake by *Vicia faba* L. *Plant Soil* 92: 431–441.

Smith, A. (1985). Polyamines. *Annu. Rev. Plant Physiol.* 36: 117–143.

Thomas, H. (1984). Effects of drought on growth and competitive ability of perennial ryegrass and white clover. *J. Appl. Ecol.* 21: 591–602.

Thompson, J. E., Legge, R. L., and Barber, R. F. (1987). The role of free radicals in senescence and wounding. *New Phytol.* 105: 317–344.

Tissera, P., and Ayres, P. G. (1986). Transpiration and the water relations of faba bean *(Vicia faba)* infected by rust *(Uromyces viciae-fabae). New Phytol.* 102: 385–395.

Tissera, P., and Ayres, P. G. (1988). Hydraulic conductance and anatomy of roots of *Vicia faba* L. plants infected by *Uromyces viciae-fabae* (Pers.) Shroet. *Physiol. Mol. Plant Pathol.* 32: 192–207.

Walmsley, L., Ashmore, M. R., and Bell, J. N. B. (1980). Adaptation of radish *(Raphanus sativus* L.) in response to continuous exposures to ozone. *Environ. Pollut.* A23: 165–177.

Walters, D. R., and Ayres, P. G. (1981). Growth and branching pattern of roots of powdery mildew–infected barley. *Ann. Bot. N. S.* 47: 159–162.

Walters, D. R., and Ayres, P. G. (1982). Water movement through root systems excised from healthy and mildewed barley: Relationship with phosphate transport. *Physiol. Plant Pathol.* 20: 275–284.

Walters, D. R., and Ayres, P. G. (1983). Changes in nitrogen utilization and enzyme activities associated with CO_2 exchange in healthy leaves of powdery mildew–infected barley. *Physiol. Plant Pathol.* 23: 447–459.

Walters, D. R., Wilson, P. W. F., and Shuttleton, M. A. (1985). Relative changes in levels of polyamines and activities of their biosynthetic enzymes in barley infected with the powdery mildew fungus *Erysiphe graminis* D.C. ex Merat J. f. sp. *hordei* Marchal. *New Phytol.* 101: 695–705.

Whitmore, M. E. (1982). A study of the effects of SO_2 and NO_2 pollution on grasses, with special reference to *Poa pratensis* L. Ph.D. dissertation, University of Lancaster.

Williams, P. F. (1975). Growth of broad beans infected by *Botrytis fabae. J. Hortic. Sci.* 50: 415–424.

Williams, P. F. (1978). Growth of broad beans infected by *Uromyces viciae-fabae. Ann. Appl. Biol.* 90: 329–334.

Williams, G. M., and Ayres, P. G. (1981). Effects of powdery mildew and water stress on CO_2 exchanges in uninfected leaves of barley plants. *Plant Physiol.* 68: 527–530.

Wolfe, M. S. (1984). Trying to understand and control powdery mildew. *Plant Pathol.* 33: 451–466.

12

Plant Stress and Insect Herbivory: Toward an Integrated Perspective

Clive G. Jones   **James S. Coleman**

I. Integrating Plant Stress and Insect Herbivory

Plants are simultaneously exposed to a diversity of abiotic environmental factors, such as drought, flooding, mineral nutrient deficiencies or imbalances, shading, temperature extremes, and air pollution, all of which can directly reduce dry matter accumulation (Chapin *et al.*, 1987). These same stresses can also alter the suitability of the plant to insect herbivores and, therefore, affect the degree or distribution of insect herbivory or both (Rhoades, 1983; White, 1984). Insect herbivory itself can also reduce plant dry matter accumulation (Crawley, 1983; Belsky, 1986), and so the net effect of stress has two components: (1) direct effects and (2) indirect effects due to changes in herbivory.

Integration of both components is currently limited by our ability to predict the effects of the environment on herbivory. The paradigm that stress increases herbivory and the mechanisms hypothesized to explain this paradigm (Rhoades, 1979; White, 1984) are not generally supported by the data in the literature. Therefore, assumptions underlying the patterns, causes, and consequences of these relationships require reevaluation. This chapter outlines a conceptual framework for predicting insect responses to stressed plants and integrating these interactions with direct effects of stress. The framework is based on understanding both the physiological responses of plants to stress and the differential sensitivity of insects to stress-induced changes in their host plants.

A. Paradigms, Hypotheses, and Patterns

It is generally argued that stress results in increased plant suitability to insect herbivores, improved herbivore performance, increased abundance, and increased herbivory (White, 1969, 1984; Mattson and Addy, 1975; Rhoades, 1979, 1983; Alstad *et al.*, 1982; Dohman, 1985; Fuhrer, 1985; Hain and Arthur, 1985). Two hypotheses have postulated mechanisms underlying this paradigm. White (1969, 1984) postulates that plant stress results in increased concentrations of mobile nitrogen in leaves, particularly amino acids. Nitrogen is often a limiting resource to insects (McNeill and Southwood, 1978; Mattson, 1980). Increased nitrogen availability to the herbivore results in increased insect growth, accelerated development, increased survivorship and fecundity, and thus increased abundance and herbivory. Rhoades (1979) also postulates that stress has a "net positive effect on fitness of herbivores." He proposes, however, that these net effects arise from an imbalance between nutritional quality, particularly increased nitrogen, and plant defensive chemicals. Concentrations of certain chemicals, such as alkaloids and cyanogenic glycosides, increase with stress, whereas others, particularly tannins, resins, and essential oils, decrease. Changes in defensive chemicals could be explained by assuming "that plants generally possess two or more defensive systems of differing cost. Under stress, plants compensate by decreasing their commitment to costly defenses and increasing their commitment to less costly, but less effective defenses" (Rhoades, 1979, p. 24). Consequently, increased nutritional quality combined with a decrease in defensive efficacy results in improved herbivore performance.

The literature, however, does not generally support the paradigm that abiotic stress results in positive effects on insect herbivores and increased herbivory; neither are the hypothesized mechanisms generally substantiated. Changes in insect herbivore abundance are, indeed, often positively correlated with the field environmental conditions under which plants are growing. For example, insect and mite outbreaks are associated with drought, flooding, mineral nutrient deficiencies, temperature extremes, shading, ice damage, lightning, forest fires, damage during cultural practices, plant exposure to gamma irradiation, pesticide application (which in some cases results in increases rather than the expected decreases in herbivore abundance), herbicide treatment, and air pollution (reviewed in White, 1969, 1984; Mattson and Addy, 1975; Rhoades, 1979, 1983; Tingey and Singh, 1980; Alstad *et al.*, 1982; Hughes and Laurence, 1984; Lechowicz, 1987; Mattson and Haack, 1987a,b; Heinrichs, 1988). It seems likely that the above paradigm has arisen from the need to explain these positive field correlations. These correlations do not necessarily mean, however, that changes in host plants are responsible because insect abundance may increase as a result of direct environmental effects on the insect, such as a more favorable climate or a decrease in the abundance of predators, parasites, and diseases (Mattson and Haack, 1987a,b).

In fact, experimental studies show that one or more insect specific performance attributes (host finding, feeding and oviposition behaviors, consumption, growth, development, survivorship, and fecundity), overall insect population growth, or the degree of herbivory can increase, decrease, or not change at all when plants are stressed (e.g., Louda, 1986; MacGarvin *et al.*, 1986; Jones and Coleman, 1988; Coleman and Jones, 1988a,b; reviewed in Larsson, 1989). Most of the examples cited by White (1984) and Rhoades (1979) to substantiate their hypothesized mechanisms come from studies in which the effects of stress on plant chemistry were examined independently from the effects on insect herbivores. Many of the chemical studies do indeed show that soluble nitrogen concentrations increase in stressed plants (Stewart and Lahrer, 1980) or that plant defensive chemicals increase or decrease in concentration (Gershenzon, 1984). Nevertheless, the relatively few studies simultaneously examining both plant chemistry and insect responses show that insect performance can change without the predicted chemical changes; that the predicted chemical changes can occur but insects may not respond; and that chemical changes occur and insects respond, but neither change in the predicted manner (see Section III; also Hughes *et al.*, 1983; Kimmerer and Potter, 1987; Larsson, 1989).

We contend that the diversity of effects of plant stress on insect herbivores observed in the literature stems from two independent sources of variation: (1) variation in plant responses to stress and (2) variation in the sensitivity of insects to stress-induced changes in their host plants. These sources of variation are rarely explicitly recognized, and this is why the paradigm and hypotheses are at odds with the data. A conceptual framework that explicitly recognizes the sources and causes of variation in responses of both plants and insects, and effectively interrelates these two components, may help to resolve these conflicts.

II. A Conceptual Framework for Integration

Our conceptual framework has two major components that deal with the two sources of variation (Fig. 1). The phytocentric model addresses the consequences of plant physiological adjustments to the environment on the biochemical, anatomical, and physiological characteristics of leaves. The exploiter model addresses the causes of differential sensitivity of insect herbivores, in terms of individual performance and population growth, to changes in suites of leaf characteristics. The two model components interact via stress-induced changes in leaf characteristics, which then feed back to the plant via changes in leaf consumption.

A. Plant Responses to Stress: The Phytocentric Model

1. Control of Plant Responses   Plants acquire carbon via leaves and nutrients and water via roots. These resources are partitioned to different plant parts, such as roots or shoots, and allocated to biochemical fates associated with the processes of resource acquisition and partitioning themselves (i.e., photosynthesis, uptake, and transport) and the functions of growth, defense, repair, maintenance, storage, senescence, and reproduction (Mooney, 1972). These processes and functions are controlled by intrinsic and extrinsic factors (see Fig. 1).

Plant genotype limits rates of acquisition, sets priorities for partitioning and allocation, and determines the operation of functions such as defense. For example, inherently fast-growing species tend to allocate a greater proportion of their resources to growth and may show more plasticity in allocation of resources to defense than slow-growers (Coley *et al.*, 1985), perhaps because leaf longevity, hence leaf value and requirements for defense, decreases as inherent growth rate increases (McKey, 1979). Furthermore, fast-growing species may synthesize small amounts of mobile secondary metabolites with high turnover rates and metabolic costs, such as phenolic glycosides, terpenoids, and alkaloids. Slow growers may synthesize relatively large amounts of immobile secondary compounds that accumulate as leaves age, such as tannins and lignins, compounds proposed to have low turnover rates and metabolic costs (Coley *et al.*, 1985).

Plant ontogeny and phenology modify the capacity and demand for

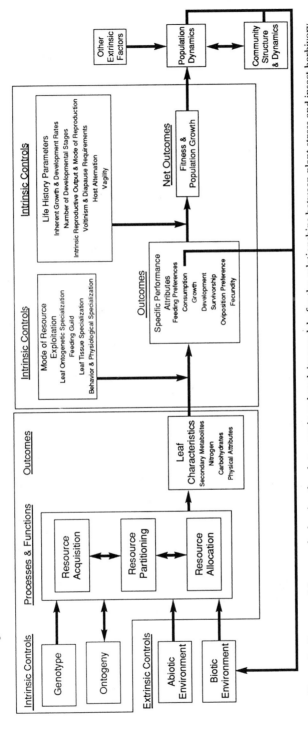

Figure 1 Conceptual framework, showing the phytocentric and exploiter models, for the relationships between plant stress and insect herbivory.

resource acquisition as well as the priorities for partitioning and allocation as plants grow and mature. For example, indeterminately growing tree species tend to use resources acquired within a growing season for growth in that season, whereas determinately growing species tend to use stored resources from the previous season; young saplings tend to allocate most resources to shoot growth, whereas mature trees allocate some resources to reproduction (Kozlowski, 1971a,b).

The environment is a critical influence. Light, soil nutrients, and moisture levels directly determine the amount of resources available for acquisition, and other environmental variables affect the plant's capacity to acquire and transport resources (Mooney, 1972; Chapin *et al.*, 1987; Chapter 3). Environmental factors such as drought, flooding, temperature extremes, ice, lightning, fire, or air pollution can also damage plant structure. Cellular or tissue damage not only reduces resource capture or inhibits transport but also increases the demands for resource allocation to defense and repair. These processes restore structural integrity, seal off damaged tissues, and reduce the risk of pathogen attack following damage (Mooney, 1972; Cruickshank, 1980). To cope with an everchanging environment, plants make short- and long-term physiological adjustments to maintain functional and structural homeostasis (Bloom *et al.*, 1985; Chapin *et al.*, 1987; Szianawski, 1987; Chapters 5, 6). Within limits set by their genotype, ontogeny, and present internal pool of resources, plants attempt to increase acquisition of the resource that is currently most limiting, restore integrity of structure, and minimize the impact of structural changes on plant functions, such that the plant maintains a carbon–nutrient balance around some optimum value (Bryant *et al.*, 1983; Chapin *et al.*, 1987). For example, plants exposed to carbon limitation often increase partitioning to shoots (Bloom *et al.*, 1985; Szianawski, 1987), which increases carbon gain relative to nutrient acquisition via the production of more leaf area or photosynthetic machinery. Plants exposed to nutrient or water limitation often allocate more resources to root growth (Hunt and Nicholls, 1986; Robinson, 1986), which increases surface area for absorption of water and nutrients. Carbon–nutrient balance has important consequences to leaves and can be used to predict changes in the quality and quantity of secondary metabolites (Bryant *et al.*, 1983), nitrogen, and other leaf characteristics.

2. Consequences to Leaves   Plant adjustments change the biochemical, anatomical, and physiological characteristics of leaves. The phytocentric model predicts changes in suites of leaf characteristics based on (1) the type, severity, duration, and combination of stresses, i.e., how carbon–nutrient balance is affected; (2) the location and extent of co-occurring damage; (3) the availability of internal resources that can be repartitioned and reallocated; and (4) genotypic, ontogenetic, and phenologic attributes of inherent growth rate, growth habit, and maturity. Here we illustrate the use of the phytocentric model to predict changes in leaf characteristics important to insect herbivores (secondary metabolites, nitrogen, carbohydrates, and leaf physical characteristics) when plants are stressed by shading, air pollution, mineral nitrogen deficiencies, and drought.

a. Secondary Metabolites   When carbon gain is limited relative to nutrient availability, as during shading, the resource availability hypothesis (Bryant *et al.*, 1983), which is based on carbon–nutrient balance, predicts a decline in allocation to carbon-based secondary metabolites (Bryant *et al.*, 1985; Bryant, 1987), while allocation to nitrogen-based secondary metabolites increases (Bryant *et al.*, 1985). On the other hand, when plants are subjected to mineral nitrogen deficiencies, allocation to carbon-based secondary compounds should increase, while production of nitrogen-based secondary compounds should decrease (Bryant *et al.*, 1985, 1987). We use the concept of inherent growth rate (Coley *et al.*, 1985; see Section II,A,1) to refine the above predictions. Under stress, fast growers should show extensive plasticity in secondary metabolite production because allocation to growth is a higher priority. In slow growers, allocation to secondary compounds is a high priority, and therefore, secondary metabolite production should be relatively unchanged under stress.

These predictions appear to be well substantiated. For example, low light reduces the concentration of phenolics in fast-growing alder (Bryant *et al.*, 1987) and willow (Waring *et al.*, 1985; Larsson *et al.*, 1986; Bryant, 1987), as does relative carbon surplus caused by high light and low mineral nutrients in willow (Waring *et al.*, 1985). Nitrogen deficiencies reduce production of nitrogen-based alkaloids in some fast growers and increases carbon-based compounds (Gershenzon, 1984). Low light has relatively little effect on phenolic production in the slow-growing shrub *Diplaucus aurantiacus* (Lincoln and Mooney, 1984). Furthermore, slow growers from resource-poor environments do not usually produce nitrogen-based secondary compounds, whereas nitrogen deficiencies result in relatively small increases in the production of tannins and the like (Bryant *et al.*, 1983; Coley *et al.*, 1985).

Air pollution reduces plant carbon gain and decreases the carbon:nitrogen ratio of plants (Guderian *et al.*, 1985; Lechowicz, 1987). Production of carbon-based secondary compounds should be reduced and nitrogen-based compounds should increase. In cottonwood leaves, mobile carbon-based phenolic glycosides decrease at ozone doses that do not cause visible injury (Jones and Coleman, 1989), but other plants show increased phenolic concentrations and increased activities of phenolic-related enzymes (phenylammonia lyase, polyphenol oxidase) (Keen and Taylor,

1975; Tingey *et al.*, 1975, 1976; Hurwitz *et al.*, 1979). These plants often have visibly injured tissue, however (Hughes and Laurence, 1984; Guderian *et al.*, 1985), so production of secondary compounds in air pollution–stressed plants is probably also related to the degree of cellular damage caused by the pollutant since structural damage often results in deposition of phenolic compounds or lignification of the cell walls around the damaged area. In fact, cottonwood leaves do show increased deposition of phenolics with ozone exposure, despite decreased concentrations of mobile phenol glycosides (Jones and Coleman, 1989). Damage may also result in the induction of biosynthesis or transport of carbon- or nitrogen-based secondary metabolites in the undamaged portions of damaged leaves and other leaves (Coleman and Jones, 1991).

Drought-induced changes in secondary metabolites are complex because drought affects the acquisition of both carbon and nutrients as well as resource transport; it also alters the concentration of secondary compounds via changes in leaf water content. Thus, secondary metabolite concentrations can both increase and decrease (Gershenzon, 1984) as a function of the duration and severity of stress [cf. the dual-defense hypothesis of Rhoades (1979); see Section I,A].

Emission of leaf volatiles changes with environmental conditions but does not appear to be related to carbon–nutrient balance. Air pollution, nutrient stress, and water stress all increase emissions of ethylene from leaves (Darrall, 1984; Pell and Puenté, 1986; Dunn *et al.*, 1986; Kimmerer and MacDonald, 1987), even when the stress is relatively mild (Kimmerer and Kozlowski, 1982).

b. Nitrogen   Many environmental stresses induce the mobilization of nitrogen in plants and increase the amounts of amino acids, imino acids, poly- and diamines, and low-molecular-weight polypeptides (Stewart and Larher, 1980; Smith, 1984; Erickson and Dashek, 1982; Grill *et al.*, 1980; Craker and Starbuck, 1972; Rhodes, 1987). Stress also affects the total concentration and relative availability of nitrogenous compounds in plants. Plants in shaded and polluted air exhibit decreased carbon : nitrogen ratios (Bryant *et al.*, 1983, 1985) and are relatively rich in nitrogenous compounds (Larcher, 1980; Lechowicz, 1987; Reich, 1987). Mild water and nutrient stresses usually result in increased carbon : nitrogen ratios in plants and decreased total nitrogen concentrations in leaves (Alberte and Thornber, 1977; Bradford and Hsiao, 1982). Amino acid composition can also change with drought and nutrient availability (Stewart and Larher, 1980). For example, phloem sap of fertilized rice plants contained aspartic acid and glutamine as dominant amino acids, whereas unfertilized plants were dominated by glycine and asparagine (Hayashi and Chin, 1986). Air pollution stress can also change amino acid composition (Godzik and Linskens, 1974), increasing the levels of serine and glycine (Heath, 1984).

c. Carbohydrates   Plants exhibit extensive diurnal variation in carbohydrate concentrations, and any stress-induced effects may be minor in comparison (Huber, 1983; Gordon, 1986; Geiger, 1986; Huber *et al.*, 1986; Chapters 1, 5). For example, the dry weight of cottonwood leaves varies by as much as 30% during a single day because photosynthate is stored as starch during the light period and translocated out of leaves as sucrose during the dark (Dickson, 1987). Most reported changes in leaf carbohydrates resulting from air pollution are much less than this (Koziol, 1984). Nevertheless, decreases in plant carbon gain result in increased carbon sink pressure by shoots relative to roots (Geiger, 1986), the retention of larger amounts of carbohydrates in leaves (Ho, 1976; Fox and Geiger, 1984; Geiger and Fondy, 1985; Shaw *et al.*, 1986), and elevated starch:sucrose ratios during the light period (Noyes, 1980; Teh and Swanson, 1982; Stitt *et al.*, 1984; Huber *et al.*, 1986). Mild drought and nitrogen deficiencies result in preferential allocation of carbohydrates to roots, with a relative decrease in the amount of carbohydrate retained in leaves, and decreased starch:sucrose ratios (Huber *et al.*, 1986; Geiger, 1986).

Air pollution and severe water stress reduce membrane integrity, changing the osmotic balance of plant tissues (Tingey and Taylor, 1982; Heath, 1984; Morgan, 1984) and causing sugars to leak onto the leaf surface (Rist and Lorbeer, 1984; Mattson and Haack, 1987a).

d. Physical Characteristics   A reduction in the plant carbon:nitrogen ratio reduces carbon available for fibers, lignin, and cell walls as carbon is shunted to roots (Larcher, 1980). Shade and air pollution often decrease leaf specific weight (Clough *et al.*, 1979; Larcher, 1980; Gulmon and Chu, 1981) and fiber content (Larcher, 1980); shade leaves are usually thinner and bigger than sun leaves (Larcher, 1980; Kimmerer and Potter, 1987). Increased leaf area and reduced flexibility of carbon partitioning can result in decreased densities of vascular bundles, trichomes, and spines (Larcher, 1980; Campbell, 1986). Increased carbon:nitrogen ratios, as a result of nutrient deficiency or mild water stress, produce the opposite effects. Carbon is shunted into the production of fibers, lignin, and cell wall material (Radin and Parker, 1979; Larcher, 1980; Bradford and Hsiao, 1982; Bell, 1984), and leaf specific weight increases (Larcher, 1980; Gulmon and Chu, 1981; Graham, 1983; Ayres, 1984). Leaves tend to be smaller, and the density of vascular bundles and anatomical defenses is higher (Bradford and Hsiao, 1982; Campbell, 1986).

Plants with reduced carbon:nitrogen ratios tend to have lower transpiration rates and higher leaf water content, the opposite of that found for plants with increased carbon:nitrogen ratios (Larcher, 1980; Gulmon and Chu, 1981; Bradford and Hsiao, 1982; Kramer, 1983). Carbon limitation also reduces the amount of leaf allocation to lipid production, resulting in reduced cuticle thickness (Baker, 1974; Wilkinson and Kasperbauer, 1972;

Larcher, 1980), which may be exacerbated by air pollution–induced degradation of leaf cuticles (Guderian *et al.*, 1985; Crossley and Fowler, 1986; Shelvey and Koziol, 1986). On the other hand, a relative surplus of carbon in nutrient- or drought-stressed plants results in increased lipid production and a thicker cuticle but apparently no change in cuticular composition (Baker, 1974; Wilkinson and Kasperbauer, 1972; Shelvey and Koziol, 1986).

Reflectance characteristics of leaves also change with stress. For example, water stress increases leaf emission of infrared radiation because of increased leaf temperature (Reicosky *et al.*, 1985). Mild water and nutrient stresses also increase concentrations of pigments such as flavonoids and anthocyannins (Gershenzon, 1984), increasing leaf absorbance of ultraviolet and increasing reflectance in the yellow–orange–red wavelengths. In contrast, shaded plants are usually cooler, reflect less infrared (Larcher, 1980), produce fewer carbon-based pigments (Gershenzon, 1984), and exhibit reduced reflection of yellow–orange–red wavelengths and reduced ultraviolet absorbance.

B. Insect Responses to Stressed Plants: The Exploiter Model

1. Determinants of Insect Sensitivity to Changes in Leaf Characteristics
Biochemical, anatomical, and physiological characteristics of leaves are important determinants of insect herbivore behavior, growth, development, survival, and fecundity. These plant characteristics include nitrogen and water content; toughness and thickness; leaf color and surface components, such as glands and hairs; volatile, surface, and internal secondary metabolites; turgor; and sink–source relations (Beck, 1965; Rosenthal and Janzen, 1979; Scriber and Slansky, 1981; Scriber, 1984; Coleman, 1986; Juniper and Southwood, 1986). Some plant characteristics are commonly correlated with insect performance. For example, growth, development, survival, and fecundity are often greatest on leaves high in nitrogen, with adequate water, low leaf toughness, and low concentrations of secondary metabolites (Feeny, 1976; Rhoades and Cates, 1976; Fox and McCauley, 1977; McNeill and Southwood, 1978; Scriber, 1978; Mattson, 1980). Also critically important, however, are the idiosyncrasies arising from taxonomic constraints on when and how insects feed and develop, insect size, and behavioral and physiological adaptations of herbivores to their particular host plants (Rosenthal and Janzen, 1979; Scriber and Feeny, 1979; Jones, 1983).

The exploiter model (see Fig. 1) recognizes that the specific performance attributes of individual insects, as well as resultant population growth, are determined by interactions between various leaf characteristics and the differential sensitivity of insect species to these characteristics. A given set of leaf characteristics in a stressed plant is therefore unlikely to result in uniform responses for all insects. We postulate that insects encounter and respond to suites of leaf characteristics, and different insect species can respond differently to the same or to a different suite of characteristics. The net effect of changes in leaf characteristics on the population growth of a given insect species is therefore determined by (1) when, where, and how the insect feeds on the plant; (2) how changes in leaf characteristics affect particular behavioral and physiological attributes of the insect and thus, herbivore performance attributes; and (3) how these various performance attributes interact to determine population growth.

The model has two subcomponents. The first predicts the sensitivity of specific performance attributes of different types of insects (host finding, feeding preference, consumption, growth, development, survival, oviposition preference, and fecundity) based on how the insects feed and on their degree of behavioral and physiological specialization. We term this subcomponent the mode of resource exploitation. Within a given insect species, various specific performance attributes can be differentially affected by the same or different leaf characteristics. The second subcomponent of the model therefore predicts how changes in specific performance attributes interact to determine overall population growth (abundance) based on an understanding of insect life-history parameters.

a. Mode of Resource Exploitation Four different aspects of the ways insect herbivores exploit plants are postulated to be important.

(1) Leaf Ontogenetic Specialization This identifies the particular leaf developmental stage(s) used, such as unexpanded, expanding, fully expanded, or senescing leaves. For example, thrips and budworms have complete or partial development in buds. Many gall-forming insects oviposit in unexpanded leaves, with their larval stages developing in galls in expanding leaves. A number of chewing insects feed on young leaves (e.g., beetles, Lepidoptera), whereas others specialize on older leaves (e.g., some Lepidoptera). Some defoliating insects necessarily feed on more than one leaf developmental stage, and aphids can feed on both young, expanding leaves and senescing leaves (e.g., Jones, 1983; Coleman and Jones, 1988b). Leaf ontogenetic specialization may well be influenced by nutritional constraints, since young or senescing tissues are often higher in nitrogen than old leaves (Raupp and Denno, 1983). Insect developmental constraints, however, may require consumption of both young and old leaves because the insect has a long developmental period relative to availability of young leaves. Environmental constraints on the breaking of diapause may mean that the insect hatches or becomes active late in the season. Constraints imposed by the availability of the most suitable leaf resources due to growth habit of the plant may also be important (Coleman, 1986). Indeterminately growing plants continually flush leaves, and thus leaves of the appropriate developmental stage may be continuously available in limited quantities

throughout the growing season. On the other hand, determinate growers tend to flush synchronously, with entire cohorts of leaves aging synchronously.

(2) Feeding Guild Insects do not all feed in the same manner, even if they use the same leaf age. Insects may gall, mine, roll, suck, crush, and suck up the cell contents, or chew. For example, many phytophagous Diptera are gall formers or miners, as are many sawfly larvae (Hymenoptera). Many Lepidoptera chew leaves, but some species mine or roll leaves. Small Coleoptera commonly crush and suck up cell contents, and the Aphidae are suckers. Feeding guild is primarily determined by ancestral taxonomic constraints, such as the basic design of mouth parts or body size.

(3) Leaf Tissue Specialization This identifies the specific leaf tissues ingested, such as portions of entire leaves, leaf margins or centers, intervein areas, particular cell layers, single cell contents, intracellular fluids, phloem, or tissues specially modified by the insect, such as nutritive cell layers of galls. Leaf tissue specialization is probably determined by the ancestral taxonomic constraints that influence both leaf ontogenic specialization and feeding guild but also by constraints on body size associated with growth and development. Thus, for example, early instars of leaf chewers may tend to feed in intervein areas or may avoid consuming hairs, spines, or glandular trichomes that may be effective barriers against larger insects; large chewers, on the other hand, including the same species at later developmental stages, may consume entire portions of leaves. Similarly, small mining insects may feed on just the spongy mesophyll layers of leaves, while later developmental stages or larger miners may consume all tissues except the epidermis and endodermis, and these insects are therefore particularly sensitive to changes in leaf thickness or developmental rates (e.g., Faeth *et al.*, 1981; Kimmerer and Potter, 1987). Sucking insects do not appear constrained to feed on one type of tissue; different species may feed in galls, on single cells, intracellular fluids, phloem, and even xylem fluids. Nevertheless, these insects may well be sensitive to leaf turgor and sink–source relationships.

These three aspects of mode of resource exploitation can be used to identify which leaf stages and tissues should be considered in regard to stress modification, reducing the list of potentially important characteristics to manageable length. For example, a phloem feeder on senescing leaves will likely respond to a different set of leaf characteristics than a chewer of entire portions of fully expanded leaves or a gall former that oviposits on unexpanded leaves with a feeding stage that uses nutritive cells in galls of expanding leaves. Major differences in mode of resource exploitation, or even subtle differences — such as a sucker that feeds on cells, intracellularly or in phloem — may be important reasons for differences in the performance of insects on stressed plants.

(4) Behavioral and Physiological Specialization Variation in the behavioral and physiological responses of insects is probably responsible for substantial differences in the insects' sensitivity to changes in leaf characteristics. Many insects are relatively specialized, being restricted to one or a few closely related plant species. A large number, however, particularly agricultural and forest pests, are generalists, exploiting species across plant families. Adaptations associated with host range are postulated to be important determinants of insect responses to stressed plants.

Plant characteristics are important visual, physical or chemical cues that function as host-finding, feeding, oviposition or dispersal attractants, repellents, stimulants, deterrents, and arrestants (Dethier, 1954, 1980; Jermy, 1966). These behavioral cues may be unique to the host plant, such as many internal secondary metabolites or surface resins, glandular trichomes, pubescence, or hairs; or more ubiquitous signals, such as surface waxes, color, internal sugars, or amino acids (Juniper and Southwood, 1986). Cues that function as attractants or stimulants to one insect may be repellents or deterrents to another insect, even if growth or survivorship on the plant is adequate (e.g., Gupta and Thorsteinson, 1960; Jermy and Szentesi, 1978). The same insect may use different cues for host finding (e.g., color or volatiles), feeding preference (e.g., toughness, sugars, amino acids, or secondary metabolites), oviposition preference (e.g., waxes, sugars, and secondary metabolites), and dispersal (e.g., amino acids), and it is rare that a single physical or chemical attribute is the sole stimulus. A mixture of cues is frequently used, with the overall response of the insect being determined by the balance of positive and negative cues (Papaj and Prokopy, 1986).

Many of the plant characteristics affecting insect behavior also influence growth, development, survivorship, and fecundity. Most insect herbivores have similar nutritional requirements for nitrogen, water, lipids, carbohydrates, steroids, vitamins, and inorganic ions (Dadd, 1973). Leaf water content and nitrogen are often positively correlated with insect growth (Scriber and Slansky, 1981), and nitrogen is frequently a limiting resource, since its concentration is relatively low in plants compared with insects (McNeill and Southwood, 1978). Not all forms of nitrogen (e.g., protein nitrogen versus amino acids) are used with the same efficiency by all insects, however, and this efficiency varies considerably (Scriber and Slansky, 1981). Plants also contain secondary metabolites that can be toxic, inhibit growth and development, and reduce the fecundity of some insects but not others (Rosenthal and Janzen, 1979) because of the variation in both the chemical's biological activity and the capacity of the insect to eliminate, sequester, or detoxify these compounds (Brattsten, 1979; Duffey, 1980). Compounds toxic to one insect may have no effect on another or may even be used as nutritional substrates by another insect (e.g., Bernays and Woodhead, 1982).

The exploiter model takes into account the variation in behavioral and physiological sensitivities of insects based on their degree of physiological specialization.

Specialists tend to use host-specific cues, sometimes in conjunction with ubiquitous signals, as positive indicators of host suitability (Dethier, 1980; Jermy, 1984). Such insects have relatively specialized receptors for detecting specific host chemicals, and they may have a limited capacity to detect non-host chemicals, particularly secondary metabolites. Consequently, specialists should show increased attraction or stimulation with increases in the concentration of chemical cues that can occur in slightly to moderately stressed plants and decreased attraction or stimulation with decreased concentrations. Similarly, specialist insects should show increased or decreased positive responses over a relatively narrow range of cue concentrations; in other words, their behavioral plasticity is low. Although receptors may be relatively specialized, it is likely that there is some capacity to detect qualitative changes in the composition of cues, particularly the presence of chemically similar components that are not normally present in high concentrations. Changes in quality that are associated with severe stress may therefore fail to elicit positive responses. Because specialists have relatively nonplastic behavior, compensatory mechanisms — such as increased consumption on nutritionally inadequate hosts — are unlikely to be present. Thus, a lack of positive cues, which decreases feeding, combined with decreased host nutritional quality, will contribute to decreased growth, development, survival, and fecundity. Furthermore, cue requirements for oviposition are likely to be relatively strict, and decreases in the concentration of these cues, or their absence, should result in marked decreases in oviposition, particularly at moderate to severe levels of plant stress.

Growth, development, or fecundity of specialists is postulated to increase (or decrease) with increased (or decreased) foliar nitrogen. But because specialists are likely to have limited physiological plasticity, they may respond to changes in plant nitrogen over a fairly narrow concentration range, such as those found in slightly to moderately stressed plants. Marked increases in foliar nitrogen associated with severe stress may not result in any greater increase in performance than that found at moderate increases (e.g., growth may be asymptotic at moderate foliar nitrogen levels). Specialists may also show a limited capacity to deal with the changes in the quality of nitrogen in the host that can occur with severe stress. Performance will therefore tend to correlate positively with concentrations of particular forms of nitrogen, rather than total foliar nitrogen. Substantial increases in amino acid content, without corresponding increases in soluble protein, may not result in any increased performance if the herbivore uses protein more efficiently than amino acids.

Specialists will tend to be well adapted to processing the secondary chemicals found in their host plants. Adaptations for detoxifying, sequestering, or eliminating these components may have little or no apparent cost in terms of decreased growth, development, survival, or fecundity. Thus increases in secondary chemical concentrations may have little adverse effect on performance, provided the changes remain within the range of the insects' capacity to process these chemicals (i.e., low to moderate plant stress). In certain circumstances, concentrations of these chemicals may correlate positively with performance, if the compounds are used as nutritional substrates. On the other hand, specialists should have only limited capacity to process plant secondary metabolites that do not normally occur in high concentrations. Thus, marked shifts in plant allelochemical composition, which may occur with severe stress, may adversely affect performance.

Generalists, in contrast to specialists, tend to use more-or-less ubiquitous signals as positive cues, and plant species–specific characteristics often function as negative indicators of host suitability (Jermy, 1984). Such insects have relatively nonspecific receptors that respond positively to such compounds as sugars or amino acids and a broad capacity to detect a range of secondary metabolites functioning as deterrents or repellents. Consequently, generalists should show increased or decreased attraction or stimulation with increases or decreases in the concentration of ubiquitous components, respectively, and increased or decreased repulsion or deterrence with increases or decreases of a wide range of secondary metabolites, respectively. Because these insects have substantial plasticity of behavior, they should show these responses over a broad range of concentrations in slightly to severely stressed plants. Because of this plasticity, qualitative changes, such as those occurring in severely stressed plants, are unlikely to play a critical role. Compensatory feeding, i.e., increased consumption on nutritionally inadequate hosts, is likely to be present (Mattson, 1980). When it is, consumption may be decoupled from growth, development, and fecundity: these do not change because increased consumption offsets decreased nutritional quality. Furthermore, many generalists do not have oviposition behaviors that are strictly regulated by the host plant (e.g., gypsy moth), and so effects on oviposition behavior may not be marked at most levels of plant stress.

The growth, development, or fecundity of generalists may increase (or decrease) over a wide range of changes in foliar nitrogen because the physiological processes associated with nitrogen use are likely to be plastic. Insect performance and foliar nitrogen may show a linear positive relationship with no asymptote at nitrogen concentrations found in slightly, moderately, or severely stressed plants or unstressed plants. Furthermore, their physiological plasticity should confer a relatively broad capacity to deal with changes in plant nitrogen quality, such that performance may correlate well with total foliar nitrogen, rather than the concentration of specific forms of foliar nitrogen at all levels of stress.

Generalists are well adapted for processing a wide range of plant second-ary metabolites and should be relatively insensitive to the qualitative com-position of plant allelochemicals. On the other hand, detoxification or elimination of these compounds may have a cost, with performance de-creasing at high total concentrations of such compounds. (Sequestration appears to be rare in generalists [Jones *et al.,* 1988].) Furthermore, be-cause generalists have a broad capacity for processing many allelochemi-cals, they may also have a limited capacity to process any given chemical efficiently. Particularly toxic or inhibitory compounds in a given plant may be processed inefficiently, leading to greater decreases in performance compared with a specialist.

b. Life-History Parameters   The various specific performance attributes of insect herbivores can be differentially affected by changes in stressed plants (see Section I,A). The same proportional magnitude of change in a given specific performance attribute, such as growth rate or fecundity, will not necessarily translate into the same degree of change in abundance for two different insect species. Furthermore, within a species, different spe-cific performance attributes can change to different degrees and in differ-ent directions. For example, fecundity may decrease while growth stays unchanged (e.g., Coleman and Jones, 1988a), or consumption could in-crease while survivorship decreases, with quite different consequences to abundance. These patterns can arise because different species of insects have different life histories (Denno and Dingle, 1981), and it is life-history parameters that determine the net outcome, in this case abundance, of interactions among changes in specific performance attributes. Failure to take into account major, or even subtle, differences in life-history parame-ters may lead to predictions that are opposite to the observed patterns.

The exploiter model incorporates life-history parameters. Some of these parameters are markedly constrained by taxonomy, but within closely related taxa, life histories can differ in important, subtle ways. The model recognizes six key parameters. Although we examine these parameters one at a time, the model requires that they be combined.

(1) Inherent Growth and Development Rates   Insects with rapid inherent growth or development on stressed plants than are slow growers or rapid changes in population abundance following increases or decreases in growth for development on stressed plants than are slow growers or developers. Fast growers or developers will reach reproductive maturity more rapidly, and therefore the potential for the abundance of their populations to change within and between years will be greater. Dynamic variation in host-plant suitability caused by changes in the intensity and duration of stress, together with plant adjustments to stress, are more likely to translate into fluctuations in abundance for fast growers or developers, primarily because their generation time is shorter.

(2) Number of Developmental Stages   Insects with relatively few developmental stages, or developmental stages that are very similar in their trophic requirements (all stages use the same resource), should show less variation in abundance than insects with many stages, or divergent trophic requirements at each stage, primarily because spatial and temporal variation in host quality because of stress has a higher probability of having an effect.

(3) Intrinsic Reproductive Output and Mode of Reproduction   Given the same proportional change in fecundity, insects with intrinsically high rates of reproductive output (many offspring per clutch or many clutches per female) are likely to show more variation in abundance on stressed plants than insects with low reproductive output. Insects with females that require only a single mating to produce multiple clutches or that reproduce parthenogenetically, producing offspring continuously (e.g., many aphids), are more likely to show variation in abundance than insects for which multiple matings are necessary simply because mate finding is less likely to constrain any changes in female reproductive potential. Insects with long-lived females may show substantial variation in abundance because stress-induced increases (or decreases) in adult-female longevity coupled with increases (or decreases) in fecundity will translate into substantially greater (or less) reproductive output.

(4) Voltinism and Diapause Requirements   Increases in growth or development that are not accompanied by increases in survival or fecundity are likely to increase abundance only if the insect is multivoltine (has more than one generation per year). In fact, increases in growth and development rates may result in decreased abundance in subsequent generations for univoltine insects (those with a single generation per year) that lack flexibility in their diapause (overwintering) requirements because insects that mature or reach the overwintering stage too early or too late in the season may experience decreases in overwintering survivorship. Flexibility in diapause requirements may also be critical for multivoltine insects. If the insect can overwinter at different stages, or, as in the tropics, if diapause is unnecessary, then increases in abundance because of an increase in the number of generations will be carried forward into the subsequent season. If, however, diapause requirements are inflexible, then changes in abundance in the subsequent season will depend on whether the extra generation(s) produced on stressed hosts reach an appropriate overwintering stage before the end of the season. If an extra generation completes development, then abundance in the next season should increase, but if development is incomplete, overwintering survivorship should decrease markedly, offsetting any increases from the extra generation produced in the previous season.

(5) Host Alternation   A number of insects have sexual generations on one host and asexual generations on another, often unrelated, host (e.g., many

aphids). Here, abundance on one or another host plant will depend on which host (or hosts) is stressed and whether stress affects specific performance attributes in the same or different ways on different hosts. It is probable that only one host will be stressed at a time. These unrelated plants (for aphids, often a herbaceous annual and a woody perennial) will probably respond differently to stress, even if both are exposed to the same environment. Furthermore, sexual and asexual generations may well respond differently to stress. Consequently, insects with obligate host alternation may well show more variation in abundance on stressed plants than insects that do not alternate hosts.

(6) Vagility Insects show different propensities for aggregation and dispersal, and this has a number of important consequences. Increases in abundance of aggregating and nondispersing species are more likely to lead to intraspecific competition and subsequent decreases in survival and reproductive output than in dispersing species. Some species may have density-dependent dispersal (e.g., some plant hoppers); in these species, increases in survival or reproduction on one host may not lead to increases in abundance on that host simply because of increased emigration.

III. Predicting Insect Responses to Stressed Plants

The effects of environmental stress on plant interactions with a given herbivore can have only three net outcomes in terms of either insect abundance or plant damage: an increase, decrease, or no change. However, as one would expect from the multiple components of the phytocentric and exploiter models, the components can interact in a very large number of ways to produce the same or different outcomes. The most parsimonious way to integrate these components is to use the appropriate model components to predict sequentially (1) particular leaf characteristics, (2) specific insect performance attributes, and (3) overall population growth. Space does not permit us to illustrate even a small fraction of all the possible combinations of stresses and insects; instead, we illustrate our approach with some real and hypothetical examples.

A. Air Pollution, Shading, Cottonwood, and Its Insect Herbivores

Exposure of inherently fast-growing, indeterminate saplings of clones of eastern cottonwood *(Populus deltoides)* to a single acute dose of ambient ozone (200 ppm for 5 hr), resulted in a reduction in photosynthetic rates over 72 hr that was accompanied by a corresponding reduction in the relative growth rates (RGR) of shoots compared with charcoal-filtered-air controls (Jones, Coleman and Wait, unpublished data). These responses suggested that ozone exposure caused carbon stress. Although younger leaves were not visibly injured, nocturnal leaf respiration rates were unchanged relative to controls, despite the reduction in shoot RGR. This indicated that the decrease in respiration associated with growth was accompanied by increases in respiration associated with repair of cellular damage (Jones, Coleman and Wait, unpublished data). Thus, ozone exposure probably caused both carbon stress and damage, as is known from other studies (e.g., Tingey *et al.*, 1975, 1976; Reich, 1987).

Changes in leaf biochemical characteristics were compatible with these physiological and growth responses. Leaf physical characteristics (specific weight, area, water content, and so on) were unchanged, as would be expected from such a short-term, low-dose exposure (Jones and Coleman, unpublished data). There were, however, a number of biochemical changes in leaves. The major form of carbon-based, mobile secondary metabolites, phenol glycosides, decreased in concentration by about 30% (Jones and Coleman, 1989). This would be expected with carbon stress to a fast-growing plant with a high priority for growth (see Section II,A,2). In addition, the proportion of leaf carbon and nitrogen in low-molecular-weight, polar, mobile forms increased, and phenolics bound to cell walls, absent in controls, appeared (Jones and Coleman, 1989). These changes would be expected with cellular damage, phenolics being deposited to seal off damaged cells, and carbon and nitrogen being mobilized for repair. Both of these responses are known for other plants exposed to oxidant injury.

Bioassays with a leaf-chewing beetle, *Plagiodera versicolora,* showed that the following changes in specific performance attributes occurred: Beetles preferred and consumed more of the ozone-exposed plant material but showed no change in growth rates; consequently, relative growth rates (growth per unit food ingested) decreased. Development and survivorship were unchanged. Beetles preferred to oviposit on plants grown in charcoal-filtered air, even if they had been previously reared on control plants (Jones and Coleman, 1988; Coleman and Jones, 1988a).

The changes in specific performance attributes are compatible with both the leaf biochemical changes and with certain aspects of this insect's mode of resource exploitation. *P. versicolora* is a specialist on the Salicaceae (willows and poplars; Raupp and Denno, 1983). It prefers expanding leaves at the sink–source transition—the same leaves that showed the greatest degree of biochemical alteration by ozone (Coleman, 1986; Jones and Coleman, 1988; Coleman and Jones, 1988a). This insect feeds on cells by crushing and sucking the fluids (larvae) and by ingesting the fluids and some of the tissues (adults)—the same cells that have been affected by ozone. Feeding preference appears to be regulated by phenol glycosides and sugars (Tahvanainen *et al.*, 1985). Some phenol glycosides and sugars must be present to stimulate feeding, but high concentrations of phenol glycosides are inhibitory, and lower concentrations are stimulatory. We suggest that the beetle's increased preference for ozone-exposed plants was due to increased stimulation caused by the ozone-induced reductions in phenol glycoside concentrations (carbon stress), which may also have been responsible for the increased consumption. We suspect that the reduction in relative growth rates was due to a decrease in the quality of food because of phenolic deposition on the cells. This might reduce the availability of soluble protein-nitrogen because of the formation of phenolic–membrane complexes, particularly since this insect ingests primarily cell fluids. The fact that growth, development, and survivorship did not change suggests that compensatory consumption, perhaps facilitated by the stimulation of feeding by reduced phenol glycosides, was sufficient to offset decreases in food quality.

We do not yet understand the mechanisms responsible for the beetle's decreased preference for oviposition on ozone-exposed plants, although it is possible that this is due to release of a repellent. The change in oviposition preference, however, is entirely responsible for the reduced fecundity because females first reared on control plants exhibit the same subsequent reduction in fecundity on ozone-treated plants (Jones and Coleman, 1988; Coleman and Jones, 1988a).

The overall effect on beetle population growth indicates the critical importance of life-history parameters. Overall, beetle populations on ozone-exposed plants are likely to decline because the effects on fecundity are equivalent to the loss of a whole generation of beetles (Coleman and Jones, 1988a). This beetle is multivoltine with flexible diapause requirements, having 3–4 generations per year in New York. Consequently, reductions in oviposition preference lead to decreases in fecundity that could be carried over into a decrease in population density in the next generation. Thus, despite increased preference and consumption and unchanged growth, development, and survivorship, reductions in fecundity should result in marked decreases in population growth.

The changes in both specific performance attributes and population growth have important implications for plant damage. Ozone-exposed plants will receive increased damage initially because beetles are stimulated to feed. Furthermore, ozone-exposed plants next to controls experience more damage, and the controls less damage, than separated plants because beetles will move onto ozone-exposed plants. Subsequently, however, ozone-exposed plants on their own would experience less damage because of a reduction in fecundity, while controls with ozone-exposed plants would receive more damage because beetles would move onto these plants to oviposit (Jones and Coleman, 1988).

In contrast to the beetle, exactly the same ozone treatment had no effect on the survivorship, reproduction, or population growth of a host-specific aphid, *Chaitophorus populicola* (Coleman and Jones, 1988b). This demonstrates that the same changes in the plant do not result in uniform responses of all insect herbivores. Although the reasons for this difference in sensitivity are as yet unknown (because it is hard to know why the aphid failed to respond), it may well be due to the differences in the mode of resource exploitation between the two insect species. The aphid prefers exactly the same leaf developmental age as the beetle (Jones and Coleman, 1988; Coleman and Jones, 1988b), but the aphid feeds on phloem, not on cellular fluids. It is possible at these low ozone doses that phloem loading and composition were not altered, most of the biochemical changes being localized in cells, so from the aphid's point of view, no real changes occurred.

We can further exemplify our approach by speculating on the likely effects of ozone exposure with another insect that feeds on cottonwood. The chrysomelid beetle, *Chrysomela scripta*, is also a salicaceous specialist whose larvae and adults feed on the same tissues of the same leaf developmental ages in a similar manner to *P. versicolora*. In contrast to *P. versicolora*, however, *C. scripta* is strongly stimulated to feed by increasing concentrations of phenol glycosides. This is partly because this insect benefits from the phenol glycosides for its own chemical defenses, using the sugar cleaved by hydrolysis of the glycosides as an energy source (Pasteels *et al.*, 1983). We would therefore predict that ozone-induced reductions in phenol glycosides would result in decreased feeding preference, consumption, and growth — the opposite of our findings for *P. versicolora*.

We can similarly speculate that shading cottonwoods would produce different effects on the plant and these herbivores than ozone air pollution. For example, short-term shading is likely to reduce carbon-acquisition but, unlike ozone, not to cause cellular damage. Consequently, phenol glycosides should decrease in concentration, but phenolic deposition should not occur, and release of an oviposition repellent against *P. versicolora* may not occur. Here we would expect stimulation of *P. versicolora* feeding and inhibition of *C. scripta* feeding, as with ozone. Increases in relative plant nitrogen content because of shading (see Section II,A,2), however, might be expected to increase food quality, leading to increases in *P. versicolora* growth and survivorship, no reduction in fecundity (perhaps even an increase), and therefore an increase in abundance. Increases in nitrogen might also offset reductions in growth of *C. scripta*. The aphid, feeding on phloem, may also experience increased nitrogen concentrations, which could result in increased reproduction and abundance.

B. Other Examples

Two other examples in the literature can further illustrate the validity of our approach. Hughes *et al.* (1981, 1982, 1983) showed that fumigation of soybeans with sulfur dioxide resulted in increased feeding and growth of the Mexican bean beetle *Elpilachna varivestis*. A specialist on the Fabaceae, this insect is stimulated to feed primarily by sugars. Apparently, sulfur dioxide fumigation inhibits phloem loading, causing an increase in leaf sugar content (Teh and Swanson, 1982). The increased feeding provoked by the increased sugars is apparently sufficient to account for increased beetle growth.

In another case, Kimmerer and Potter (1987) showed that survivorship of the holly leaf miner is greater on sun leaves than on shade leaves. This is due to an increase in the number of cell layers (thickness) of sun leaves caused by increased leaf carbon, which results in more food for larvae and less chance that the mine will collapse.

IV. Toward Further Integration

A number of additional factors must be taken into account if we are to move toward integration of stress and herbivory.

1. *Insect species versus consumer communities*   All plants have communities of insects, and thus the exploiter model must be applied to each member of the community. Focusing on a single, dominant herbivore is unsatisfactory because stress has the potential to make previously rare insects abundant. Insect herbivores can also interact with each other directly, via inter- or intraspecific interference or facilitation, and indirectly, via herbivore-induced modifications in plant suitability (i.e., damage-induced changes) (Coleman and Jones, 1991; Tallamy and Raupp, 1991). Most important, insect folivores are not the only members of the consumer community. Above- and belowground plant pathogens and mutualists and belowground herbivores can cause or ameliorate stress (Barbosa *et al.*, 1991), and their interactions with the plant can also be modified by abiotic stress (Jones, 1991; Jones and Last, 1991). Stress to plants can even modify the characteristics of senescent leaves, thereby altering leaf decomposition (Findlay and Jones, 1990) and the return of nutrients to the plant. Although all of these factors complicate consideration of stress effects on insect herbivores, in principle, the phytocentric and exploiter perspectives, with suitable modifications to the models, could also be applied to these interactions.

2. *Extrinsic factors affecting insect herbivores*   Clearly, plant characteristics, and insect sensitivity to changes in these characteristics, are by no means the sole determinants of insect herbivore performance and population growth. Predators, parasites, and diseases of insects, together with direct environmental effects on insects, can all directly affect insect growth, development, survivorship, or fecundity. Furthermore, complex, indirect interactions occur between host plants, insects, their

biotic mortality agents, and the abiotic environment. For example, many predators and parasites use host-plant cues to find prey (Price *et al.*, 1980), and these cues may be altered by plant stress. The nutritional or allelochemical status of the host plant can play a major role in determining insect mortality from parasites and diseases via modifications in the suitability of the insect as a host (Barbosa *et al.*, 1991), and again, such interactions may well be modified by plant stress. Environmental conditions, such as leaf-surface microclimate, are also important to some herbivores (Willmer, 1986), and stress-induced changes in these conditions can also play a critical role. Although these factors add a further degree of complexity to stress–plant–insect interactions, future modifications to the existing conceptual framework should take these factors into account.

3. *Feedbacks between stress and herbivory*   Abiotic stress results in plant adjustments that modify leaf characteristics and the degree and distribution of herbivory. This in turn causes further plant adjustments which may exacerbate (or even ameliorate) the degree of stress to the plant, and so on, in a continuous, dynamic series of interactions. It is unlikely that we can gain an integrated understanding of stress–plant–herbivore interactions without explicitly recognizing these feedbacks, and this could be explored by iterating the phytocentric and exploiter models through a number of cycles.

4. *Multiple stresses and stress dynamics*   Plants are rarely, if ever, exposed to a single stress or source of damage at a time. Stress and damage can also vary markedly in their duration and intensity. Hence, relative effects on carbon versus nutrient and water acquisition, as well as the demands for resource allocation to defense and repair, will also vary. A potential solution to this problem would be to develop the phytocentric model to deal with such conditions. This will require a substantial increase in the empirical data base, particularly addressing the additive, synergistic, or antagonistic nature of mixtures of stresses and damage, as well as a better understanding of the temporal nature of plant responses.

Given the complexity of interactions between plant stress and insect herbivory, plus the factors outlined above, how can progress toward integration best be achieved? We believe that attention must be focused on two types of net effects of these interactions (Jones, 1991): first, net effects on herbivore performance and population growth and second, net effects of stress and changes in herbivory on plant dry matter accumulation. It is likely that four simultaneous approaches will be necessary to achieve such integration.

First, direct experimental tests of model components and predictions (see Section III,A) will permit us to evaluate whether the various components of the phytocentric and exploiter models adequately predict changes in both leaf characteristics and insect sensitivity to these changes. The use of model systems, such as cottonwood and its insect herbivores, will permit studies under conditions in which confounding effects can be minimized. Key plant characteristics, such as inherent growth rate, growth habit, and genotypically defined resistance traits, can be carefully selected and controlled, and insect herbivores with particular modes of resource exploitation and life-history parameters can be chosen. Second, we need to do a more thorough, quantitative analysis of the extant data to ascertain whether variables in the phytocentric and exploiter models produce a clearer understanding of pattern than the existing paradigm and hypotheses. Analyses that clearly separate sources of variation in the responses of plants from those caused by variation in the sensitivity of different insect herbivores will be critical here. Third, we need more empirical data, particularly on the responses of plants to multiple stresses and the temporal patterns of plant response. Last, we need to develop more sophisticated models that refine predictions; permit iteration; and incorporate factors such as other consumers and the predators, parasites, and diseases of insects.

V. Summary and Conclusion

Plants are exposed to a diversity of abiotic environmental factors that can cause stress. Stress can also alter insect herbivore performance and change the degree and distribution of damage. Considerable evidence indicates that abiotic stress can change insect behavior, growth, development, survival, and fecundity, but consistent patterns are difficult to discern in the literature. The paradigm that plant stress increases insect herbivore performance and abundance and the mechanistic explanations that underly this paradigm are not supported by the evidence. Our conceptual framework recognizes that plant stress–induced variation in insect herbivory is due to both variation in plant responses to stress and variation in insect sensitivity to changes in stressed plants. Two models address these two sources of variation.

The phytocentric model predicts stress-induced changes in leaf anatomical, physiological, and biochemical characteristics from a whole-plant physiological understanding of stress-induced changes in the carbon–nutrient balance of plants and plant adjustments to these changes. Alterations in carbon, nutrient, and water acquisition; the presence of cellular and tissue damage; and the genotypic and ontogenetic traits of plants, such as inherent growth rate and growth habit, appear to be the most critical components for predicting many leaf characteristics.

Variation in the sensitivity of insects is predicted by the exploiter model, which takes into account differences in insect modes of resource exploitation and life-history parameters. Information on the mode of resource exploitation, particularly leaf ontogenetic and tissue specialization and feeding guild, can be used to identify which leaf characteristics are the most critical. Knowledge of the degree of host-plant specialization can be used to predict the behavioral and physiological sensitivity of the insect and, hence, specific performance attributes. Integrated insect life-history parameters —particularly inherent growth and development rates, number of developmental stages, intrinsic reproductive output and mode of reproduction, voltinism and diapause, host alternation, and vagility—predict how changes in specific performance attributes will translate into changes in abundance.

Since the net effect on either insect abundance or plant damage will be the outcome of all of these interacting components, predicting insect abundance or damage on stressed plants will require the development of specific predictions for particular species of insects that can be tested experimentally with model systems. We have exemplified the utility of the models using data from a few studies, including ours. Validation of these specific models will permit generalizations that could then be evaluated from the literature. Finally, further integration requires that we recognize that plants have consumer communities with numerous insects, pathogens, and mutualists; insect herbivores are affected by factors other than the plant; feedbacks between stress and herbivory are likely to occur; and plants are rarely exposed to a single stress of uniform duration and intensity.

We believe that the development of a truly integrated approach is more likely to lead to reliable predictions than are the current, oversimplistic generalizations. While we clearly have a long way to go toward achieving a truly integrated understanding of the relationships between plant stress and insect herbivory, we hope that the conceptual framework presented here will facilitate such efforts.

Acknowledgments

We thank the editors, Steward Pickett, and Vera Krischik for critical review. Concepts presented here were developed during research funded by NSF (BSR-8516679 to C. G. J. and W. H. Smith; BSR-8817519 to C. G. J.), the Mary Flagler Cary Charitable Trust (C. G. J. and a Cary Fellowship to J. S. C.), the Electrical Power Research Institute (via H. Mooney to J. S. C. and for travel to the symposium), the Andrew W. Mellon Foundation (via F. Herbert Bormann to J. S. C.), and Fakri Bazzaz (support to J. S. C.). This is a contribution to the program, Institute of Ecosystem Studies, the New York Botanical Garden.

References

Alberte, R. S., and Thornber, J. P. (1977). Water stress effects on the content and organization of chlorophyll in mesophyll and bundle-sheath chloroplasts of maize. *Plant Physiol.* 59: 351–353.

Alstad, D. N., Edmunds, G. F., Jr., and Weinstein, L. H. (1982). The effect of air pollutants on insect populations. *Annu. Rev. Entomol.* 27: 369–384.

Ayres, P. G. (1984). The interaction between environmental stress injury and biotic disease physiology. *Annu. Rev. Phytopathol.* 22: 53–75.

Baker, E. A. (1974). The influence of environment on leaf wax deposition in *Brassica oleracea* var. *gemmifera. New Phytol.* 73: 955–966.

Barbosa, P., Krischik, V. A., and Jones, C. G., eds. (1991). "Microbial Mediation of Plant–Herbivore Interactions." Wiley, New York.

Beck, S. D. (1965). Resistance of plants to insects. *Annu. Rev. Entomol.* 10: 207–232.

Bell, A. A. (1984). Morphology, chemistry, and genetics of *Gossypium* adaptation to pests. *Recent Adv. Phytochem.* 18: 197–230.

Belsky, A. J. (1986). Does herbivory benefit plants? A review of the evidence. *Am. Nat.* 127: 870–892.

Bernays, E. A., and Woodhead, S. (1982). Plant phenols utilized as nutrients by a phytophagous insect. *Science* 216: 201–203.

Bloom, A. J., Chapin, F. S., III, and Mooney, H. A. (1985). Resource limitation in plants—an economic analogy. *Annu. Rev. Ecol. Syst.* 16: 363–392.

Bradford, K. J., and Hsiao, T. C. (1982). Physiological responses to moderate water stress. *In* "Encyclopedia of Plant Physiology, New Series" (O. L. Lange, P. S. Nobel, C. B. Osmond, and H. Ziegler, eds.), Vol. 12B, pp. 263–324. Springer-Verlag, Berlin.

Brattsten, L. B. (1979). Biochemical defense mechanisms in herbivores against plant allelochemicals. *In* "Herbivores: Their Interaction with Secondary Plant Metabolites" (D. A. Rosenthal and D. H. Janzen, eds.), pp. 199–270. Academic Press, New York.

Bryant, J. P. (1987). Feltleaf willow–snowshoe hare interactions: Plant carbon–nutrient balance and floodplain succession. *Ecology* 68: 1319–1327.

Bryant, J. P., Chapin, F. S., III, and Klein, D. R. (1983). Carbon–nutrient balance of boreal plants in relation to vertebrate herbivory. *Oikos* 40: 357–368.

Bryant, J. P., Chapin, F. S., III, Reichardt, P. B., and Clausen, T. (1985). Adaptation to resource availability as a determinant of chemical defense strategies in woody plants. *Recent Adv. Phytochem.* 19: 219–237.

Bryant, J. P., Chapin, F. S., III, Reichardt, P. B., and Clausen, T. P. (1987). Response of winter chemical defense in Alaskan paper birch and green alder to manipulation of plant carbon–nitrogen balance. *Oecologia* 72: 510–514.

Campbell, B. M. (1986). Plant spinescence and herbivory in a nutrient-poor ecosytem. *Oikos* 47: 168–172.

Chapin, F. S., III, Bloom, A. J., Field, C. B., and Waring, R. H. (1987). Plant responses to multiple environmental factors. *Bioscience* 37: 49–57.

Clough, J. M., Alberte, F. S., and Teeri, J. A. (1979). Photosynthetic adaptation of *Solanum culcamara* L. to sun and shade environments. *Plant Physiol.* 64: 25–30.

Coleman, J. S. (1986). Leaf development and leaf stress: Increased susceptibility associated with sink–source transition. *Tree Physiol.* 2: 289–299.

Coleman, J. S., and Jones, C. G. (1988a). Plant stress and insect performance: Cottonwood, ozone and a leaf beetle. *Oecologia* 76: 57–61.

Coleman, J. S., and Jones, C. G. (1988b). Acute ozone stress on eastern cottonwood (*Populus deltoides*, Bartr.) and the pest potential of the aphid, *Chaitophorus populicola* Thomas (Homoptera: Aphidae). *Environ. Entomol.* 17: 207–212.

Coleman, J. S., and Jones, C. G. (1991). A phytocentric perspective of phytochemical induction by herbivores. *In* "Phytochemical Induction by Herbivores" (D. Tallamy and M. J. Raupp, eds.). Wiley, New York.

Coley, P. D., Bryant, J. P., and Chapin, F. S., III. (1985). Resource availability and plant antiherbivore defense. *Science* 230: 895–899.

Craker, L. E., and Starbuck, J. S. (1972). Metabolic changes associated with ozone injury of leaves. *Can. J. Plant Sci.* 52: 589–597.

Crawley, M. J. (1983). "Herbivory." Blackwell, Oxford.

Crossley, A., and Fowler, D. (1986). The weathering of Scots pine epicuticular wax in polluted and unpolluted air. *New Phytol.* 103: 207–218.

Cruickshank, I. A. M. (1980). Defenses triggered by the invader. *In* "Plant Disease" (J. Horsfall and E. B. Cowling, eds.), Vol. V, pp. 247–267. Academic Press, New York.

Dadd, R. H. (1973). Insect nutrition: Current developments and metabolic implications. *Annu. Rev. Entomol.* 18: 382–419.

Darrall, N. M. (1984). Biochemical diagnostic tests for the effect of air pollution on plants. *In* "Gaseous Air Pollutants and Plant Metabolism" (M. J. Koziol and F. R. Whatley, eds.), pp. 333–349. Butterworth, London.

Denno, R. F., and Dingle, H., eds. (1981). "Insect Life-History Patterns: Habitat and Geographic Variation." Springer-Verlag, New York.

Dethier, V. G. (1954). Evolution of feeding preferences in phytophagous insects. *Evolution* 8: 33–54.

Dethier, V. G. (1980). Evolution of receptor sensitivity to secondary plant substances with special reference to deterrents. *Am. Nat.* 115: 45–66.

Dickson, R. E. (1987). Diurnal changes in leaf chemical constituents and ^{14}C partitioning in cottonwood. *Tree Physiol.* 3: 157–172.

Dohman, G. P. (1985). Secondary effects of air pollution: Aphid growth. *Environ. Pollut. Ser. A Ecol. Biol.* 39: 227–234.

Duffey, S. S. (1980). Sequestration of plant natural products by insects. *Annu. Rev. Entomol.* 18: 447–477.

Dunn, J. P., Kimmerer, T. W., and Nordin, G. L. (1986). Attraction of the two-lined chestnut borer, *Agrilus bilineatus* (Weber) (Coleoptera: Buprestidae), and associated borers to volatiles of stressed white oak. *Can. Entomol.* 118: 503–509.

Erickson, S. S., and Dashek, W. V. (1982). Accumulation of foliar soluble protein in sulphur-dioxide-stressed *Glycine max* c.v. "Essex" and *Hordium vulgare* cvs. "Proctor" and "Excelsior" seedlings. *Environ. Pollut. Ser. A Ecol. Biol.* 28: 89–108.

Faeth, S. H., Mopper, S., and Simberloff, D. (1981). Abundances and diversity of leaf-mining insects on three oak host species: Effects of host-plant phenology and nitrogen content of leaves. *Oikos* 37: 238–251.

Feeny, P. P. (1976). Plant apparency and chemical defense. *Recent Adv. Phytochem.* 10: 1–40.

Findlay, S., and Jones, C. G. (1990). Exposure of cottonwood plants to ozone alters subsequent leaf decomposition. *Oecologia* 82: 248–250.

Fox, L. R., and Macauley, B. J. (1977). Insect grazing on *Eucalyptus* in response to variation in leaf tannins and nitrogen. *Oecologia* 29: 145–162.

Fox, T. C., and Geiger, D. R. (1984). Effects of decreased net carbon exchange on carbohydrate metabolism in sugar beet source leaves. *Plant Physiol.* 76: 763–768.

Fuhrer, E. (1985). Air pollution and the incidence of forest insect problems. *Z. Angew. Entomol.* 99: 371–377.

Geiger, D. R. (1986). Processes affecting carbon allocation and partitioning among sinks. *In* "Phloem Transport" (J. Cronshaw, W. J. Lucas, and R. T. Giaquinta, eds.), pp. 375–388. Alan R. Liss, New York.

Geiger, D. R., and Fondy, B. R. (1985). Responses of export and partitioning to internal and environmental factors. *In* "Regulation of Sources and Sinks in Crop Plants" (B. Jeffcoat, A. F. Hawkins, and A. D. Stead, eds.), pp. 177–194. Monogr. 12, British Plant Growth Regulator Group, Bristol.

Gershenzon, J. (1984). Changes in the levels of plant secondary metabolites under water and nutrient stress. *Recent Adv. Phytochem.* 18: 273–320.

Godzik, S., and Linskens, H. F. (1974). Concentration changes in free amino acids in primary bean leaves after continuous and interrupted sulfur dioxide fumigation and recovery. *Environ. Pollut. Ser. A Ecol. Biol.* 7: 25–38.

Gordon, A. J. (1986). Diurnal patterns of photosynthate allocation and partitioning among sinks. *In* "Phloem Transport" (J. Cronshaw, W. J. Lucas, and R. T. Giaguinta, eds.), pp. 499–517. Alan R. Liss, New York.

Graham, R. D. (1983). Effects of nutrient stress on susceptibility of plants to disease, with particular reference to the trace elements. *Adv. Bot. Res.* 10: 222–297.

Grill, D., Esterbauer, H., Schaner, M., and Felgitsch, C. (1980). Effect of sulfur dioxide on protein-SH in needles of *Picea abies. Eur. J. For. Pathol.* 10: 263–267.

Guderian, R., Tingey, D. T., and Rabe, R. (1985). Effects of photochemical oxidants on plants. *In* "Air Pollution by Photochemical Oxidants" (R. Guderian, ed.), pp. 129–333. Springer-Verlag, Berlin.

Gulmon, S. L., and Chu, C. C. (1981). The effects of light and nitrogen on photosynthesis, leaf characteristics, and dry matter allocation in the chaparral shrub, *Diplacus aurantiacus. Oecologia* 49: 207–212.

Gupta, P. D., and Thorsteinson, A. J. (1960). Food plant relationships of the diamondback moth (*Plutella maculipennis* Curt.). *Entomol. Exp. Appl.* 3: 241–250.

Hain, F. P., and Arthur, F. H. (1985). The role of atmospheric deposition in the latitudinal variation of fraser fir mortality caused by the balsam woolly adelgid, *Adelges piceae* Ratz. (Hemiptera: Adelgidae): A hypothesis. *Z. Angew. Entomol.* 99: 145–152.

Hayashi, H., and Chin, M. (1986). Effect of nitrogen nutrition and light on the chemical composition of rice phloem sap. *In* "Phloem Transport" (J. Cronshaw, W. J. Lucas, and R. T. Giaquinta, eds.), pp. 465–468. Alan R. Liss, New York.

Heath, R. L. (1984). Air pollutant effects on biochemicals derived from metabolism: Organic, fatty, and amino acids. *In* "Gaseous Air Pollutants and Plant Metabolism" (M. J. Koziol and F. R. Whatley, eds.), pp. 275–290. Butterworth, London.

Heinrichs, E. A., ed. (1988). "Plant Stress–Insect Interactions." Wiley, New York.

Ho, L. C. (1976). The relationship between the rates of carbon transport and of photosynthesis in tomato leaves. *J. Exp. Bot.* 27: 87–97.

Huber, S. C. (1983). Relation between photosynthetic starch formation and dry weight partitioning between the shoot and root. *Can. J. Bot.* 61: 2709–2716.

Huber, S. C., Kerr, P. S., and Kalt-Torres, W. (1986). Biochemical control of allocation of carbon for export and storage in source leaves. *In* "Phloem Transport" (J. Cronshaw, W. J. Lucas, and R. T. Giaquinta, eds.), pp. 355–367. Alan R. Liss, New York.

Hughes, P. R., and Laurence, J. A. (1984). Relationship of biochemical effects of air pollutants on plants to environmental problems: Insect and microbial interactions. *In* "Gaseous Air Pollutants and Plant Metabolism" (M. J. Koziol and F. R. Whatley, eds.), pp. 361–377. Butterworth, London.

Hughes, P. R., Potter, J. E., and Weinstein, L. H. (1981). Effects of pollutants on plant–insect interactions: Reactions of the Mexican bean beetle to sulfur dioxide–fumigated pinto beans. *Environ. Entomol.* 10: 741–744.

Hughes, P. R., Potter, J. E., and Weinstein, L. H. (1982). Effects of air pollutants on plant–insect interactions: Increased susceptibility of greenhouse-grown soybeans to the Mexican bean beetle after plant exposure to sulfur dioxide. *Environ. Entomol.* 11: 173–176.

Hughes, P. R., Dickie, A. I., and Penton, M. A. (1983). Increased success of the Mexican bean beetle on field-grown soybeans exposed to sulfur dioxide. *J. Environ. Qual.* 12: 565–568.

Hunt, R., and Nicholls, A. D. (1986). Stress and coarse control of root–shoot partitioning in herbaceous plants. *Oikos* 47: 149–158.

Hurwitz, B., Pell, E. J., and Sherwood, R. T. (1979). Status of coumesterol and 4',7-dihydroxyflavone in alfalfa foliage exposed to ozone. *Phytopathol.* 69: 810–813.

Jermy, T. (1966). Feeding inhibitors and food preference in chewing phytophagous insects. *Entomol. Exp. Appl.* 9: 1–12.

Jermy, T. (1984). Evolution of insect–host plant relationships. *Am. Nat.* 124: 609–630.

Jermy, T., and Szentesi, A. (1978). The role of inhibitory stimuli in the choice of oviposition site by phytophagous insects. *Entomol. Exp. Appl.* 24: 458–471.

Jones, C. G. (1983). Phytochemical variation, colonization and insect communities: The case of bracken fern, *Pteridium aquilinum* L. (Kuhn). *In* "Variable Plants and Herbivores in Natural and Managed Systems" (R. F. Denno and M. S. McClure, eds.), pp. 513–537. Academic Press, New York.

Jones, C. G. (1991). Interactions among insects, plants and microorganisms: A net-effects perspective on insect performance. *In* "Microbial Mediation of Plant–Herbivore Interactions" (P. Barbosa, V. A. Krischik, and C. G. Jones, eds.), pp. 7–35. Wiley, New York.

Jones, C. G., and Coleman, J. S. (1988). Plant stress and insect behavior: Cottonwood, ozone, and the feeding and oviposition preference of a beetle. *Oecologia* 76: 51–56.

Jones, C. G., and Coleman, J. S. (1989). Biochemical indicators of air pollution effects in trees: Unambiguous signals based on secondary metabolites and nitrogen in fast-growing species? *In* "Biologic Markers of Air Pollution Stress and Damage in Forests" (National Research Council), pp. 261–273. National Academy Press, Washington, D. C.

Jones, C. G., and Last, F. T. (1991). Ectomycorrhizae and trees: Implications for aboveground herbivory. *In* "Microbial Mediation of Plant–Herbivore Interactions" (P. Barbosa, V. A. Krischik, and C. G. Jones, eds.), pp. 65–103. Wiley, New York.

Jones, C. G., Whitman, D. W., Silk, P. J., and Blum, M. S. (1988). Diet breadth and insect chemical defenses: A generalist grasshopper and a general hypothesis. *In* "Chemical Mediation of Coevolution" (K. C. Spencer, ed.), pp. 417–512. Academic Press, San Diego.

Juniper, B., and Southwood, T. R. E., eds. (1986). "Insects and the Plant Surface." Edward Arnold, London.

Keen, N. T., and Taylor, O. C. (1975). Ozone injury in soybeans: Isoflavonoid accumulation is related to necrosis. *Plant Physiol.* 55: 731–733.

Kimmerer, T. W., and Kozlowski, T. T. (1982). Ethylene, ethane, acetaldehyde, and ethanol production by plants under stress. *Plant Physiol.* 69: 840–847.

Kimmerer, T. W., and MacDonald, R. C. (1987). Acetaldehyde and ethanol biosynthesis in leaves of plants. *Plant Physiol.* 84: 1204–1209.

Kimmerer, T. W., and Potter, D. A. (1987). Nutritional quality of specific leaf tissues and selective feeding by a specialist leaf miner. *Oecologia* 71: 548–551.

Koziol, M. J. (1984). Interactions of gaseous pollutants with carbohydrate metabolism. *In* "Gaseous Air Pollutants and Plant Metabolism" (M. J. Koziol and F. R. Whatley, eds.), pp. 252–273. Butterworth, London.

Kozlowski, T. T. (1971a). "Growth and Development of Trees," Vol. 1: Seed Germination, Ontogeny, and Shoot Growth. Academic Press, New York.

Kozlowski, T. T. (1971b). "Growth and Development of Trees," Vol. 2: Cambial Growth, Root Growth, Reproductive Growth. Academic Press, New York.

Kramer, P. (1983). "Water Relations of Plants." Academic Press, New York.

Larcher, W. (1980). "Plant Physiological Ecology." Springer-Verlag, Berlin.

Larsson, S. (1989). Stressful times for the plant stress–insect performance hypothesis. *Oikos* 56: 277–283.

Larsson, S., Wiren, A., Lundgren, L., and Erisson, T. (1986). Effects of light and nutrient stress on leaf phenolic chemistry in *Salix dasyclados* and susceptibility to *Galerucella lineola*. *Oikos* 47: 205–210.

Lechowicz, M. J. (1987). Resource allocation by plants under air pollution stress: Implications for plant–pest–pathogen interactions. *Bot. Rev.* 53: 281–300.

Lincoln, D. E., and Mooney, H. A. (1984). Herbivory on *Diplacus aurantiacus* shrubs in sun and shade. *Oecologia* 64: 173–176.

Louda, S. M. (1986). Insect herbivory in response to root cutting and flooding stress on a native crucifer under field conditions. *Acta Oecologia* 7: 37–53.

MacGarvin, M., Lawton, J. H., and Heads, P. A. (1986). The herbivorous insect communities of open and woodland bracken: Observations, experiments and habitat manipulations. *Oikos* 47: 135–148.

McKey, D. (1979). The distribution of secondary compounds within plants. *In* "Herbivores: Their Interaction with Secondary Plant Metabolites" (G. A. Rosenthal and D. H. Janzen, eds.), pp. 55–133. Academic Press, New York.

McNeill, S., and Southwood, T. R. E. (1978). The role of nitrogen in the development of insect–plant relationships. *In* "Biochemical Aspects of Plant and Animal Coevolution" (J. B. Harborne, ed.), pp. 77–98. Academic Press, London.

Mattson, W. J. (1980). Herbivory in relation to plant nitrogen content. *Annu. Rev. Ecol. Syst.* 11: 119–162.

Mattson, W. J., and Addy, N. D. (1975). Phytophagous insects as regulators of forest primary production. *Science* 190: 515–522.

Mattson, W. J., and Haack, R. A. (1987a). The role of drought stress in provoking outbreaks of phytophagous insects. *In* "Insect Outbreaks: Ecological and Evolutionary Perspectives" (P. Barbosa and J. Schultz, eds.), pp. 365–407. Academic Press, Orlando.

Mattson, W. J., and Haack, R. A. (1987b). The role of drought in the outbreaks of plant-eating insects. *Bioscience* 37: 110–118.

Mooney, H. A. (1972). The carbon balance of plants. *Annu. Rev. Ecol. Syst.* 3: 315–346.

Morgan, J. M. (1984). Water stress in higher plants. *Annu. Rev. Plant Physiol.* 35: 299–348.

Noyes, R. D. (1980). The comparative effects of sulfur dioxide on photosynthesis and translocation in bean. *Physiol. Plant Pathol.* 16: 73–79.

Papaj, D. R., and Prokopy, R. J. (1986). Phytochemical basis of learning in *Rhagoletis pomonella* and other herbivorous insects. *J. Chem. Ecol.* 12: 1125–1143.

Pasteels, J. M., Rowell-Rahier, M., Braekman, J. C., and Dupont, A. (1983). Salicin from host plant as a precursor of salicylaldehyde in defensive secretion of chrysomeline larvae. *Physiol. Entomol.* 8: 307–314.

Pell, E. J., and Puente, M. (1986). Emission of ethylene by oat plants treated with ozone and simulated acid rain. *New Phytol.* 103: 709–715.

Price, P. W., Bouton, C. E., Gross, P., McPheron, B. A., Thompson, J. N., and Weis, A. E. (1980). Interactions among three trophic levels: Influence of plants on interactions between insect herbivores and natural enemies. *Annu. Rev. Ecol. Syst.* 11: 41–65.

Radin, J. W., and Parker, L. L. (1979). Water relations of cotton plants under nitrogen deficiency. I. Dependence on leaf structure. *Plant Physiol.* 64: 495–498.

Raupp, M. J., and Denno, R. F. (1983). Leaf age as a predictor of herbivore distribution and abundance. *In* "Variable Plants and Herbivores in Natural and Managed Systems" (R. F. Denno and M. S. McClure, eds.), pp. 91–125. Academic Press, New York.

Reich, P. B. (1987). Quantifying plant responses to ozone: A unifying theory. *Tree Physiol.* 3: 63–91.

Reicosky, D. C., Smith, R. C. G., and Meyer, W. S. (1985). Foliage temperature as a means of detecting stress of cotton subjected to a short-term water-table gradient. *Agric. For. Meterol.* 35: 192–203.

Rhoades, D. F. (1979). Evolution of plant chemical defenses against herbivores. *In* "Herbivores: Their Interaction with Secondary Plant Metabolites" (G. A. Rosenthal and D. H. Janzen, eds.), pp. 3–54. Academic Press, New York.

Rhoades, D. F. (1983). Herbivore population dynamics and plant chemistry. *In* "Variable Plants and Herbivores in Natural and Managed Systems" (R. F. Denno and M. S. McClure, eds.), pp. 155–220. Academic Press, New York.

Rhoades, D. F., and Cates, R. G. (1976). Toward a general theory of plant antiherbivore chemistry. *Recent Adv. Phytochem.* 10: 168–213.

Rhodes, D. (1987). Metabolic responses to stress. *In* "The Biochemistry of Plants," vol. 12: Physiology of Metabolism (D. D. Davies, ed.), pp. 201–241. Academic Press, London.

Rist, D. L., and Lorbeer, J. W. (1984). Ozone-enhanced leaching of onion leaves in relation to lesion production by *Botrytis cinerea*. *Phytopathology* 74: 1217–1220.

Robinson, D. (1986). Compensatory changes in the partitioning of dry matter in relation to nitrogen uptake and optimal variations in growth. *Ann. Bot.* 58: 841–848.

Rosenthal, G. A., and Janzen, D. H., eds. (1979). "Herbivores: Their Interaction with Secondary Plant Metabolites." Academic Press, New York.

Scriber, J. M. (1978). The effects of larval feeding specialization and plant growth form on the consumption and utilization of plant biomass and nitrogen: An ecological consideration. *Entomol. Exp. Appl.* 24: 494–510.

Scriber, J. M. (1984). Food plant suitability. *In* "Chemical Ecology of Insects" (W. Bell and R. Cardé, eds.), pp. 159–202. Chapman and Hall, London.

Scriber, J. M., and Feeny, P. P. (1979). Growth of herbivorous caterpillars in relation to feeding specialization and to growth form of their food plants. *Ecology* 60: 829–850.

Scriber, J. M., and Slansky, F., Jr. (1981). The nutritional ecology of immature insects. *Annu. Rev. Entomol.* 26: 183–211.

Shaw, A. F., Grange, R. I., and Ho, L. C. (1986). The regulation of source-leaf assimilate compartmentation. *In* "Phloem Transport" (J. Cronshaw, W. J. Lucas, and R. T. Giaquinta, eds.), pp. 391–398. Alan R. Liss, New York.

Shelvey, J. D., and Koziol, M. J. (1986). Seasonal and SO_2-induced changes in epicuticular wax of ryegrass. *Phytochemistry* 25: 415–420.

Smith, T. A. (1984). Putrescine and inorganic ions. *Recent Adv. Phytochem.* 18: 7–54.

Stewart, G. R., and Larher, F. (1980). Accumulation of amino acids and related compounds in relation to environmental stress. *In* "The Biochemistry of Plants," vol. 5: Amino Acids and Derivatives (B. J. Miflin, ed.), pp. 609–635. Academic Press, New York.

Stitt, M., Herzog, B., and Helot, H. W. (1984). Control of photosynthetic sucrose synthesis by fructose 1,6-biphosphate. *Plant Physiol.* 75: 548–553.

Szianawski, R. K. (1987). Plant stress and homeostasis. *Plant Physiol. Biochem.* 25: 63–72.

Tahvanainen, J., Julkunen-Tiitto, R., and Kettunen, J. (1985). Phenolic glycosides govern the food selection pattern of willow-feeding leaf beetles. *Oecologia* 67: 52–56.

Tallamy, D., and Raupp, M. J., eds. (1991). "Phytochemical Induction by Herbivores." Wiley, New York.

Teh, K. H., and Swanson, C. A. (1982). Sulfur dioxide inhibition of translocation in bean plants. *Plant Physiol.* 69: 88–92.

Tingey, D. T., and Taylor, G. E. Jr. (1982). Variation in plant response to ozone: A conceptual model of physiological events. *In* "Effects of Gaseous Air Pollutants in Agriculture and Horticulture" (M. H. Unsworth and D. P. Ormrod, eds.), pp. 113–138. Butterworth, London.

Tingey, D. T., Fites, R. C., and Wickliff, C. (1975). Activity changes in selected enzymes from soybean leaves following ozone exposure. *Physiol. Plant.* 33: 316–320.

Tingey, D. T., Fites, R. C., and Wickliff, C. (1976). Differential foliar sensitivity of soybean cultivars to ozone associated with differential enzyme activities. *Physiol. Plant.* 37: 69–72.

Tingey, W. M., and Singh, S. R. (1980). Environmental factors influencing the magnitude and expression of resistance. *In* "Breeding Plants Resistant to Insects and Diseases" (F. G. Maxwell and P. R. Jennings, eds.), pp. 87–114. Wiley, New York.

Waring, R. H., McDonald, A. J. S., Larsson, S., Ericcson, T., Wiren, A., Ericcson, A., and Lohammar, T. (1985). Differences in chemical compositions of plants grown at constant relative growth rates with stable mineral nutrition. *Oecologia* 66: 157–160.

White, T. C. R. (1969). An index to measure weather-induced stress of trees associated with outbreaks of psyllids in Australia. *Ecology* 50: 905–909.

White, T. C. R. (1984). The abundance of invertebrate herbivores in relation to the availability of nitrogen in stressed food plants. *Oecologia* 63: 90–105.

Wilkinson, R. E., and Kasperbauer, M. J. (1972). Epicuticular alkane content of tobacco as influenced by photoperiod, temperature, and leaf age. *Phytochemistry* 11: 1273–1280.

Willmer, P. (1986). Microclimatic effects on insects at the plant surface. *In* "Insects and the Plant Surface" (B. Juniper and T. R. E. Southwood, eds.), pp. 65–80. Edward Arnold, London.

III

Plant Growth Forms

13

Annual Plants: Potential Responses to Multiple Stresses

F. A. Bazzaz   S. R. Morse

I. Introduction

Plants with an annual life history constitute a heterogeneous group unified by a set of traits revolving around reproduction. By definition, annual-plant reproduction is sexual, occurs once, and must take place within the same year that vegetative growth began. These life-history traits, however, fail to completely circumscribe groups recognized as annual taxa. The putative annual species *Poa annua,* for example, comprises extreme variations in life history, including perennial ecotypes as well as individuals of annual ecotypes that can live up to ten years (Law *et al.,* 1977). All other characteristics "common" to annual plants also span a spectrum, and hence every potential generality about annuals has exceptions. Nevertheless, the recurrence of the annual life history within certain ecological settings suggests that there are suites of environmental factors that favor the maintenance of an annual habit. Furthermore, similarities among phylogenetically distinct groups within a given community indicate that certain traits enable annuals to persist in habitats where perennials fail to grow and reproduce. The intent of this chapter is to explore how the restriction of the vegetative generation to one year, in addition to other features common to annual plants, may facilitate or limit the ability of annuals to survive in the face of multiple abiotic and biotic stresses.

II. Annuals: Definition, Distribution, and Evolutionary History

Annuals as a group vary widely in growth form (rosette, erect, twining), nutrition (autotrophic or parasitic), population regulation (see Symonides, 1988), and the habitats they exploit (agricultural fields to extreme deserts). In general, they are herbaceous and flower only once (semelparous). Biennials and some perennials, however, also share these traits. The distinguishing feature of annual plants is that the active (postlatent) phase of their life is limited to a single year's growth and a single reproductive episode. Within annuals as a group, the time to reproductive maturity is extremely variable, ranging from ten days in some *Brassica* species to months for many crop species. Individual longevity, however, is not necessarily limited to one year. With seed dormancy, annuals can be very long-lived; only germination, growth, establishment, and reproduction are restricted to one year.

Annuals make up approximately 10% of world's flora (Raunkiaer, 1934), but they are not equally represented in all regions. They are notably lacking in alpine and arctic zones, rare in the aseasonal tropics, and disproportionately represented in the desert regions of the world (Raunkiaer, 1934; Danserau, 1957; Billings and Mooney, 1968). For example, 60% of the Saharo–Sindian vegetation of Palestine are annual taxa (Orshan, 1953). With very few exceptions, annuals occur in open, sunny habitats that either experience or are created by frequent disturbances; these include early successional habitats, desert systems dominated by perennials in which annuals sporadically flourish, and annual crop systems. The first two habitats are characterized by substantial temporal and spatial variation in soil moisture availability, temperature, relative humidity, and soil nutrient content (Rice, 1987). In crop systems, in contrast, human intervention attempts to reduce variation through the manipulation of soil water and nutrient content.

Life-history theory predicts that the accelerated reproductive cycle of annual plants relative to other vascular plants will have two consequences: as the age of reproduction shifts to earlier in the life cycle (1) birth rate increases, and thus the intrinsic rate of natural increase (r) of a population increases (Cole, 1954), and (2) a negative correlation exists between future survivorship and current fecundity. These two characteristics will favor the presence and persistence of annuals in unpredictable environments (Schaffer, 1974). Thus, characteristics such as seasonal drought, high temperatures, and frequent disturbances are invoked as strong selective factors favoring the annual habit. The predominance of annuals in deserts and regions with Mediterranean climates (30–60% of the flora), where the establishment and growth of plants is strongly affected by the great seasonal and yearly variation in rainfall and the extreme fluctuations in temperature, supports the notion that drought has been a primary selective force favoring the annual habit (Raven and Axelrod, 1978; Orshan, 1953; Mulroy and Rundel, 1977).

Although annuals are able to persist in habitats that restrict the establishment of perennials because of extreme and unpredictable environmental conditions, the coexistence of annuals and perennials in many regions requires further explanation. Unfavorable growing conditions for part of the year do not preclude perennials. Rather, these conditions favor selection of some dormant structure, such as protected buds, that enables perennials to survive conditions detrimental to herbaceous growth. Consequently, the limit of perennial distribution is probably set by demographic factors, such as juvenile mortality patterns. When growing conditions become so extreme that the probability of survival to the second year becomes very small, then selection for annual-dominated communities will occur. Where conditions are less extreme, perennials are more likely to prevail. For example, annuals dominate in the Sahara, whereas perennials dominate in the environmentally less extreme Sonoran–Chihauhan deserts. The diversity of annuals in the latter desert is nevertheless still very high (Went, 1949; Koller, 1972; Whittaker, 1975).

The phylogenetic distribution of annuals is limited. Annuals are unknown in gymnosperms and virtually absent among pteridophytes (Klekowski, 1967; Cousens, 1988). Annual species arose relatively late in the evolution of terrestrial plants and have been found only within the angiosperms (Niklas *et al.*, 1985). Furthermore, annual life histories are not evenly spread throughout flowering plants, but rather have originated numerous times from perennial ancestors within evolutionarily advanced plant families (see Stebbins, 1965, 1974). Associated with the origination of an annual life history are reproductive traits not found among biennials, perennial herbs, shrubs, or trees. These traits include the lowest base chromosome numbers and the highest frequency of autogamy among flowering plants (Grant, 1975). Both extreme chromosome complements and breeding systems are presumed to have facilitated the proliferation of annual taxa within those families in which the life history initially arose. Short generation times, together with population fragmentation because of environmental stresses, may also have led to rapid speciation in certain annual clades (Lewis, 1972; Bartholomew *et al.*, 1973; cf. Gottlieb, 1974).

Annual plants are of special interest in the study of environmental stresses for several reasons, in particular because the majority of agricultural crops on which humans depend heavily for food are annual. These include maize, wheat, soybean, and sunflower. Annuals are also weeds and important colonists of naturally and human-disturbed habitats that can permanently modify plant communities, such as the present-day Californian grassland. Because of their short lifespan, both wild and cultivated annual plants also provide good experimental systems for assessing the impact of environmental stresses on individuals throughout their ontogeny from germination to reproduction and on the potential evolutionary effect of novel stresses, through changes in population size and structure over time.

III. Potential for Annuals to Respond to Stress

Annual plants must set seed in the year they germinate if they are to replace themselves in the next generation. Because seed set is the pivotal constraint of an annual life history, the following analysis emphasizes the effects of stresses on correlates of fitness such as fecundity and biomass. We address both individual and population responses because the former reveals the plasticity of a response through the ontogeny of an individual, while population-level responses reflect the ecological and evolutionary potential of a species to respond to stresses.

We will not discuss the physiological literature that uses annuals primarily as a model system for plant physiology. This body of literature has demonstrated that the potential physiological responses of a plant's vegetative and reproductive tissues appear to be shared among most life-forms because of the biochemical and physiological similarities among all vascular plants. Thus annuals, like perennials, can alter nutrient and water acquisition, carbon gain, growth, and reproduction in response to a myriad of environmental conditions (Fitter and Hay, 1988; Grime, 1979; Nobel, 1983; Radosevich and Holt, 1984).

We will instead focus on the importance and limitations of the suite of traits listed in Table I. These traits, common to many annuals, are important for maximizing the probability of survival and reproduction within a single growing season (traits 3–5) or are necessary physiological and environmental correlates for precocious reproduction (traits 6–9). Each

Table I   Features of Annual Plants

---

1. The vegetative and reproductive phases of a genet together persist for less than a year.
2. Replacement of individuals between generations is only by seed.
3. Breeding systems facilitating outcrossing are common. (Yet autogamy is largely restricted to annuals.)
4. A genet may last as a seed for one to many years.
5. Germination of seeds is typically linked to environmental cues indicating that a major resource is temporarily not limiting.
6. Plants tend to grow in open habitats.
7. They are usually entirely herbaceous.
8. They tend to have high photosynthetic capacity and high relative growth rates.
9. The proportion of net assimilation allocated to reproduction ranges from 15% to 40% (compared with 1% to 20% for perennials).

---

trait may also constrain the ability of annuals to persist in certain habitats. To examine the positive and negative implications of these traits for the persistence of annuals, the following discussion considers annuals at three different levels: the individual, the population, and the community.

A. The Annual as an Individual

Although environmental cues for germination (light, rainfall, photoperiod, fire, etc.) may enable annuals to avoid conditions that are detrimental to herbaceous growth, the assumption that these plants therefore do not experience stresses is invalid. First, as a seed, a genet may experience considerable variation in the environment. Second, the environment at the time of germination is not a fail-safe predictor of the environment later, so germination cues alone cannot guarantee later survival. Pre-reproductive mortality in annual plants may be attributed to many factors, including competition for light and nutrients, drought, disease, frost, herbivory, and toxins. Thus it is safe to conclude that annuals experience multiple stresses within their vegetative and latent phases.

Individuals may experience excesses and shortages of resources on very short time scales, ranging from several times in one day to many times in their life cycle. Resource imbalances may also limit plant growth and reproduction even if the levels of individual factors are not in themselves stressful (discussion in Bloom *et al.*, 1985; Chapin *et al.*, 1987). The sensitivity of plants to resource imbalances and the speed with which they adjust to them are important components of plant strategies. These strategies are manifest in the flexibility of allocation and partitioning patterns. Both may be mechanisms for balancing resources for maximum growth and precocious flowering.

Flexibility in partitioning patterns differs among annuals. Determinate species, for example, show an abrupt transition between vegetative and reproductive growth and are unable to revert to vegetative growth. Senescence of determinate annuals appears to be developmentally controlled; with the onset of flowering, root growth soon ceases because of the preferential allocation of carbohydrates to reproduction. Indeterminate annuals, in contrast, continue vegetative growth after flowering begins and continue to flower until a catastrophic event ends their life (see Harper, 1977). The sensitivity of a species to unpredictable acute stresses should vary as a function of these developmental differences. Because determinate annuals end vegetative growth when reproduction starts, they are more vulnerable than indeterminate annuals to acute stresses during floral initiation, anthesis, and seed filling (Schulze, 1982). Indeterminate annuals, in contrast, may abort flowers during acute stresses but then resume growth and flower initiation when favorable conditions return. We also predict that chronic stresses are more likely, on average, to severely limit determinate annuals if they fail to reach the critical size necessary for the onset of flowering. This hypothesis has not, however, been directly explored.

A study comparing the performance of an annual–perennial pair of *Polygonum* species across a wide range of a resource gradients suggests that the response of annuals versus perennials to multiple factors may only be distinctive when comparing a previously established perennial with an annual started from seed (Zangerl and Bazzaz, 1983; Lee *et al.*, 1986). When both the annual and perennial were started from seed, both exhibited similar patterns of responses to a range of resource gradients even though the annual always grew faster and partitioned less of its biomass to belowground parts. When the annual was compared with the perennial started from a rootstock, however, the similarity of response patterns between the two species broke down. Only the annuals exhibited a significant change in response to crossed gradients of light and nutrients, or when grown with neighbors. These results suggest that the response patterns of annuals and perennials to multiple factors may be similar during the first year of establishment. As the perennial grows, however, the presence of underground reserves decouples the effects of an imbalance between resources such as nutrients and light. These resource imbalances in the environment are probably compensated for by drawing on internal resource reserves and may give the perennial, over its lifetime, a higher competitive ability. Thus, as the perennial *Polygonum* matures, an ontogenetic shift appears to take place, from tracking resource heterogeneity through rapid partitioning shifts to buffering growth from resource heterogeneity through the use of stored reserves.

Some annual plants can store reserves. Without the opportunity to postpone reproduction to another growing season, however, the manifestation of buffering through storage in annuals is distinct from perennials: storage buffers reproductive potential rather than biomass accumulation. *Hemizonia luzulifolia*, for example, stores carbohydrate early when the growing season is favorable for carbon gain and then reallocates these carbohydrates to reproduction when flowering begins during the summer drought (see Chiariello and Roughgarden, 1984). In this species, storage decouples reproductive allocation from the unpredictable summer growing conditions, much as rates of biomass accumulation in perennials are buffered from resource heterogeneity.

The response of annuals to stress factors may be highly modified by the general vigor of the plant and its stage of development. Overall response of old-field annuals to fixed resource gradients varies as the plant changes in size, develops different types of tissues, and reallocates internal resources (Bazzaz, 1987). The probability of survivorship of plants confronted with similar acute stresses, such as temperature shock, also changes with ontogeny (Parrish and Bazzaz, 1985). Vegetative plants of *Abutilon*, for example, survive heat shock equally well when grown at ambient and at twice ambient concentrations of carbon dioxide. Reproductive individuals, in contrast, show differential survivorship as a function of carbon dioxide concentrations. Flowering plants grown under ambient carbon dioxide consistently survived, whereas plants grown under high carbon dioxide concentrations consistently died after the heat shock treatment (Bazzaz, unpublished data).

Flexibility of partitioning in annuals is linked to high relative growth rates (Bradshaw, 1965) because a small ratio of total biomass to net assimilation rate permits a faster developmental response to changes in the environment. Many crops and annual weeds are able to change the proportional allocation of biomass in response to environmental stresses (Baker, 1974). *Polygonum pensylvanicum*, for example, responds to rooting volume with a significant change in the allometric relationship between vegetative and reproductive biomass, even when the total amount of nutrients available remains constant (Fig. 1). At the largest soil volume, this species allocates more dry weight per gram to reproductive tissues (McConnaughay and Bazzaz, unpublished data). Proportional root partitioning may also change on very short time scales. The onset of tissue-water deficits in soybean, for example, induces changes in root partitioning within 12 hr (Meyer and Boyer, 1981).

Limiting resources, such as nonmobile cations, may alter the growth of plants with typically high leaf turnover because nutrients are not available at the apical meristem (Harper and Sellek, 1987). When the deficiency is not severe enough to cause meristem death, the recovery from such an acute stress can be rapid in plants with potentially high relative growth rates.

One potentially limiting consequence of the large allocation to leaf area required for high relative growth rates is that less energy can be allocated to defensive compounds (Coley *et al.*, 1985) and other maintenance activities (Chapin *et al.*, 1987). This trade-off may make annuals more vulnerable to biotic stresses such as pathogens and herbivores. A model evaluating the trade-offs between allocation to nitrogen-based defense and growth indi-

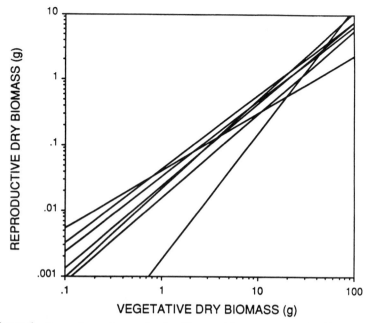

Figure 1 Regressions of reproductive biomass (g) versus vegetative biomass (g) for *Polygonum pensylvanicum* individuals grown in seven different soil volumes. Analyses show that the only significantly different slope is the one for the largest soil volume. [McConnaughay and Bazzaz, unpublished data).]

cates that the cost of defense, measured in terms of potential leaf production, increases as the maximum photosynthetic rate increases (Gulmon and Mooney, 1986). With the diversion of even a small percentage of nitrogen into defense chemicals, compound growth rate rapidly declines and ultimately will limit productivity and seed set. This decline in growth may affect fecundity in ephemerals if they fail to flower before the onset of deleterious growing conditions and in old-field annuals as a result of a decline in competitive ability.

The sensitivity of annuals to air pollutants such as sulfur dioxide (SO_2), nitrogen oxides (NO_x), and ozone (O_3) also results in part from their high relative growth rates, high conductances, and photosynthetic rates. Reductions in biomass accumulation and reproduction are caused mainly by direct interference with carbon fixation inside the leaves (see articles in Winner *et al.*, 1985 and this volume; Reinert, 1984; Reich, 1987). Thus the higher the conductance, the more sensitive a species. Further reductions in photosynthesis, growth, and seed production may then occur with a reduction in leaf stomatal conductance in response to the damage incurred from gaseous pollutants. This physiological feedback will reduce the entry of these gases into the mesophyll (Winner and Mooney, 1980) but may prevent an annual from reaching minimum critical size for flowering.

The photosynthetic and growth responses of annuals to various combinations of sulfur dioxide, nitrogen dioxide, and carbon dioxide also vary as a function of photosynthetic pathway. C_3 and C_4 annual species from an old-field community, for example, responded to SO_2 and CO_2 in complex ways (see Carlson and Bazzaz, 1982). In general, SO_2 caused a reduction in the growth of C_3 plants at 300 ppm CO_2 but not at higher concentrations. In contrast, SO_2 reduced the growth of C_4 species at high CO_2 levels but increased growth at 300 ppm CO_2. Thus, C_3 species were more sensitive to SO_2 than were the C_4 species at ambient CO_2 levels, but that sensitivity was reversed at high CO_2 levels. This differential response is attributed to differences in the responsiveness between C_3 and C_4 plants to ambient CO_2 concentrations. Because carbon fixation of C_3 plants tends to increase with increasing ambient CO_2 concentrations, the growth of the C_3 plants was stimulated with elevated CO_2 concentrations. The increase in carbon fixation also enabled the C_3 plants to compensate for reductions in photosynthesis because of damage by SO_2. The latter compensation probably did not occur in the C_4 species because a constant carboxylation rate is maintained irrespective of ambient CO_2 concentration (Carlson and Bazzaz, 1982).

In addition to eliciting differential responses with respect to carbon-fixation pathway, gaseous pollutants in various combinations together may have unpredictable effects (Carlson and Bazzaz, 1986; Carlson, 1983). For example, SO_2 (0.6 ppm) reduced photosynthesis more than NO_2 applied at the same concentration. When SO_2 was supplied together with NO_2, however, photosynthesis was reduced more than it was by SO_2 alone. Thus, there appears to be a negative synergistic interaction of the two pollutants. By raising the CO_2 concentration, however, the reduction in photosynthetic rates caused by SO_2 fumigation alone and together with NO_2 was alleviated.

B. Annuals in a Population

The response of populations to a wide range of multiple resources can be quantified by estimating the response (niche) breadth in the field or experimentally in the greenhouse. Measures of response breadth may include survivorship, biomass production and allocation, and seed production. In extensive studies with old-field annuals (see Bazzaz, 1987), a strong correlation was consistently found between the degree of fluctuation in a given environmental factor in the field and the breadth of response to that factor. Results from experimental studies designed to elucidate the response of annuals to excesses and shortages of resources and to neighbors (competitors?) also demonstrate a similar correlation.

Results from single-factor studies indicate that the population responses of annuals are (1) broad: each species tolerates a wide range of resource levels, and this breadth of response is seen in recruitment, survivorship, growth, and reproduction, although not to the same degree for all environmental factors or equally for all species; (2) highly overlapping: species therefore may compete strongly on large portions of these gradients; (3) different primarily in response breadth for specific gradients: having a broad response to one resource does not necessarily mean that a species has broad responses on all gradients; (4) sensitive to competition: under heterospecific competition some annuals show a dramatic shift in their response on the environmental gradients, whereas others do not.

Although some progress has been made in understanding how popula-

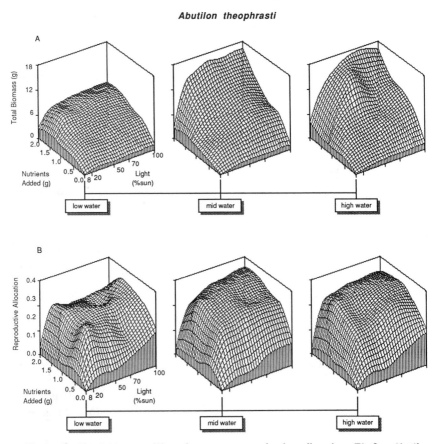

Figure 2 Total biomass (A) and percent reproductive allocation (B) for *Abutilon theophrasti* on a three-dimensional gradient of light, nutrients, and water. [Bazzaz, Crabtree, and Buckley (unpublished data).]

tions respond to single stresses, remarkably little is known about the combined effect of more than two resource states or, more important, the response of annuals to conditions that typify their ecological setting. Annuals flourish in highly heterogeneous environments that vary not only in absolute quantities of many resources but also in the pattern of resource availability relative to the life cycle. The consequences of this resource heterogeneity for plant performance, however, has only begun to be examined experimentally. We recently completed a study in which three annual species were grown under 75 resource states generated by crossing moisture, nutrient, and light gradients. The moisture gradient consisted of three treatment levels, and the nutrient and light gradients spanned five resource levels. Figures 2 and 3 illustrate the performance of two of the

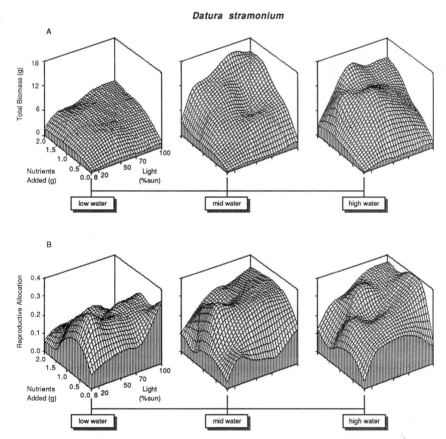

Figure 3 Total biomass (A) and percent reproductive allocation (B) for *Datura stramonium* on a three-dimensional gradient of light, nutrients, and water. [Bazzaz, Crabtree, and Buckley (unpublished data).]

species, *Abutilon theophrasti* and *Datura stromonium*, for all resource combinations; the three plots for each species together represent the three-dimensional niche volume.

The response surfaces of total biomass suggest that *Abutilon* has a very broad niche. The response surfaces of the intermediate and high water levels were nearly indistinguishable. For these two water levels, biomass changed constantly as a function of nutrient availability. Below 50% sunlight, however, the nutrient response was suppressed. At the lowest water level, in contrast, peak biomass was significantly reduced and the accumulated biomass was nearly constant for all resource states. The significant light-by-nutrient interaction was detected only when light levels dropped to less than 20% full sun. Reproductive allocation patterns in response to light and nutrient levels were also indistinguishable at intermediate and high water levels. The isoclines at 20% light and 0.5 g nutrients define a broad plateau of constant percent allocation. At the lowest water level, the plateau is triangular because of a negative interaction at the highest light and nutrient levels.

For *Datura*, the response pattern for biomass accumulation across the water gradient was very similar to that of *Abutilon*. Reproductive allocation, however, was much more complex. Plants in the low-water treatment consistently reproduced less with high nutrients, irrespective of the light level. Maximum allocation occurred at the lowest nutrient and highest light level. At mid- and high water levels, however, reproductive allocation increased with both nutrient and light level, reaching a peak allocation at maximum light and nutrient levels. The contingency of responses on a given resource level, represented in these complex surfaces, indicates that we cannot readily assess the response of plants to multiple stresses from the analysis of three independent environmental gradients. Yet such a complex and labor-intensive experiment is truly only a first step. Another layer of complexity will need to be added if we are to examine the responses to the daily and seasonal incongruities among resources that we know exist in the microenvironment of annual plants.

In the field, the population structure and dynamics of any species is the best measure we have of how a species responds to multiple stresses. For annuals with no seed dormancy, severe stresses that reduce reproductive fitness may become a potent selective force because of very short generation times. Species with the potential for extended seed dormancy, however, will be buffered from any strong selective force that is not constant and unidirectional because many genotypes can persist as seed through selective bottlenecks (Templeton and Levin, 1979). Recruitment of *Bromus tectorum* from the seed bank, for example, is stimulated by irregular patterns of rainfall throughout the year in western Washington. Mortality among these seedling cohorts, however, has multiple causes ranging from drought and frost to disease (Mack and Pyke, 1984).

Novel or extreme stresses, such as soil contaminants and gaseous pollutants, may also unmask variation within the population, variation on which natural selection will act (Schmalhausen, 1949; Coleman *et al.*, 1990). Growth of radish plants exposed to SO_2 fumigation, for example, show no significant change in mean biomass accumulation but do show significant increases in variance with increased fumigation. This increased variability may result from the differential resistance to SO_2 among genotypes or the differential disruption of developmental homeostasis among genotypes. In either case, the increased variance in response to chronic stresses provides a route by which rapid selection of traits may occur, given the necessary genetic link with the expressed variability (Fisher, 1930; Coleman *et al.*, 1990).

The rapidity of potential response within a population will vary not only as a function of the genetic variability within a population and the presence of seed dormancy, but also with the relative fecundity of individuals within a population. One might predict that determinate and indeterminate annuals will differ substantially in this regard. The proportion of individuals of an indeterminate annual that would contribute to the next generation is predicted to be higher than in determinate annuals if each starts to flower before the onset of density-dependent mortality (Hickman, 1975).

Density-dependent regulation of fecundity and population size may also result in a rapid change in the genetic structure of the population if the asymmetry in size of individuals is genetically linked. Although it is not clear whether size hierarchies in general are mediated primarily by chance events (Weiner and Thomas, 1986) or are linked to heritable traits (Gottlieb, 1977), it is probable that strong competitive interactions, in combination with rapid generation time, were the selective forces that gave rise to weeds that mimic the life cycle of annual crop systems (Baker, 1974).

The vulnerability of an individual and a population to either chronic or acute stresses will also depend on whether the species is outcrossing or autogamous. The survival or extinction of an outcrossing plant population, for example, depends on the synchrony of flowering, in addition to vegetative and reproductive tolerance to a multitude of stresses. The first limitation is not relevant for selfing species, and instead, the ability to maximize biomass or plant size may be of primary importance. This simple example suggests that if we are to examine the vulnerability of annuals to stresses, it will be critical to examine the effects of multiple stresses with respect to plant breeding system.

Although heavy-metal tolerance is one of the best examples of selective divergence (Antonovics *et al.*, 1971; Bradshaw, 1952), it is quite unlikely that evolution will be controlled solely by one environmental factor except in those extreme and atypical situations where a single factor is spatially distinct and temporally constant (Bazzaz and Sultan, 1987; Sultan, 1987). Instead it is expected that plant populations will be buffered from selection by their plastic responses to the marked heterogeneity they perceive on daily and seasonal scales. When these plastic responses to several environmental factors are integrated by the plant, the overall response of the plant is not necessarily predictable from the response to each resource or stress. Even in the case of heavy-metal pollution, where the physiological response to each metal is understood (review in Clijsters and Van Assche, 1985), the reduction in growth when both cadmium and lead are present is greater than would be predicted from the responses to each metal applied individually. Thus, multiple stresses together will certainly tend to act as important, yet often unpredictable, selective forces. The challenge that remains nearly untouched is finding a powerful means of studying multiple factors with additive, synergistic, or antagonistic effects that can vary across genotypes and across different stages of the life cycle of the same individual.

C. Annuals in a Community

Annual plants occur in a spectrum of environments ranging from unproductive deserts to temperate deciduous forests; this spectrum includes highly modified and disturbed systems, such as agricultural fields and industrial waste sites, as well as transitional or successional communities. All of the annual habitats within these communities, while different in the mean and range of their environmental factors, share two attributes: large fluctuations and a high degree of unpredictability in environmental factors. At large scales, such as in deserts, climate patterns determine the environmental heterogeneity; within dense stands, on the other hand, disturbance creates spatial heterogeneity such as that exploited by old-field annuals. While the source of unpredictability varies from community to community, the ecological role annuals play in these systems is the same — the opportunist. A striking example of opportunism is fire-adapted annuals, which may remain in the seed bank from 30 to 100 years and flourish only when a temporarily open and nutrient-rich habitat is created with the burning of the perennial vegetation that otherwise usurps the majority of the light and nutrients (Trabaud, 1987; Keeley and Keeley, 1989).

In the most extreme deserts, such as the Sahara, annuals use environmental resources that may be available as infrequently as once every 50 years. These annuals quickly germinate, grow, and bloom after adequate rain but may again remain dormant in the soil for many years (Shreve, 1951; Noy-Meir, 1973). Thus persistence of desert annuals is mediated through temporal resource variability and seed dormancy. Coexistence of desert annuals with perennials is also mediated through dormancy. In this instance, however, the habitats used by each of the life-forms are distinct from a phytocentric view. The distribution of perennials appears to be dictated by competition for deep water stores (Ehleringer, 1984; Fowler, 1986), whereas the annuals effectively exploit shallow water stores when available. With substantial rainstorms, the annuals germinate in the high-light environment between the widely dispersed perennials.

Diversity patterns of annuals suggest that the number of taxa in a community is correlated with the severity of competition among all species within the community. Within productive communities, annual species diversity is relatively low (e.g., Odum, 1969; Bazzaz, 1975). Only a few, highly competitive taxa can successfully colonize the repeatedly disturbed or early successional microhabitats. Communities with characteristically low productivity (e.g., arid and semiarid regions), in contrast, have notably diverse annual floras (e.g., Went, 1949; Koller, 1972; Whittaker, 1975). This diversity primarily reflects the repeated origination of short-lived taxa that are able to persist under conditions where extreme variation in resource availability is abiotically controlled.

The responses of annuals to growth conditions are as diverse as the types of communities in which they are found. Desert annuals, for example, are widely dispersed and tend to be limited most by moisture and nutrients and least by light (Ehleringer, 1985). Old-field annuals, in contrast, grow on fertile soils with ample moisture and become light limited as dense canopies develop through the growing season (Bazzaz, 1984; Tilman, 1988). Consequently, the light environment for individual leaves of annuals in these two communities are very different: all leaves except those in the uppermost part of the canopy become light limited. In the desert annuals, older leaves still receive much light, whereas the leaves of old-field annuals become progressively shaded by the canopy developing above them. Correlated with these differences in light environment are physiological differences: leaves of the light-unlimited annuals of Death Valley have higher photosynthetic capacity, higher specific leaf weight, and higher nitrogen content than do the light-limited old-field annuals (Mooney *et al.*, 1981). Not surprisingly, these characteristics change little with leaf age for the Death Valley annuals but greatly for old-field annuals (Fig. 4).

Perennial plants are often thought to be more competitive for resources than annual plants because annuals allocate proportionally more to reproduction and less to stems and roots, structures required for resource acquisition (Fig. 5). Experimental studies with closely related annuals and perennials explicitly designed to examine specific aspects of the response to multiple stresses are unfortunately very limited. Several studies of closely related annual–perennial pairs verify the generality of the inherent differences in allocation patterns but do not examine differences in competitive ability (Gaines *et al.*, 1974; Pitelka, 1977; Primack, 1979). One recent study with *Hypochoeris glabra* and *H. radicata*, which are similar in growth form, tested assumptions regarding competitive ability and life history and found no difference between the two taxa in competitive ability, despite the typical differences in allocation pattern between the annual and the peren-

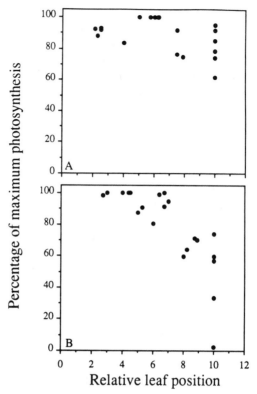

Figure 4 Percentage of maximum photosynthesis (area basis) in relation to relative leaf position for desert (A) and old-field annuals (B). Oldest leaves sampled were scaled to 10. [After Mooney *et al.*, (1981).]

nial (Fone, 1989). In rare instances, annuals can maintain themselves in competition with woody perennials through habitat dominance. *Impatiens capensis,* for example, will persist in the understory of deciduous forests for years because its rapid growth and canopy closure prevents the establishment of trees (Winsor, 1983).

Persistence of annuals in abandoned agricultural fields is limited by both temporal and spatial resource availability. The annuals flourish in fields where moisture, light, and temperature are extremely variable relative to the adjacent perennial community (Rice, 1987; Fig. 6). Steep temperature profiles may also develop in these open fields, and these in turn significantly affect plant performance (e.g., Regehr and Bazzaz, 1976). Annuals of successional environments can eventually be lost from an area through competition for light with invading perennial species (see Bazzaz, 1984; Tilman, 1988). Persistence of these annual species is mediated through dispersal among disturbed habitats and seed dormancy.

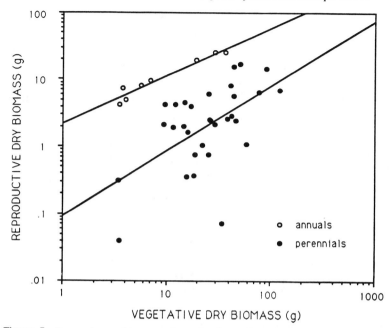

Figure 5 Regressions of log-transformed values of reproductive biomass (g) versus (aboveground) vegetative biomass for annual and perennial British grass species. [McConnaughay, Thomas, Morse, and Bazzaz (unpublished), figure based on data of Wilson and Thompson (1989).]

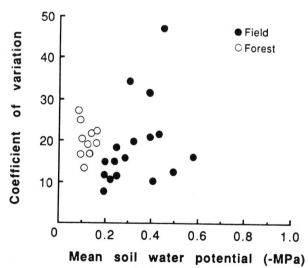

Figure 6 Variability in soil moisture in a field and deciduous forest in the midwestern United States. [After Rice (1987).]

D. Seed Dormancy: A Key (Confounding?) Factor

The ability of annuals to exploit the variation of resource levels within different communities is intimately linked to seed dormancy because the seed is the only possible dormant state. Although seed dormancy is not an obligate stage in the development of the sporophyte (Sussex, 1978), the existence of an arrested stage is common, if not typical. With dormancy, seeds disperse before germinating, individuals can persist for many years under a wide range of environmental conditions, and the probability of survival after germination increases when germination is cued by environmental conditions optimal for subsequent growth (Simpson, 1978).

Although seed longevity may buffer annual populations against environmental unpredictabilities, the reservoir of seeds in the soil does not guarantee persistence. Extinction or persistence of a population depends on the congruence between the type of seed produced and the spatial–temporal patterns of environmental heterogeneity. Human modification of fire frequency, for example, may lead to the extinction of fire-dependent annuals because the frequency of favorable years may no longer overlap with seed longevity (Trabaud, 1987; Keeley and Keeley, 1989).

The degree of specialization to temporal variation, in turn, is limited by both within- and between-genotype variation of seed phenotypes (see Amen [1968] and Simpson [1978] for extensive reviews). Within-genotype differences in seed longevity and seed dormancy are correlated with many factors, including developmental position, seed heteromorphism (Fenner, 1985), environmental conditions at the time of seed development (e.g., Cresswell and Grime, 1981), and modification of dormancy through time (Bewley and Black, 1978). Models of the response of seed-bank annuals to temporal variation also suggest that specialization of germination responses to particular environmental conditions may be further hindered if the magnitude and direction of selection on non-seedbank traits vary from season to season (Brown and Venable, 1986). Clearly, if we are to understand the responses of annual plants to stresses (either chronic or acute) and the consequences for plant population dynamics, then the effects of multiple stresses on this particular developmental stage, in conjunction with the response of non-seed traits, may prove to be an important interface to explore.

IV. Conclusion

Annuals as a group vary greatly in time to reproductive maturity, growth form, population regulation, and the habitats they exploit. The only unifying trait of this diverse group is a vegetative generation time of approximately one year or less. This contracted lifespan, together with potentially high relative growth rates and seed dormancy, enable annuals to persist on resources available on spatial and temporal scales that ultimately make the resources unavailable to perennials. High metabolic rates enable individual annual plants to respond rapidly, both physiologically and morphologically, to variable microenvironmental conditions throughout ontogeny, while environmental sensitivity of germination allows annuals to track between-year variation in resource availability. Although these features correlated with the annual habit contribute to precocious reproduction, they also define the sensitivity or vulnerability of individuals to stresses through development. The outcome of this tension between enhanced survival to reproduction and increased vulnerability to a single stress, however, remains largely unexplored.

Through evolutionary time, plants can respond to stresses, novel or otherwise, in three possible ways: (1) genetic change in response to selection, (2) no detectable genetic change, and (3) extinction. Although annual plants have the potential for very short generation times, and thus the ability to respond rapidly to strong selective forces, both seed dormancy and phenotypic plasticity during ontogeny may provide a strong buffer against genetic change and extinction. Thus, the relative importance of each of the three potential responses of annual plants is only partially dictated by the pivotal necessity of their life history—seed production within a single year. Perhaps only the simultaneous examination of both seed dormancy and phenotypic plasticity during ontogeny will provide the critical information on how population dynamics change in the face of new combinations of novel and old stresses and will ultimately shed more light on the evolutionary forces that shape the annual life history.

Acknowledgments

We thank D. Ackerley, D. Buckley, J. Coleman, K. McConnaughay, and P. Wayne for comments and references.

References

Amen, R. D. (1968). A model of seed dormancy. *Bot. Rev.* 34: 1–31.
Antonovics, J., Bradshaw, A., and Turner, R. G. (1971). Heavy-metal tolerance in plants. *Adv. Ecol. Res.* 7: 1–85.
Baker, H. G. (1974). The evolution of weeds. *Annu. Rev. Ecol. Syst.* 5: 1–24.
Bartholomew, B., Eaton, L. C., and Raven, P. H. (1973). *Clarkia rubicunda:* A model of plant evolution in semi-arid regions. *Evolution* 27: 505–517.
Bazzaz, F. A. (1975). Plant species diversity in old-field successional ecosystems in southern Illinois. *Ecology* 56: 485–488.
Bazzaz, F. A. (1984). Demographic consequences of plant physiological traits: Some case

studies. *In* "Perspectives on Plant Population Biology" (R. Dirzo and J. Sarukhan, eds.), pp. 324–346. Sinauer, Sunderland, Massachusetts.

Bazzaz, F. A. (1987). Experimental studies on the evolution of niche in successional plant populations: A synthesis. *In* "Colonization, Succession, and Stability" (A. J. Gray, M. J. Crawley, and P. J. Edwards, eds.), pp. 245–272. Blackwell, Oxford.

Bazzaz, F. A., and Sultan, S. E. (1987). Ecological variation and the maintenance of plant diversity. *In* "Differentiation Patterns in Higher Plants" (K. M. Urbanska, ed.), pp. 69–93. Academic Press, Orlando.

Bewley, J. D., and Black, M. (1978). "Physiology and Biochemistry of Seeds in Relation to Germination. II: Viability, Dormancy and Environmental Control." Springer-Verlag, Berlin.

Billings, W. D., and Mooney, H. A. (1968). The ecology of arctic and alpine plants. *Biol. Rev.* 43: 481–529.

Bloom, A. J., Chapin, F. S., III, and Mooney, H. A. (1985). Resource limitation in plants—an economic analogy. *Annu. Rev. Ecol. Syst.* 16: 363–392.

Bradshaw, A. D. (1952). Populations of *Agrostis tenuis* resistant to lead and zinc poisoning. *Nature* 109: 1098.

Bradshaw, A. D. (1965). Evolutionary significance of phenotypic plasticity in plants. *Adv. Genet.* 13: 115–155.

Brown, J. S., and Venable, D. L. (1986). Evolutionary ecology of seed-bank annuals in temporally varying environments. *Am. Nat.* 127: 31–47.

Carlson, R. W. (1983). Interaction between SO_2 and NO_2 and their effects on photosynthetic properties of soybean *(Glycine max)*. *Environ. Pollut. Ser. A Ecol. Biol.* 32: 11–38.

Carlson, R. W., and Bazzaz, F. A. (1982). Photosynthetic and growth response to fumigation with SO_2 at elevated CO_2 for C_3 and C_4 plants. *Oecologia* 54: 50–54.

Carlson, R. W., and Bazzaz, F. A. (1986). Plant response to SO_2 and CO_2. *In* "Sulfur Dioxide and Vegetation" (H. A. Mooney, W. E. Winner, and R. A. Goldstein, eds.), pp. 313–331. Stanford Univ. Press, Stanford, California.

Chapin, F. S., III, Bloom, A. J., Field, C. B., and Waring, R. H. (1987). Plant responses to multiple environmental factors. *Bioscience* 37: 49–57.

Chiariello, N., and Roughgarden, J. (1984). Storage allocation in seasonal races of an annual: Optimal versus actual allocation. *Ecology* 65: 1290–1301.

Clijsters, H., and Van Assche, F. (1985). Inhibition of photosynthesis by heavy metals. *Photosynth. Res.* 7: 31–40.

Cole, L. C. (1954). The population consequences of life history phenomena. *Quart. Rev. Biol.* 29: 103–137.

Coleman, J. S., Mooney, H. A., and Winner, W. E. (1990). Anthropogenic stress and natural selection: Variability in radish biomass accumulation increases with increasing SO_2 dose. *Can. J. Bot.* 68: 102–106.

Coley, P. D., Bryant, J. P., and Chapin, F. S., III. (1985). Resource availability and plant anti-herbivore defense. *Science* 230: 895–899.

Cousens, M. I. (1988). Reproductive strategies of pteridophytes. *In* "Plant Reproductive Ecology: Patterns and Strategies" (J. Lovett Doust and L. Lovett Doust, eds.), pp. 307–328. Oxford Univ. Press, New York.

Cresswell, E. G., and Grime, J. P. (1981). Induction of a light requirement during seed development and its ecological consequences. *Nature* 291: 583–585.

Dansereau, P. (1957). "Biogeography—An Ecological Perspective." Ronald Press, New York.

Ehleringer, J. (1984). Intraspecific competitive effects on water relations, growth, and reproduction in *Encelia farinosa*. *Oecologia* 63: 153–158.

Ehleringer, J. (1985). Annuals and perennials of warm deserts. *In* "Physiological Ecology of North American Plant Communities" (B. F. Chabot and H. A. Mooney, eds.), pp. 162–180. Chapman and Hall, New York.

Fenner, M. (1985). "Seed Ecology." Chapman and Hall, London.

Fisher, R. A. (1930). "The Genetical Theory of Natural Selection." Clarendon, Oxford.

Fitter, A. H., and Hay, R. K. (1988). "Environmental Physiology of Plants," 2nd ed. Academic Press, London.

Fone, A. L. (1989). Competition in mixtures of the annual *Hypochoeris glabra* and perennial *H. radicata. J. Ecol.* 77: 484–494.

Fowler, N. (1986). The role of competition in plant communities in arid and semiarid regions. *Annu. Rev. Ecol. Syst.* 17: 89–110.

Gaines, M. S., Vogt, K. J., Hamrick, J. L., and Caldwell, J. (1974). Reproductive strategies and growth patterns in sunflowers *(Helianthus). Am. Nat.* 108: 889–894.

Gottlieb, L. D. (1974). Genetic stability in a peripheral isolate of *Stephonomaria exigua* ssp. *carotifera* that fluctuates in population size. *Genetics* 76: 551–556.

Gottlieb, L. D. (1977). Genotypic similarity of large and small individuals in a natural population of the annual plant *Stephanomeria exigua* ssp. *coronaria* (Compositae). *J. Ecol.* 65: 127–134.

Grant, V. (1975). "Genetics of Flowering Plants." Columbia Univ. Press, New York.

Grime, P. (1979). "Plant Strategies and Vegetation Processes." Wiley, Chichester.

Gulmon, S. L., and Mooney, H. A. (1986). Cost of defenses and their effects on plant productivity. *In* "On the Economy of Plant Form and Function" (T. J. Givnish, ed.), pp. 681–698. Cambridge Univ. Press, New York.

Harper, J. L. (1977). "Population Biology of Plants." Academic Press, New York.

Harper, J. L., and Sellek, C. (1987). The effects of severe mineral nutrient deficiencies on the demography of leaves. *Proc. R. Soc. Lond. B Biol. Sci.* 232: 137–157.

Hickman, J. C. (1975). Environmental unpredictability and plastic allocation strategies in the annual *Polygonum cascadense* (Polygonaceae). *J. Ecol.* 63: 689–701.

Keeley, J. E., and Keeley, S. C. (1989). Allelopathy and the fire-induced herb cycle. *Nat. Hist. Mus. Los Angel. Cty. Sci. Ser.* 34: 65–72.

Klekowski, E. J., Jr. (1967). Observations of pteridophyte life cycles: Relative lengths under cultural conditions. *Am. Fern J.* 57: 49–51.

Koller, D. (1972). Environmental control of seed germination. *In* "Seed Biology" (T. Kozlowski, ed.), pp. 1–101. Academic Press, New York.

Law, R., Bradshaw, A. D., and Putwain, P. D. (1977). Life history variation in *Poa annua. Evolution* 31: 233–246.

Lee, H. S., Zangerl, A. R., Garbutt, K., and Bazzaz, F. A. (1986). Within- and between-species variation in response to environmental gradients in *Polygonum pensylvanicum* and *Polygonum virginianum. Oecologia* 68: 606–610.

Lewis, H. (1972). The origin of endemics in the California flora. *In* "Taxonomy, Phytogeography and Evolution" (D. H. Valentine, ed.), pp. 179–189. Academic Press, London.

Mack, R. N., and Pyke, D. A. (1984). The demography of *Bromus tectorum:* The role of microclimate, grazing, and disease. *J. Ecol.* 72: 731–748.

Meyer, R. F., and Boyer, J. S. (1981). Osmoregulation in soybean seedlings having low water potential. *Planta* 151: 482–489.

Mooney, H. A., Field, C. B., Gulmon, S. L., and Bazzaz, F. A. (1981). Photosynthetic capacity in relation to leaf position in desert versus old-field annuals. *Oecologia* 50: 109–112.

Mulroy, T. W., and Rundel, P. W. (1977). Annual plants: Adaptations to desert environments. *Bioscience* 27: 109–114.

Niklas, K. J., Tiffney, B. H., and Knoll, A. H. (1985). Patterns in vascular plant diversification: An analysis at the species level. *In* "Phanerozoic Diversity Patterns: Profiles in Macro Evolution" (J. W. Valentine, ed.), pp. 97–128. Princeton Univ. Press, Princeton, New Jersey.

Nobel, P. S. (1983). "Biophysical Plant Physiology and Ecology." W. H. Freeman, San Francisco.

Noy-Meir, I. (1973). Desert ecosystems: Environment and producers. *Annu. Rev. Ecol. Syst.* 4: 25–51.

Odum, E. P. (1969). The strategy of ecosystem development. *Science* 164: 262–270.

Orshan, G. (1953). Note on the application of Raunkiaer's system of life forms in arid regions. *Palest. J. Bot.* 6: 120–122.

Parrish, J. A. D., and Bazzaz, F. A. (1985). Ontogenetic niche shifts in old-field annuals. *Ecology* 66: 1296–1302.

Pitelka, L. F. (1977). Energy allocation in annual and perennial lupines (*Lupinus:* Leguminosae). *Ecology* 58: 1055–1065.

Primack, R. B. (1979). Reproductive effort in annual and perennial species of *Plantago* (Plantaginaceae). *Am. Nat.* 114: 51–62.

Radosevich, S. R., and Holt, J. S. (1984). "Weed Ecology." Wiley, New York.

Raunkiaer, C. (1934). "The Life Forms of Plants and Statistical Plant Geography." Clarendon, Oxford.

Raven, P. H., and Axelrod, D. I. (1978). Origins and relationships of the California flora. *Univ. Calif. Publ. Bot.* 72.

Regehr, D. L., and Bazzaz, F. A. (1976). Low-temperature photosynthesis in successional winter annuals. *Ecology* 57: 1297–1303.

Reinert, R. A. (1984). Plant responses to air pollutant mixtures. *Annu. Rev. Phytopathol.* 22: 421–442.

Reich, P. B. (1987). Quantifying plant response to ozone: A unifying theory. *Tree Physiol.* 3: 63–91.

Rice, S. A. (1987). Environmental variability and phenotypic flexibility in plants. Ph.D. dissertation, University of Illinois, Champaign-Urbana.

Schaffer, W. M. (1974). Optimal reproductive effort in fluctuating environments. *Am. Nat.* 108: 783–790.

Schmalhausen, I. I. (1949). "Factors of Evolution: The Theory of Stabilizing Selection." Univ. of Chicago Press, Chicago.

Schulze, E.-D. (1982). Plant life forms and their carbon, water, and nutrient relations. *In* "Encyclopedia of Plant Physiology, New Series" (O. L. Lange, P. S. Nobel, C. B. Osmond, and H. Ziegler, eds.), Vol. 12B, pp. 615–676. Springer-Verlag, Berlin.

Shreve, F. (1951). Vegetation of the Sonoran Desert. *Carnegie Inst. Washington Publ.* 591.

Simpson, G. M. (1978). Metabolic regulation of dormancy in seeds — a case history of the wild oat (*Avena fatua*). *In* "Dormancy and Developmental Arrest: Experimental Analysis in Plants and Animals" (M. E. Clutter, ed.), pp. 168–220. Academic Press, New York.

Stebbins, G. L. (1965). The probable growth habit of the earliest flowering plants. *Ann. Mo. Bot. Gard.* 52: 1463–1469.

Stebbins, G. L. (1974). "Flowering Plants: Evolution Above the Species Level." Harvard Univ. Press, Cambridge, Massachusetts.

Sultan, S. E. (1987). Evolutionary implications of phenotypic plasticity in plants. *Evol. Biol.* 21: 127–178.

Sussex, I. (1978). Dormancy and development. *In* "Dormancy and Developmental Arrest: Experimental Analysis in Plants and Animals" (M. E. Clutter, ed.), pp. 297–301. Academic Press, New York.

Symonides, E. (1988). Population dynamics of annual plants. *In* "Plant Population Ecology" (A. J. Davy, M. J. Hutchings, and A. R. Watkinson, eds.), pp. 221–248. Blackwell, Oxford.

Templeton, A. R., and Levin, D. A. (1979). Evolutionary consequences of seed pools. *Am. Nat.* 114: 232–249.

Tilman, D. (1988). "Plant Strategies and the Structure and Dynamics of Plant Communities." Princeton Univ. Press, Princeton, New Jersey.

Trabaud, L. (1987). Fire and survival traits of plants. *In* "The Role of Fire in Ecological Systems" (L. Trabaud, ed.), pp. 65–89. S. P. B. Academic, The Hague.

Weiner, J., and Thomas, S. C. (1986). Size variability and competition in plant monocultures. *Oikos* 47: 211–222.

Went, F. W. (1949). Ecology of desert plants. II. The effect of rain and temperature on germination and growth. *Ecology* 30: 26–38.

Whittaker, R. H. (1975). "Communities and Ecosystems." Macmillan, New York.

Wilson, A. M. and Thompson K. (1989). A comparative study of reproductive allocation in 40 British grasses. *Functional Ecology* 3: 297–302.

Winner, W. E., and Mooney, H. A. (1980). Ecology of SO₂ resistance. II. Photosynthetic changes of shrubs in relation to SO_2 absorption and stomatal behavior. *Oecologia* 44: 296–302.

Winner, W. E., Mooney, H. A., and Goldstein, R. A., eds. (1985). "Sulfur Dioxide and Vegetation." Stanford Univ. Press, Stanford, California.

Winsor, J. (1983). Persistence by habitat dominance in the annual *Impatiens capensis* (Balsaminaceae). *J. Ecol.* 71: 451–466.

Zangerl, A. R., and Bazzaz, F. A. (1983). Responses of an early and a late successional species of *Polygonum* to variations in resource availability. *Oecologia* 56: 397–404.

14

Dryland Herbaceous Perennials

S. J. McNaughton

I. Drylands and Herbaceous Perennials

Herbaceous perennial plants are characterized by unlignified aboveground tissues and adult lifespans of many years. The aboveground organs often live only a limited time, but subterranean and soil-surface organs persist from growing season to growing season, and the plants themselves survive for many growing seasons. Herbaceous perennials inhabit more extreme

Response of Plants to Multiple Stresses. Copyright © 1991 by Academic Press, Inc. All rights of reproduction in any form reserved.

terrestrial environments than any other vascular plants, ranging from the highest altitudes to the highest latitudes and from desert to tropical forest understory (Schulze, 1982). They reach their greatest abundance, however, in arid to subhumid climates as a major component of the dryland ecosystems commonly classified as rangelands.

Widespread and economically important (Williams *et al.*, 1968; Stoddart *et al.*, 1975; World Resources Institute and International Institute for Environment and Development, 1988), rangelands are used for grazing by domestic livestock, for wildlife habitat, and for watershed management. More than a third of Earth's land surface is too dry for precipitation-supported agriculture. Rangelands comprise this area as well as areas unsuitable for cultivation because of other properties of the physical environment, such as erratic or low precipitation; poor soil development, with considerable unconsolidated material in surface layers; rough or steep topography; and a wide variety of edaphic limitations ranging from salinity to micronutrient deficiencies. Africa and Asia encompass major expanses

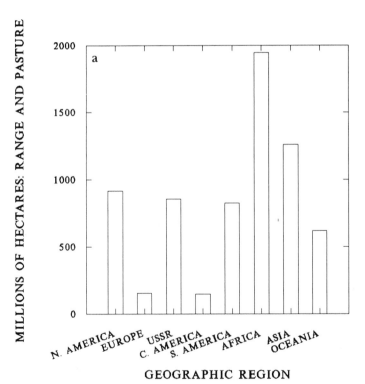

Figure 1 Geographic distribution of drylands. (a) Absolute area in different regions in millions of hectares; (b) area as a percentage of total land.

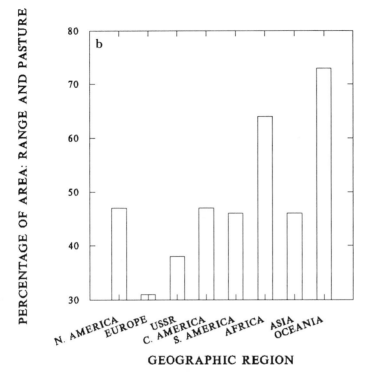

PERCENTAGE OF AREA: RANGE AND PASTURE

GEOGRAPHIC REGION

Figure 1 *(continued).*

of rangeland, although the greatest percentage of rangeland is found in Oceania (Fig. 1). Although no strict climatic bounds can be established for drylands—they depend on the balances among the amount and seasonal distribution of precipitation, evaporative demand, and soil properties—annual precipitation is typically less than 100 cm, and potential evaporation exceeds precipitation for substantial periods of the year. Rangeland vegetation is dominated by herbaceous perennials, sometimes with an arborescent overstory.

The principal perennial herbaceous plants of dryland habitats are (1) members of the Poaceae, grading from Kranz-anatomy species with C_4 photosynthesis in more arid areas to species with C_3 photosynthesis in more mesic sites; (2) a wide variety of C_3 dicots, including a conspicuous representation of Asteraceae and Fabaceae; (3) shallow-rooted plants with CAM photosynthesis and a succulent growth form; and (4) semishrubs with chlorophyllous stems able to photosynthesize when leaves are shed (Fig. 2; Parker, 1968; Lange *et al.*, 1976). Although the succulents are often drought tolerant, the other plant types are physiologically active mainly

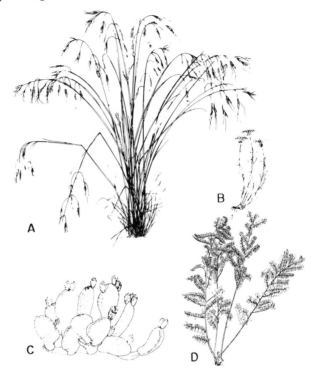

Figure 2 Common growth forms of herbaceous perennials in dryland ecosystems. (a) Grass (*Stipa pulchra,* Poaceae); (b) dicot (*Aster stenomeres,* Asteraceae); (c) succulent (*Opuntia* spp., Cactaceae); and (d) semishrub (*Amorpha canescens,* Fabaceae). Drawings not to scale.

during wet periods and dormant during seasonal drought. A major adaptation of herbaceous perennials to dryland habitats is storage of reserves in belowground organs during periods, often brief, of active growth; rapid mobilization of those reserves to generate aboveground biomass at the onset of growth; and death of aboveground tissues during dry periods (Schulze, 1982).

No single physiological or morphological syndrome unifies the herbaceous perennials of dryland ecosystems. Rather, a variety of evolutionary responses has characterized these plants. A comparison of nearby, similar-sized individuals of a C_4 grass, a CAM succulent, and a C_3 semishrub revealed strikingly different physiological patterns (Nobel and Jordan, 1983). Daily peak water potentials indicated greatest water stress in the grass, least in the CAM plant, and intermediate levels in the semishrub. Daily total transpiration per unit of foliage surface was least for the succulent; the grass transpired 20 times more water, the semishrub 40 times. Daily water loss per plant, however, was lowest for the succulent and semishrub, which each transpired about a fifth of the total water lost by the grass. A major difference among the growth forms was the capacitance, or storage capacity, of the transpiration stream. Diurnal changes in water storage could account for only 4 min of transpiration by the grass, 7 min for the semishrub, and 16 hr for the succulent plant. Thus, CAM plants exploit arid environments by having high capacitances and low stomatal conductances, thereby buffering their tissues from drought. Both grasses and semishrubs are capable of extracting water at lower soil water potentials, but they have a much more ephemeral growth strategy than CAM species.

An important element of niche partitioning and the coexistence of growth forms in dryland ecosystems is different rooting patterns (Weaver, 1954). For example, shrubs and other woody plants may obtain water from deep soil pools inaccessible to more shallowly rooted herbaceous plants. And many herbaceous perennial dicots are more deeply rooted than the grasses of dryland ecosystems. Recent evidence, however, indicates that some of the water obtained by deep roots is lost into drier, shallow soil layers during the night (Richards and Caldwell, 1987). While this "hydraulic lift" could provide a more favorable water balance for deeply rooted plants, particularly after several cloudy days that suppress transpiration, it also provides the potential for water parasitism by shallowly rooted species.

An additional mechanism of species coexistence in dryland perennial herbaceous species is temporally distinct growth periods, often related to photosynthetic pathways (Kemp and Williams, 1980). Plants with growth concentrated during cool seasons typically have C_3 photosynthesis, whereas warm-season species assimilate by the C_4 pathway. The temperatures at which thermal stress is translated into irreversible inhibition of quantum yield, as indicted by chlorophyll fluorescence tests, are lower for species that grow during cool periods than for those growing during warm seasons (Monson and Williams, 1982).

Although a consideration of stress in dryland ecosystems naturally centers around water shortage, a wide variety of other factors can also cause stress, including temperature; high radiant energy loads; nutritional shortage; elemental toxicity; and such biological factors as herbivores, pathogens, and competitors. In dryland ecosystems, water shortage is an obvious stress, yet to the extent that plants occupying such environments have evolved mechanisms to alleviate that dominant stress, other factors assume increasing importance.

II. Water Loss versus Carbon Gain

The fundamental agent of energy storage and transmission in ecosystems is carbon-to-carbon bonds. These bonds are the first stable, storable, and transmittable form of energy derived from photosynthesis. The energy of carbon-to-carbon bonds can be translocated within the plant, affecting energy and growth–allocation patterns, and transmitted through the food chain.

Conversion of light energy into carbon-to-carbon bond energy is inevitably linked to loss of water. Indeed, on a molar basis, water loss is considerably greater than carbon gain. This is because the diffusion gradient from water vapor in the leaf to the atmosphere is steeper than the gradient in carbon dioxide concentration from the atmosphere to the leaf. Moreover, the diffusion pathway of water is spatially shorter since it lacks a significant mesophyll component, and the radius of the carbon dioxide molecule is greater than that of water. For an idealized "leaf" at 20° C and 70% relative humidity, water loss is 20 times greater than carbon gain; at 50° C and 10% relative humidity, water loss on a molar basis is more than 400 times greater than carbon gain (Raschke, 1976). Depletion of water supplies by plants thus coincides with gaining energy, and the imbalance toward water loss is substantial. Three of the common interacting stresses in dryland environments are demonstrated by this example: water stress based on the balance between soil pools and atmospheric demand, thermal stress based on ambient temperature and radiant heat load, and limits to carbon assimilation due to the interaction of water and thermal stress.

Photosynthesis, desiccation from an imbalance between water loss and supply, and both thermal and radiant heat loads on aerial tissues are intimately related (Gates, 1968). Plants have evolved a stomatal response to atmospheric humidity that conserves water as the vapor-pressure gradient from leaf to air increases (Lange et al., 1971). Rather than reacting directly to plant water status, the stomata respond in a feedforward manner (Cowan, 1977; Farquhar, 1978), closing before plant water deficits develop to significant levels. This mechanism helps protect the plant against water deficits that would otherwise arise and may allow significantly more efficient use of soil water supplies than if stomatal control were coupled solely to plant water status (Schulze, 1986). This conservation mechanism may be particularly important in the arid to semiarid habitats of herbaceous perennials (Schulze et al., 1987), where combinations of low midday atmospheric humidities and variable soil water stores are common.

It has long been known that species from dryland ecosystems often have smaller and more dissected leaves than congeneric species from more mesic habitats (Clements, 1907), thereby increasing convective heat flux with the atmosphere (Gates et al., 1968). There is also, however, a countervailing balance between leaf size and the dissipation of heat load by transpiration (Smith, 1978). When water supplies are adequate, large-leaved species have higher leaf conductances and transpiration rates. This generates leaf temperatures that are often substantially lower than air temperatures, and these broad-leaved species have lower optimum photosynthetic temperatures than fine-leaved species (Smith, 1978).

The thermotolerance of photosynthesis is typically higher in perennials than in annuals of dryland habitats (Downton *et al.*, 1984), which likely reflects the more stable growth patterns of perennials. Succulents appear to be particularly resistant to thermal damage of photosynthetic processes.

Plants with different photosynthetic pathways differ in their sensitivity to atmospheric humidity and the resultant gradients in water vapor pressure from leaf to air. In one study, raising humidity to decrease the vapor-pressure gradient to about half of ambient resulted in daily totals of net photosynthesis that were 17% higher in a C_3 species, whereas C_4 plants were insensitive to the environmental change (Bunce, 1983). These results indicate that stomates of C_4 plants are less sensitive to a desiccating atmosphere than are C_3 plants, which would provide for greater carbon gain in low-humidity atmospheres.

Little is known about carbohydrate storage and mobilization in dryland herbaceous perennials (Schulze, 1982), and the available evidence is contradictory. For example, experiments on whole sods of *Bouteloua gracilis*, a C_4 grass, indicated that water stress favored transport of assimilate to roots, providing greater allocation to storage pools under the environmental stimulus of an impending drought (Chung and Trlica, 1980). A seemingly conflicting result was obtained with another C_4 perennial grass *(Dichantium annulatum)*, in which the percentage of total nonstructural carbohydrate —i.e., storage pools—progressively declined as water stress increased (Misra and Singh, 1981). Nevertheless, there is firm evidence that the accumulation of carbohydrate reserves in shoot bases and roots is an important element of the ability of dryland herbaceous perennials to survive dry periods (Trlica and Singh, 1979). Similarly, the establishment of seedlings of *B. gracilis* under water stresses characteristic of rangelands depends on carbohydrate reserves (Khan and Wilson, 1984). Therefore, the ability of herbaceous perennials to accumulate carbon in belowground storage pools appears to be an important element of their ability to dominate dryland ecosystems, affecting both establishment at the seedling stage and survival at the adult stage.

III. Halomorphic Soils

Water stress is exacerbated by soil-solution salt concentrations that are often high in dryland ecosystems (Brady, 1974). Halomorphic soils include saline soils, with concentrations of neutral soluble salts detrimental to plant growth, and sodic soils, with high sodium concentrations. These soil properties become an important limitation on agricultural exploitation of dryland ecosystems receiving less than about 50 cm of annual precipitation. The deleterious effects of sodium chloride and other osmotically active elements in the soil solution appear closely related to plant water status (Munns et al., 1983). Effects on carbon assimilation, cell division, and cell expansion have all been implicated in the deleterious impact of soil-solution salts. But as McCree (1986: p. 42) observed, "In general, the physiological literature has little to say on the subject of water stress in the presence of salt, despite its obvious practical importance. Probably as a matter of convenience, most physiologists continue to treat water and salt stress as separate topics for research." This reflects a general lack of attention to interactive environmental factors in both ecology and physiology (McNaughton et al., 1983; Chapin et al., 1987).

Adaptations of herbaceous perennials to halomorphic soils differ between monocots and dicots (Munns et al., 1982). Dicots typically sequester accumulated ions in the vacuole, protecting metabolic processes occurring in the cytoplasm. Some of these species respond to increasing salinity with substantial increases in growth over moderate salinity ranges. An exclusion mechanism seems more important in salt-tolerant monocots since they appear not to accumulate concentrations as high as in dicots and typically do not show a growth response over moderate salinities.

Despite the undoubted importance of salt tolerance in the dryland ecosystems dominated by herbaceous perennials, little information exists on species from those ecosystems. Much of the research has concentrated on halophytes from extreme environments, such as salt marshes and coastal regions, halophytic lower plants, and crop plants. Since dryland ecosystems are often subject to heavy rain showers alternating with intervening dry periods, as soils dry out herbaceous perennials must cope with soil solutions of escalating salinity if they are to maintain growth between showers. Then, when the next shower occurs, soil-solution salinities undoubtedly drop precipitously. Understanding the ability of herbaceous perennials to cope with these fluctuations would add considerably to our understanding of adaptation to dryland systems and the functional processes in those systems.

IV. Osmotic Adjustment

It was first documented in 1972 that vascular plants could compensate for soil drying by osmolyte accumulation (Greacen and Oh, 1972; Meyer and Boyer, 1972) and that this accumulation could allow growth to proceed with less soil water than if osmoregulation did not occur (Meyer and Boyer, 1972; see also Chapter 5). Salt stress is also associated with osmolyte accumulation (Dreier, 1983). Sugars and amino acids are major osmoregulators in vascular plants (Morgan, 1984), and proline accumulation appears to be a particularly sensitive indicator of both water stress (Stewart *et al.*, 1966; Bokhari and Trent, 1985) and salt stress (Dreier, 1983). Yancey *et al.* (1982) argue that the repeated evolutionary adoption of a narrow range of organic molecules as osmolytes is related to two factors. First, the physicochemical interactions among macromolecules, water, and solutes allow only a limited array of solutes to be compatible with maintaining macromolecular structure and function. Second, a degree of parsimony has been involved in the evolution of the organic osmolyte systems, thereby allowing a simple array of organic osmotic regulators to stabilize macromolecules over a broad range of water potentials and obviating the necessity to synthesize a variety of proteins of different structures under different degrees of water stress. Therefore, the osmolytes used by herbaceous perennials in dryland habitats are likely to be very similar, even identical, to those employed by a wide variety of other organisms.

The rate that soil water pools are depleted affects the ability of herbaceous dryland perennials to adjust osmotically. Rapid desiccation, of the type that plants grown in small pots encounter when watering stops, prevents osmotic adjustment and does not allow plants to maintain physiological competence as soils dry (Ludlow *et al.*, 1985). Tropical perennial grasses grown in small containers reached water potentials of -3.5 MPa in the youngest fully expanded leaf blades within two to four days after watering ended. Plants in the field and in large pots required three to six weeks to reach the same leaf water potential. Rapidly desiccating plants were completely unable to adjust to the rapid drying, whereas the slowly drying soils allowed plants to adjust the relation of both stomatal conductance and photosynthetic rate to leaf water potential, so they became progressively less sensitive as the drying proceeded. A tropical legume *(Macroptilium atropurpureum)* in the same experiment, known to be less drought tolerant than the C_4 grasses, did not accumulate solutes; adjusted osmotically to only a minor degree; and was therefore incapable of maintaining photosynthetic rates as the soil dried out, regardless of drying rate.

In addition to maintaining physiological competence over a broader range of environmental conditions, osmotic adjustment to water or salt stress might allow continuance of cellular expansion and growth, among the most sensitive processes to water deficit (Boyer, 1968). Studies of several perennial C_4 grasses, however, suggest that only one to four days of growth extension can be attributed to osmotic adjustment (Wilson and Ludlow, 1983a,b; Toft *et al.*, 1987). Nevertheless, this osmotic adjustment can accomplish two functions: (1) extending the lifetime of active tissues between ephemeral showers so that physiological activity and growth can be rapidly resumed if a shower does happen, and (2) extending the period of tissue preparation for drought (drought-hardening), if showers fail.

In field plots of a C_4 grass *(Cenchrus ciliaris)*, the water potential at which stressed leaves lost turgor decreased by 1 MPa because of a reduction in osmotic potential at full turgor and decreased cell wall elasticity (Wilson and Ludlow, 1983a). Water potential at full turgor was still lower for prestressed leaves after rewatering than for unstressed leaves. The cyclic wetting and drying characteristic of dryland habitats can, therefore, lead to modifications improving the ability of plant tissue to survive subsequent drought. This drought-hardening in response to wet-dry cycles can lead to greater adjustment than the single drought periods typical of many experiments (Morgan, 1984) and result in a faster recovery on rewetting than does exposure to a single dry period (Ashton, 1956).

These studies on osmotic adjustment suggest that many laboratory experiments, in which plants grown in pots are exposed to rapid drying without previous water stress, have little explanatory utility under field conditions. Alternating wet and dry cycles, slow depletion of soil water pools available to the plant, and combined water and salinity stress are typical of the environments of dryland herbaceous perennials, but they are not common features of laboratory studies.

Two major classes of dryland herbaceous perennials, members of the Fabaceae and Poaceae, have different methods of coping with water shortage (Ludlow, 1980). Legumes avoid stress to a certain extent by deep rooting and the exploitation of a greater soil volume; sensitive stomatal control of transpiration; and reducing leaf area under stress, often associated with the subsequent production of small, thick, pubescent leaves. Grasses, in contrast, are more drought tolerant, acclimating to water stress by osmotic adjustments that maintain physiological competence in the face of water shortage and preserving tissues during short periods of drought to rapidly reestablish full activity when another shower arrives.

V. Growth Balances and Adaptation to Dryland Habitats

The sensitivity of growth by cellular expansion to mild water stress can rapidly restrict canopy growth while root growth may be less affected, leading to an increase in root:shoot ratio that will maintain a more favorable water balance than if expansion growth were highly resistant to water stress (Bradford and Hsiao, 1982). Modifications of root:shoot balance, therefore, are a major aspect of the adaptation of herbaceous perennials to dryland habitats. Reductions of leaf area due to senescence under water stress were about twice as important as stomatal conductance as a regulator of water loss in unirrigated *Pennisetum typhoides* (Squire *et al.*, 1984). Both a restriction of growth in early stages of water stress and the senescence of older leaves as stress intensifies can lead to a modification of root:shoot ratios that tends to maintain water balance.

The importance of plant developmental processes as regulators of water stress is evident both in species distribution patterns in nature and the growth responses of species to water stress. Two species of *Solidago* differentially distributed along a soil-moisture gradient had similar stomatal conductances, water-use efficiencies, and stomatal responses to low leaf water potentials when grown in controlled environments (Potvin and Werner, 1983). In other words, none of the physiological responses discussed earlier in this chapter explained the species' distributions. Instead, the species from dry habitats was distinguished by a lower leaf area, more rapid phenological development, and the presence of a nonreproductive rosette growth form. Similarly, differential drought susceptibility of *Andropogon* (a C_4 grass) species from moist and dry habitats was not due to differences in osmoregulation or different stomatal sensitivity to differences in water-vapor pressure between air and leaf (Barnes, 1985). Rather, the major dissimilarities were pronounced leaf rolling and epicuticular wax accumulation in the dry-habitat plants.

Growth responses of three C_3 grasses to soil drying during the seedling phase indicated that root growth patterns in response to soil moisture influence their tolerance of moderate drought (Molyneux and Davies, 1983). One species was deep rooted when well watered but shallow rooted when less well watered; this species became severely water stressed when water was withheld. Conversely, a species whose root growth into deep soil layers was stimulated by mild water stress was much less severely water stressed when water was withheld for a longer period. Plants capable of continued root growth into deeper soil layers in drying soils have root growth rates that are considerably less sensitive to root-tip water potentials than plants with roots that cease growing as soils dry (Davies *et al.*, 1986). This implicates plant growth regulators in developmental events influencing acclimation to shortages in soil moisture (Chapters 1, 4, 5).

VI. Nutrient and Water Interactions

Decomposition and mineralization rates in dryland soils are coupled to soil water, but the relationship is not simply more water, more nutrients. Instead, dryland soils are subject to the Birch effect: a rapid burst of mineralization when soils are first wetted, followed by a decline if wetting is continuous but pulsed nutrient release if soils are alternately dried and rewetted (Birch, 1958a,b). Birch (1958a) was able to maintain decomposition pulses over 40 cycles of wetting and drying, reducing total soil organic carbon about 35%, but the magnitude of the flush declined with subsequent cycles. In situ mineralization studies in Sahelian rangelands indicated that 5–8% of the nitrogen was mineralized annually, with a strong pulse of mineralization as the wet season began (Bernhard-Reversat, 1982).

Plant water and nitrogen status are related (Hanson and Hitz, 1983). In general, shoot nitrogen content of grasses increases under mild water stress, but nitrogen fixation by legumes is inhibited. Leaf nitrate reductase activity often decreases markedly, showing a pattern similar to the inhibition of expansion growth and protein synthesis under short-term water stress (Hsaio, 1973). Under water stress, therefore, nitrate nitrogen in leaves often increases more markedly than total nitrogen. When four cultivars of *Festuca arundinacea* were subjected to water stress combined with nitrogen fertilization, Kjeldahl nitrogen of leaves increased 5.5% under moderate water stress, but foliage nitrate nitrogen increased 475% (Belesky *et al.*, 1982). In addition, fertilization increased the yield in water-stressed plots by 23%, indicating that nitrogen can reduce the effects of water stress on yield, perhaps through promoting root growth and proliferation.

VII. Grazing and Fire

Fire and grazing are two other environmental factors that often prevail in the dryland ecosystems dominated by herbaceous perennials. Although they are sometimes equated because both remove plant tissues, they are, in fact, quite distinct because of their timing with respect to plant growth. Grazers typically seek out actively growing plant tissue whenever it is available, so their impact is strongest when plants are physiologically active. Fire, in contrast, is most prevalent during the dormant season, when aboveground plant tissues of herbaceous perennials are dead, and the plants are largely impervious to its impact. Both grazing and fire, however, recycle nutrients within the system and can also lead to nutrient loss through volatilization, from microbial activity in animal wastes or heat in fires.

The accumulation of litter and standing dead tissues in unburned, ungrazed grassland has a deleterious effect on both individual plants and community primary productivity. For example, *Andropogon gerardii* on burned plots had higher photosynthetic rates, leaf conductances, leaf nitrogen concentrations, and shoot biomasses than plants on unburned plots (Knapp, 1985). During a drought year, however, the greater transpiring plant surface generated on burned plots led to greater plant water stress there. Plants on burned plots had lower osmotic potentials at full turgor and at zero turgor during the drought. Still, as the drought progressed, water-use efficiency of plants on burned plots was consistently greater than on control plots.

The accumulation of standing dead tissues also inhibits nitrogen fixation by microbes and blue-green algae on the soil surface and reduces soil temperatures, which diminishes root, soil microbe, and invertebrate activities (Knapp and Seastedt, 1986). The general effect of this is to slow a variety of ecosystem processes that support primary and secondary productivity (McNaughton *et al.*, 1988).

Herbivory in rangelands is among the most intense of that in any terrestrial ecosystem, with worldwide averages estimated at 30–40% of aboveground production consumed by large mammals and 5–15% consumed by insects (Detling, 1988). Grazers have such a major impact upon plant and ecosystem processes that it is impossible to treat them fully in this chapter (McNaughton 1983a,b; 1984; 1985). Two aspects of their impact, however, are particularly relevant.

First, light filtered through a canopy feeds back to meristems at the base of grasses to regulate tillering (Deregibus *et al.*, 1985). That is, the depletion of red wavelengths relative to far red as a canopy develops inhibits meristem activation, so the red:far red ratio serves as a signal of canopy closure that can interact with such variables as water and nutrients to determine the rate of tiller initiation and death and the activation of terminal meristems leading to taller canopies. Grazers, by opening the canopy and allowing unfiltered light to penetrate to the shoot bases, promote tillering and affect both plant growth rate and the geometry of the canopy. Second, the accumulation of nutrients in standing dead tissues, litter, and recalcitrant fractions of soil organic matter can limit nutrient supply to growing tissues (Knapp and Seastedt, 1986). Grazing tramples dead tissues and litter into the soil, where they become more accessible to decomposition; it increases soil temperature by opening the canopy; and it slows the depletion of soil moisture by reducing the transpiration surface (McNaughton *et al.*, 1988). All of these effects can promote soil microbial activity. Grazing also recycles elements in consumed plant tissues into forms that may be both directly available to the plants or more rapidly mineralized than other substrates.

Prairie dogs (*Cynomys* spp.) and bison (*Bison bison*) were extremely abundant herbivores in the Great Plains of North America, and they have profound effects upon the vegetation in areas where they still occur (Whicker and Detling, 1988a,b). In the gradient from the highly disturbed centers of prairie dog towns to the surrounding prairies, a community of unpalatable herbaceous dicots in the center changes to dwarf perennial grasses and shrubs and then to taller perennial grasses on the periphery and the surrounding prairie. Bison graze preferentially on the disturbed areas outside the central core of unpalatable forbs. Although stress would seem extreme given a variety of interacting factors including higher insolation, mound building, trampling, higher temperatures, and evaporative stresses, aboveground production on the towns is indistinguishable from the surrounding areas, and the towns appear much greener than the surrounding prairies because standing dead tissues do not accumulate in the short canopies as they do in the taller surrounding grassland.

In addition to the higher levels of aboveground herbivory on the towns, belowground herbivory is also higher. Nitrogen content of aboveground plant tissues is consistently higher on the towns, so that net nitrogen yield can be more than 50% higher than in the undisturbed prairie. The plants, either as species distinct from those in the surrounding area or as dwarf growth forms of species also common off the towns, therefore have higher nutrient-accumulating abilities and, very likely, greater photosynthetic rates per unit of aboveground tissues (Detling *et al.*, 1979), traits that allow them to maintain adequate carbon and nutrient balances in the face of multiple, severe stresses.

Thus, an important factor in herbaceous plant growth in dryland ecosystems can be the effect of grazers on both plant physiological processes and ecosystem processes (McNaughton, 1985; McNaughton *et al.*, 1988). In general, grazers accelerate physiological processes in individual plants and promote more rapid rates of nutrient cycling and tissue flux through the canopy.

VIII. Interactions among Water, Nutrients, and Defoliation in the Laboratory

An experiment with an African perennial grass *(Eustachys paspaloides)* that was designed to examine the interaction of drought and defoliation indicated that water stress developed much more slowly in defoliated plants than in controls (Toft *et al.*, 1987). The rate of stress development, as measured by predawn leaf water potential, reflected much more rapid depletion of soil water by unclipped plants (Fig. 3). Clipped plants were able to maintain turgor pressures sufficient to sustain leaf elongation for a week longer than the unclipped plants. Osmotic adjustment from solute accumulation occurred in all water-stressed plants, and clipping effects or interactions between clipping and water stress were evident for osmotic potential at full turgor, water potential when zero turgor was first reached, turgid weight divided by dry weight, and bulk modulus of elasticity (a measure of cell wall rigidity). The general effect of defoliation was to reduce the rate at which water stress developed and modify the compensating osmotic adjustments.

Experiments on short *(Kyllinga nervosa)*, medium-height *(Themeda triandra)*, and tall *(Hyperthelia dissoluta)* perennial graminoids from Africa were explicitly designed to examine interactions among defoliation, nitrogen stress, and water stress (McNaughton *et al.*, 1983; Coughenour *et al.*, 1985a,b). Plants were grown in comparable 2^4 factorial experiments where

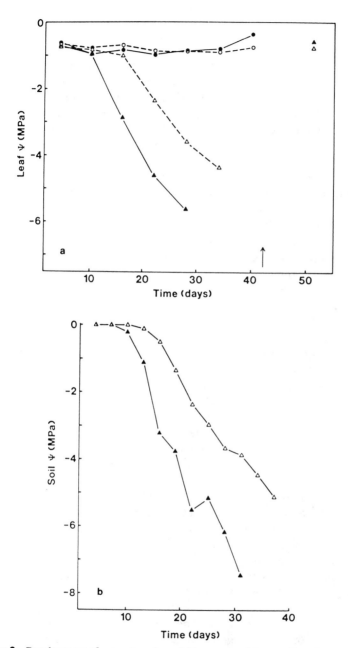

Figure 3 Development of water stress in an African perennial grass, *Eustachys paspaloides*, subject to drought and defoliation. Open symbols represent clipped plants; filled symbols, unclipped plants. (a) Time course of mean predawn water potential in leaves of well-watered plants (circles) and water-stressed plants (triangles). (b) Time course of mean soil water potential at 100 mm depth in pots containing water-stressed plants that were either clipped or unclipped. [Reprinted from Toft *et al.* (1987) with permission from CSIRO Australia.]

the variables were clipping height, clipping frequency, watering frequency, and nitrogen supply. The short-statured species was able to compensate fully for defoliation under most conditions, even overcompensating to produce a greater terminal residual biomass when clipped at a moderate height (McNaughton *et al.*, 1983). The principal mechanisms of compensation were increased activation of intercalary meristems and greater leaf elongation rates. This species was also highly sensitive to nitrogen stress, however, which was the principal environmental factor limiting most components of growth and yield. In contrast, the medium-height species was highly sensitive to clipping height, and yield was inhibited strongly by defoliation (Coughenour *et al.*, 1985b). Leaf elongation rate was stimulated by defoliation, but tillering was not affected, so there was no mechanism for increasing leaf area below the clipping height. The tall species was less severely inhibited by clipping than the medium-height grass, although it was unable to compensate to the same degree as the short species (Coughenour *et al.*, 1985a). The tall species exhibited the greatest variety of environmental controls, with some plant traits affected most by clipping, others by nitrogen stress, and others by water stress.

A principal result of these factorial experiments was that formal statistical interaction among environmental factors was relatively rare. Of 28 traits measured with statistically significant effects for the short-statured species, the major statistical effect was a main effect in 25 cases; factor interaction accounted for only 3 cases. Of 28 traits of the midheight species, none of the most significant statistical effects was due to factor interaction. Of 33 traits of the tall species, only 2 of the most significant statistical effects were due to factor interaction. Therefore, main-factor effects dominated the experiments, with higher-order interactions taking secondary importance.

Nevertheless, the conclusion that the responses of herbaceous perennials to multiple stresses are noninteractive, dominated by main effects, must be tempered by judicious evaluation of the data. For example, tillering and enhanced rates of leaf elongation were the principal responses of the short grass species that allowed it to cope with defoliation. But both of these responses had nitrogen level as the principal environmental factor, explaining 50% and 43%, respectively, of the variances in the two traits. For tillering, the second most important variable was clipping height, explaining 9% of the variance, and the next most important was an interaction between height and nitrogen, explaining 5% of the variance. Therefore, nitrogen regulated the ability of this species to compensate for defoliation. Although factor interaction in the statistical sense may be of minor importance, it is, of course, ecologically important that nutrient level regulates a plant's ability to compensate for defoliation. Similar principles emerge on close examination of the data for the other two species.

IX. Dryland Ecosystems: Are They Water Limited?

Three sets of field experiments and an agronomic pattern suggest that water may be less important as a direct limiter than as an interactive limiter of dryland ecosystems. In the early 1970s, the Grassland Biome Program of the U.S. International Biological Program applied a 2^2 factorial experiment of supplemental irrigation and nitrogen fertilization to native rangeland in Colorado (Lauenroth *et al.*, 1978; Dodd and Lauenroth, 1979). Mean annual precipitation in the location is 311 mm, about 70% of which falls during the frost-free period. Full supplementation lasted five years, with irrigation designed to maintain soil water potential less negative than -0.08 MPa and nitrogen added to maintain a difference of at least 50 kg ha^{-1} of soil mineral nitrogen. The rank of aboveground peak biomass was control $<$ nitrogen-fertilized $<$ irrigated $\ll$ fertilized and irrigated.

Nitrogen supplementation alone increased production by about two-thirds, a substantial stimulation. But irrigation increased yield almost threefold, and fertilization plus irrigation resulted in an eightfold increase. The relative abundance of various plant groups was also affected by the treatments (Table I). C_4 grasses increased substantially in response to the water treatment, and their importance doubled on the dual-supplementation plots. C_3 grasses were nowhere important; neither were cool-season forbs. Warm-season forbs increased from insignificant contributors to a maximum on dual supplementation, and semishrubs declined on all treatments, although not dramatically. Succulents diminished to minor community constituents on both supplementations involving irrigation. Much of this reduction was indirect, however, because of an outbreak of insect herbivory on irrigated plots.

Table I Effects of Water and Nitrogen Supplementation on the Balance of Growth Forms in a Semiarid Grassland

Growth form	Control	Irrigated	Fertilized	Fertilized and irrigated
Warm-season grass	23[a]	38	26	46
Cool-season grass	2	3	2	Nil
Warm-season forbs	Nil	11	1	18
Cool-season forbs	1	3	2	1
Semishrubs	47	39	38	31
Succulents	27	6	31	2

[a] Expressed as the percent of total peak biomass (g m^{-2}) after four years of treatment. [Calculated from Lauenroth *et al.* (1978).]

Trends in species composition over the course of the experiment suggest that long-term nitrogen supplementation would likely have little effect on species composition, whereas continued water supplementation would lead to replacement of the present community by legumes, and dual supplementation would result in displacement of the native flora by exotic, invading species (Dodd and Lauenroth, 1979). These conclusions suggest that the native herbaceous perennials dominating this dryland ecosystem —largely warm-season grasses and semishrubs—are nitrogen limited but can nonetheless cope effectively with the system's inherent water stress. Potentially invading exotics, in contrast, are prevented from colonizing by that water stress. Thus, nutrient supplementation would appear to increase the yield of the system without affecting community composition, whereas water supplementation, with or without nutrients, would convert the plant community into a totally different one.

Fertilization experiments in Sahelian grasslands receiving 40–60 cm of annual precipitation also indicate that the region's C_4 grasslands are nitrogen deficient (Penning de Vries et al., 1980). Fertilization of five different grassland types on soils ranging from sands to heavy clays resulted in substantial yield increments, sometimes up to application levels of 200 kg nitrogen ha^{-1}. There was some evidence of phosphorus limitation, although phosphorus was less important than nitrogen as a limiting factor.

Experimental studies of the decomposition of shrub litter in desert ecosystems suggest that the process is not limited by water (Whitford et al., 1986). When plots were irrigated by sprinkling to provide 25 mm supplementation every four weeks or 6 mm per week, the researchers found no evidence that added water, or the pattern of supplementation, affected litter decomposition. Whitford et al. (1986: p. 512) concluded that ". . . the conventional wisdom linking decomposition to rainfall in deserts is wrong."

Finally, agronomic evidence points to a geographically widespread iron deficiency in plants of the U.S. Great Plains (Clark, 1982). The neutral to alkaline soils, often derived from calcareous parent materials, produce evident iron deficiency, which increases from cooler to warmer climates. Whether such limitations affect native plants is open to question, since I know of no studies designed to examine that possibility. If the native plants tolerate this nutritional stress, understanding the mechanisms alleviating it would help clarify how herbaceous perennials dominate dryland ecosystems and be of practical importance to agriculture.

X. Conclusions

There can be little doubt that the traditional view of dryland ecosystems dominated by herbaceous perennials is correct: water shortage and all its attendant effects are important determinants of those systems. It is not as clear, however, whether species composition or balances among growth forms, productivity, and plant growth rates are directly limited by water stress in either the short or long term. In evolutionary time, the species dominating dryland ecosystems have evolved a whole suite of traits — ranging from osmotic adjustment to adjustments of shoot : root ratios — that allows them to circumvent many of the direct effects of water stress that plants adapted to other environments confront when water is lacking. Even during extended severe drought, some herbaceous perennials can survive several years in a dormant state and then reestablish themselves rapidly through vegetative growth and massive seed production when drought is broken. Major changes in species composition that may have occurred during the drought may thus be ephemeral unless some additional stress, such as overgrazing, has been superimposed (Weaver and Albertson, 1944). A better picture of the responses of dryland herbaceous perennials to stress, therefore, would come from focusing on the interactions between water stress and other factors, rather than exclusively on water stress. Nutrients, grazing, and salinity seem to be particularly important interactive factors whose study in concert would significantly increase our understanding of these widespread and economically important ecosystems.

References

Ashton, F. M. (1956). Effects of a series of cycles of alternating low and high soil water contents on the rate of apparent photosynthesis in sugar cane. *Plant Physiol.* 31: 266–274.

Barnes, P. W. (1985). Adaptation to water stress in the big bluestem–sand bluestem complex. *Ecology* 66: 1908–1920.

Belesky, D. P., Wilkinson, S. R., and Pallas, J. E., Jr. (1982). Responses of four tall fescue cultivars grown at two nitrogen levels to soil water availability. *Crop Sci.* 22: 93–97.

Bernhard-Reversat, R. (1982). Biogeochemical cycle of nitrogen in a semiarid savanna. *Oikos* 38: 321–332.

Birch, H. F. (1958a). Pattern of humus decomposition in East African soils. *Nature* 181: 788.

Birch, H. F. (1958b). Further aspects of humus decomposition. *Nature* 182: 1172.

Bokhari, U. G., and Trent, J. D. (1985). Proline concentrations in water-stressed grasses. *J. Range Manage.* 38: 37–38.

Boyer, J. S. (1968). Relationship of water potential to growth of leaves. *Plant Physiol.* 43: 1056–1062.

Bradford, K. J., and Hsiao, T. C. (1982). Physiological responses to moderate water stress. *In* "Encyclopedia of Plant Physiology, New Series" (O. L. Lange, P. S. Nobel, C. B. Osmond, and H. Ziegler, eds.), Vol. 12B, pp. 263–324. Springer-Verlag, Berlin.

Brady, N. C. (1974). "The Nature and Properties of Soils." Macmillan, New York.

Bunce, J. A. (1983). Differential sensitivity to humidity of daily photosynthesis in the field in C_3 and C_4 species. *Oecologia* 57: 262–265.

Chapin, F. S., III, Bloom, A. J., Field, C. B., and Waring, R. H. (1987). Plant responses to multiple environmental factors. *Bioscience* 37: 49–57.

Chung, H., and Trlica, M. J. (1980). ^{14}C distribution and utilization in blue grama as affected by temperature, water potential, and defoliation regimes. *Oecologia* 47: 190–195.

Clark, R. B. (1982). Iron deficiency in plants grown in the Great Plains of the U.S. *J. Plant Nutr.* 5: 251–268.

Clements, F. E. (1907). "Plant Physiology and Ecology." Holt, New York.

Coughenour, M. B., McNaughton, S. J., and Wallace, L. L. (1985a). Responses of an African tall-grass (*Hyparrhenia filipendula* stapf.) to defoliation and limitations of water and nitrogen. *Oecologia* 68: 80–86.

Coughenour, M. B., McNaughton, S. J., and Wallace, L. L. (1985b). Responses of an African graminoid (*Themeda triandra* Forsk.) to frequent defoliation, nitrogen, and water: A limit of adaptation to herbivory. *Oecologia* 68: 105–110.

Cowan, I. R. (1977). Stomatal behaviour and environment. *Adv. Bot. Res.* 4: 117–228.

Davies, W. J., Metcalfe, J., Lodge, T. A., and daCosta, A. R. (1986). Plant growth substances and the regulation of growth under drought. *Aust. J. Plant Physiol.* 13: 105–125.

Deregibus, V. A., Sanchez, R. A., Casal, J. J., and Trlica, M. J. (1985). Tillering responses to enrichment of red light beneath the canopy in a humid natural grassland. *J. Appl. Ecol.* 22: 199–206.

Detling, J. K. (1988). Grassland and savannas: Regulation of energy flow and nutrient cycline by herbivores. *In* "Concepts of Ecosystem Ecology" (L. R. Pomeroy and J. J. Alberts, eds.), pp. 131–148. Springer-Verlag, New York.

Detling, J. K., Dyer, M. I., and Winn, D. T. (1979). Net photosynthesis, root respiration, and regrowth of *Bouteloua gracilis* following simulated grazing. *Oecologia* 41: 127–134.

Dodd, J. L., and Lauenroth, W. K. (1979). Analysis of the response of a grassland to stress. *In* "Perspectives in Grassland Ecology" (N. French, ed.), pp. 43–58. Springer-Verlag, Berlin.

Downton, W. J. S., Berry, J. A., and Seemann, J. R. (1984). Tolerance of photosynthesis to high temperature in desert plants. *Plant Physiol.* 74: 786–790.

Dreier, W. (1983). Le teneur en proline et la résistance des plantes aux sels. *Biol. Plant. (Prague)* 25: 81–87.

Farquhar, G. D. (1978). Feedforward responses of stomata to humidity. *Aust. J. Plant Physiol.* 5: 787–800.

Gates, D. M. (1968). Transpiration and leaf temperature. *Annu. Rev. Plant Physiol.* 18: 211–238.

Gates, D. M., Alderfer, R., and Taylor, E. (1968). Leaf temperatures of desert plants. *Science* 159: 994–995.

Greacen, E. L., and Oh, J. S. (1972). Physics of root growth. *Nature* 235: 24–25.

Hanson, A. D., and Hitz, W. D. (1983). Whole-plant response to water deficits: Water deficits and the nitrogen economy. *In* "Limitations to Efficient Water Use in Crop Production," pp. 331–343. ASA-CSSA-SSSA, Madison, Wisconsin.

Hsiao, T. C. (1973). Plant responses to water stress. *Annu. Rev. Plant Physiol.* 24: 519–570.

Kemp, P. R., and Williams, G. J., III. (1980). Niche displacement between *Agropyron smithii* and *Bouteloua gracilis:* A study of the role of differing photosynthetic pathways in the shortgrass prairie ecosystem. *Ecology* 61: 846–858.

Khan, S. M., and Wilson, A. M. (1984). Nonstructural carbohydrate and dehydration tolerance of blue grama seedlings. *Agron. J.* 76: 637–642.

Knapp, A. K. (1985). Effect of fire and drought on the ecophysiology of *Andropogon gerrardi* and *Panicum virgatum* in a tallgrass prairie. *Ecology* 66: 1309–1320.

Knapp, A. K., and Seastedt, T. R. (1986). Detritus accumulation limits productivity of tallgrass prairie. *Bioscience* 36: 662–668.

Lange, O. L., Kappen, L., and Schulze, E.-D. (1976). "Water and Plant Life." Springer-Verlag, Berlin.

Lange, O. L., Losch, R., Schulze, E.-D., and Kappen, L. (1971). Responses of stomata to changes in humidity. *Planta* 100: 76–86.

Lauenroth, W. K., Dodd, J. L., and Sims, P. L. (1978). The effects of water- and nitrogen-induced stresses on plant community structure in a semiarid grassland. *Oecologia* 36: 211–222.

Luldow, M. M. (1980). Stress physiology of tropical pasture plants. *Trop. Grass.* 14: 136–145.

Ludlow, M. M., Fisher, M. J., and Wilson, J. R. (1985). Stomatal adjustments to water deficits in three tropical grasses and a tropical legume grown in controlled conditions and in the field. *Aust. J. Plant Physiol.* 12: 131–149.

McCree, K. J. (1986). Whole-plant carbon balance during osmotic adjustment to drought and salinity stress. *Aust. J. Plant Physiol.* 13: 33–43.

McNaughton, S. J. (1983a). Physiological and ecological implications of herbivory. *In* "Encyclopedia of Plant Physiology, New Series" (O. L. Lange, P. S. Nobel, C. B. Osmond, and H. Ziegler, eds.), Vol. 12C, pp. 657–678. Springer-Verlag, Berlin.

McNaughton, S. J. (1983b). Compensatory plant growth as a response to herbivory. *Oikos* 40: 329–336.

McNaughton, S. J. (1984). Grazing lawns: Animals in herds, plant form, and coevolution. *Am. Nat.* 124: 863–886.

McNaughton, S. J. (1985). Ecology of a grazing ecosystem: The Serengeti. *Ecol. Monogr.* 55: 259–294.

McNaughton, S. J., Wallace, L. L., and Coughenour, M. B. (1983). Plant adaptation in an ecosystem context: Effects of defoliation, nitrogen, and water on growth of an African C_4 sedge. *Ecology* 64: 307–318.

McNaughton, S. J., Ruess, R. W., and Seagle, S. W. (1988). Large mammals and process dynamics in African ecosystems. *Bioscience* 38: 794–800.

Meyer, R. F., and Boyer, J. S. (1972). Sensitivity of cell division and cell elongation to low water potentials in soybean hypocotyls. *Planta* 108: 77–87.

Misra, G., and Singh, K. P. (1981). Total nonstructural carbohydrates of one temperate and two tropical grasses under varying clipping and soil moisture regimes. *Agro-Ecosystems* 7: 213–223.

Molyneux, D. E., and Davies, W. J. (1983). Rooting pattern and water relations of three pasture grasses in drying soil. *Oecologia* 58: 220–224.

Monson, R. K., and Williams, G. J., III. (1982). A correlation between photosynthetic temperature adaptation and seasonal phenology patterns in the shortgrass prairie. *Oecologia* 54: 58–62.

Morgan, J. M. (1984). Osmoregulation and water stress in higher plants. *Annu. Rev. Plant Physiol.* 35: 299–319.

Munns, R., Greenway, H., and Kirst, G. O. (1982). Halotolerant eukaryotes. *In* "Encyclopedia of Plant Physiology, New Series" (O. L. Lange, P. S. Nobel, C. B. Osmond, and H. Ziegler, eds.), Vol. 12C, pp. 59–135. Springer-Verlag, Berlin.

Nobel, P. S., and Jordan, P. W. (1983). Transpiration stream of desert species: Resistances and capacitances for a C_3, a C_4, and a CAM plant. *J. Exp. Bot.* 147: 1379–1391.

Parker, J. (1968). Drought-resistance mechanisms. *In* "Water Deficits and Plant Growth" (T. T. Kozlowski, ed.), pp. 195–234. Academic Press, New York.

Penning de Vries, F. W. T., Krul, J. M., and van Keulen, H. (1980). Productivity in Sahelian rangelands in relation to the availability of nitrogen and phosphorus from the soil. *In* "Nitrogen Cycling in West African Ecosystems" (T. Rosswall, ed.), pp. 95–113. Royal Swedish Acad. Sci., Stockholm.

Potvin, M. A., and Werner, P. A. (1983). Water use physiologies of co-occurring goldenrods (*Solidago juncea* and *S. canadensis*): implications for natural distributions. *Oecologia* 56: 148–152.

Raschke, K. (1976). How stomata resolve the dilemma of opposing priorities. *Philos. Trans. R. Soc. Lond. B Biol. Sci.* 273: 551–560.

Richards, J. H., and Caldwell, M. M. (1987). Hydraulic lift: Substantial nocturnal water transport between soil layers by *Artemisia tridentata* roots. *Oecologia* 73: 486–489.

Schulze, E.-D. (1982). Plant life forms and their carbon, water, and nutrient relations. *In* "Encyclopedia of Plant Physiology, New Series" (O. L. Lange, P. S. Nobel, C. B. Osmond, and H. Ziegler, eds.), Vol. 12B, pp. 615–676. Springer-Verlag, Berlin.

Schulze, E.-D. (1986). Carbon dioxide and water vapor exchange in response to drought in the atmosphere and in the soil. *Annu. Rev. Plant Physiol.* 37: 247–274.

Schulze, E.-D., Robichaux, R. H., Grace, J., Rundel, P. W., and Ehleringer, J. R. (1987). Plant water balance. *Bioscience* 37: 30–37.

Smith, W. K. (1978). Temperatures of desert plants: Another perspective on the adaptability of leaf size. *Science* 201: 614–616.

Squire, G. R., Gregory, P. J., Monteith, J. L., Russell, M. B., and Singh, P. (1984). Control of water use by pearl millet *(Pennisetum typhoides). Exp. Agric.* 20: 135–149.

Stewart, C. R., Morris, C. J., and Thompson, J. F. (1966). Changes in amino acid content of excised leaves during incubation. II. Role of sugar in the accumulation of proline in wilted leaves. *Plant Physiol.* 41: 1585–1590.

Stoddart, L. A., Smith, A. D., and Box, T. W. (1975). "Range Management." McGraw-Hill, New York.

Toft, N. L., McNaughton, S. J., and Geordiadis, N. J. (1987). Effects of water stress and simulated grazing on leaf elongation and water relations of an East African grass, *Eustachys paspaloides. Aust. J. Plant Physiol.* 14: 211–216.

Trlica, M. J., and Singh, J. S. (1979). Translocation of assimilates and creation, distribution, and utilization of reserves. *In* "Arid-Land Ecosystems: Structure, Functioning, and Management" (R. A. Perry and D. W. Goodall, eds.), pp. 537–571. Cambridge Univ. Press, London.

Weaver, J. E. (1954). "North American Prairie." Johnsen Publ., Lincoln, Nebraska.

Weaver, J. E., and Albertson, F. W. (1944). Nature and degree of recovery of grassland from the great drought of 1933 to 1940. *Ecol. Monogr.* 14: 393–479.

Whicker, A. D., and Detling, J. K. (1988a). Ecological consequences of prairie dog disturbances. *Bioscience* 38: 778–785.

Whicker, A. D., and Detling, J. K. (1988b). Modification of vegetation structure and ecosystem processes by North American grassland mammals. *In* "Plant Form and Vegetation Structure" (M. J. A. Werger, P. J. M. van der Aart, H. J. During, and J. T. A. Verhoeven, eds.), pp. 301–316. SPB Academic Publ., The Hague.

Whitford, W. G., Steinberger, Y., MacKay, W., Parker, L. W., Freckman, D., Wallwork, J. A., and Weems, D. (1986). Rainfall and decomposition in the Chihuahuan desert. *Oecologia* 68: 512–515.

Williams, R. E., Allred, B. W., DeNio, R. M., and Paulsen, H. E. (1968). Conservation, development, and use of the world's rangelands. *J. Range Manage.* 21: 355–360.

Wilson, J. R., and Ludlow, M. M. (1983a). Time trends for change in osmotic adjustment and water relations of leaves of *Cenchrus ciliaris* during and after water stress. *Aust. J. Plant Physiol.* 10: 15–24.

Wilson, J. R., and Ludlow, M. M. (1983b). Time trends of solute accumulation and the influence of potassium fertilizer on osmotic adjustment of water-stressed leaves of three tropical grasses. *Aust. J. Plant Physiol.* 10: 523–537.

World Resources Institute and International Institute for Environment and Development. (1988). "World Resources 1988–89." Basic Books, New York.

Yancey, P. H., Clark, M. E., Hand, S. C., Bowlus, R. D., and Somero, G. N. (1982). Living with water stress: Evolution of osmolyte systems. *Science* 217: 1214–1222.

15

Interactive Role of Stresses on Structure and Function in Aquatic Plants

Jon E. Keeley

I. Introduction

Any environmental parameter that reduces plant growth can be termed a stress factor. In most natural environments, regardless of their productivity, plants probably do not attain maximum potential growth and therefore must be considered stressed. Stress and plant productivity are interrelated such that as one stress (e.g., water stress) is ameliorated and productivity increases, other stresses (e.g., shading) increase. In general, low-productivity (oligotrophic) environments are dominated by abiotic stresses, whereas in high-productivity (eutrophic) environments, biotic stresses dominate.

This chapter examines how stresses interact to affect the growth forms, physiology, and productivity of aquatic macrophytes in oligotrophic and eutrophic environments. These two environments were selected because they represent the extremes along a gradient of productivity. The focus is first to identify the environmental factors that are most stressful in these two habitats and then to examine how these stresses affect plant structure and function over differing time scales of exposure. These considerations can help predict how multiple stresses have interacted to influence evolutionary change in structure and function and how future changes in stresses may alter community composition.

II. Stresses in the Aquatic Environment

Aquatic habitats include freshwater seasonal pools, marshes, rivers, and lakes as well as saline environments such as interior salt lakes, estuaries, and tidal and marine areas (Sculthorpe, 1967). Both microscopic and macroscopic plants live within these habitats. Macrophytes come in a variety of growth forms, including free-floating plants (e.g., duckweed, *Lemma*) and rooted plants whose foliage emerges from the water, floats on the surface, or is entirely submerged.

Stresses faced by submerged aquatic plants in freshwater lakes include (1) limited availability of inorganic carbon; (2) limited availability of other nutrients, e.g., nitrogen and phosphorus; (3) photosynthetic inhibition due to elevated oxygen levels; (4) limited light availability; and (5) microsite instability.

For several reasons, aquatic concentrations of inorganic carbon, although usually higher than atmospheric concentrations, are often limiting to photosynthesis: carbon dioxide is not readily soluble in water (particularly at higher temperatures), its rate of diffusion in water is 10^4 times slower than in air, and not all forms of inorganic carbon (e.g., bicarbonate) are used by all aquatic plants (Raven, 1970; Wetzel and Grace, 1983).

Other nutrients, in particular nitrogen and phosphorous, may limit growth. All rooted aquatic plants obtain these nutrients from the sediment; some species may also obtain nutrients from the water by means of foliar absorption (Hutchinson, 1975; Raven, 1981).

For submerged aquatic macrophytes, it is unlikely that too little oxygen would inhibit respiratory processes because oxygen is produced during photosynthesis. Moreover, because of its low solubility in water and water's high diffusive resistance, oxygen tends to accumulate in intercellular spaces within photosynthetic tissues, from which it diffuses freely to nonphotosynthetic tissues such as roots (Raven *et al.*, 1988). When photosynthesis is rapid, however, elevated oxygen levels in photosynthetic tissues may inhibit photosynthesis via effects on photorespiration (Hough, 1974; Samishi, 1975).

Light levels are potentially limiting to the growth of aquatic macrophytes (Kirk, 1983). Water absorbs light in increasing amounts with greater depth and as turbidity from organic and inorganic suspended particles increases. In addition, because water is buoyant, aquatic macrophytes may produce structures with relatively little support, and thus the specific weight of foliage (surface area to dry weight) is typically high. Rapid growth can produce dense mats of vegetation that block out much of the light for plants beneath (Westlake, 1967; Van *et al.*, 1976).

Some stress factors are unrelated to site productivity. Aquatic environments may be stressful because the water is turbulent and the site is unstable or because the sediment is too shallow or poorly consolidated for adequate anchorage. Some sites may be unstable because of shifting sediments, which could potentially bury some plants (Hutchinson, 1975).

The importance of particular stress factors differs between oligotrophic and eutrophic lakes (Table I). Oligotrophic lakes are typically, although not always, acidic, soft-water lakes with low conductivity. Inorganic carbon, as well as other inorganic nutrients such as nitrogen and phosphorous, is low. These abiotic stresses account for the very low biomasses characteristic of oligotrophic sites. Eutrophic lakes are commonly basic, hard-water lakes with high conductivity and high levels of most inorganic nutrients. Biomass is very high, and thus the potential for competitive interference, or biotic stress, is also high.

Numerous studies have attempted to relate the trophic status of lakes, certain water chemistry characteristics, or both with species distributions (e.g., Moyle, 1945; Sculthorpe, 1967; Seddon, 1972; Hutchinson, 1975; Hellquist, 1980). Macrophytes typically cited as indicators of oligotrophic environments include species of the genus *Isoetes* (Isoetaceae), which has hundreds of species worldwide, and *Lobelia dortmanna* L. (Campanulaceae). Other macrophytes commonly found under oligotrophic conditions include species of *Littorella* (Plantaginaceae) and *Eriocaulon* (Eriocaulaceae).

Table I   Comparison of Oligotrophic and Eutrophic Lakes[a]

Parameter	Oligotrophic	Eutrophic
pH	5–7	7–9
Inorganic carbon (mol m^{-3})	0.01–0.25	2.00–5.00
Conductivity (μmho cm^{-1})	<50	>200
Inorganic N (gm m^{-3} sediment)	1–3	10–20
Plant biomass (gm dry weight m^{-2})	0.1–50.0	200–800

[a] Summarized from the literature cited in the text.

Eutrophic indicator species include *Myriophyllum spicatum* L. (Haloraga-ceae) and *Potamogeton pectinatus* L. (Potamogetonaceae), but other species in these genera, as well as *Hydrilla verticillata* Royle (Hydrocharitaceae) and *Ceratophyllum demersum* L. (Ceratophyllaceae), are also found in eutrophic habitats.

III. Effect of Stresses on Macrophyte Structure and Function

Plant responses to stress need to be viewed in the context of time. Brief exposure to stress (e.g., for seconds or minutes) may affect biochemical processes; longer exposures (e.g., weeks) may induce morphological and anatomical changes. Even longer time frames may cause localized extinc-tion of certain species and, thus, community changes. Over evolutionary time, stresses may select for structural and physiological adaptations in some species that optimize growth in the presence of conditions that would be stressful for other species. Thus limited levels of a resource may not be stressful (i.e., potential growth is not limited) to species adapted to such conditions (see also Chapters 7, 14). Conversely, an environment in which a resource is relatively abundant could nonetheless be stressful to species adapted for very rapid growth.

This chapter contrasts the structural and functional attributes of species adapted to oligotrophic and eutrophic sites and relates these to the stresses imposed by their environments. Similar comparisons have recently been made by other authors (e.g., Bowes, 1985; Boston, 1986; Chambers, 1987; Raven *et al.*, 1988).

Table II Morphological and Anatomical Comparison of Submerged Aquatic Macrophytes from Oligotrophic and Eutrophic Lakes[a]

Plant Character	Oligotrophic	Eutrophic
Indicator species	*Isoetes* spp.	*Myriophyllum spicatum*
	Lobelia dortmanna	*Potamogeton pectinatus*
Plant height (m)	0.1–0.3	1.0–2.5
Growth form	Rosette	Caulescent
Leaves		
Shape	Mostly cylindrical	Finely dissected
Surface:volume ratio	Low	High
Air space	Extensive	Limited
Chloroplast concentration	Cells adjacent to airspaces	Epidermal cells
Cuticle	Thick–moderate	Thin
Longevity	Evergreen	Winter deciduous
Root:shoot ratio	0.5–3.5	0.1–0.4

[a] Summarized from the literature cited in the text.

Structural characteristics of species such as *L. dortmanna* and *Isoetes* spp. from oligotrophic lakes differ from those of species such as *P. pectinatus* and *M. spicatum* from eutrophic habitats (Table II). The former are small rosette-forming macrophytes (Fig. 1), whereas the latter are significantly larger caulescent macrophytes (Fig. 2). One noteworthy difference between species indigenous to these two habitats is the extensive root system typical of oligotrophic species (Raven *et al.*, 1988). Leaf structure is also quite different. Oligotrophic species have thicker, bulkier leaves that are often cutinized to various degrees, compared with the much smaller, thinner, often finely dissected and weakly cutinized leaves of eutrophic species. Oligotrophic species typically have much more internal air space, contributing to their smaller leaf surface : volume ratio. Eutrophic species have chloroplasts concentrated in the epidermal cells, whereas *L. dortmanna* and species of *Isoetes* have chloroplasts distributed throughout the leaf and often concentrated around lacunae. Oligotrophic species are commonly evergreen (but cf. Keeley *et al.*, 1983), whereas eutrophic species are usually deciduous in winter.

Physiological differences are equally pronounced (Table III). Maximum photosynthetic rates, and consequently annual productivity, are substantially higher among species of eutrophic environments (Sand-Jensen and Sondergaard, 1978; Boston, 1986). Modal differences are apparent in

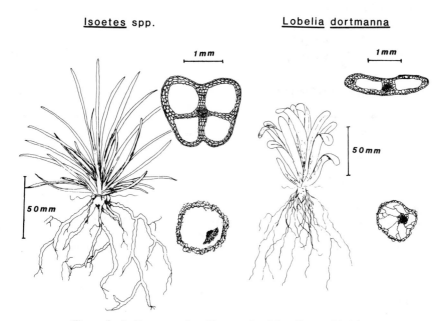

Figure 1   Indicator species of low-productivity oligotrophic lakes.

Myriophyllum spicatum   Potamogeton pectinatus

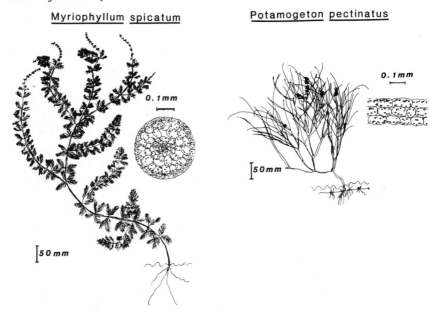

Figure 2   Indicator species of high-productivity eutrophic lakes.

mechanisms of carbon uptake and fixation. Oligotrophic species are seldom capable of bicarbonate uptake, whereas eutrophic species often are (Lucas and Berry, 1985). CAM photosynthesis is well developed in certain oligotrophic species but absent in most eutrophic species (Keeley and Morton, 1982); where present (e.g., *Hydrilla verticillata;* Holaday and

Table III   Physiological Comparison of Submerged Aquatic Macrophytes from Oligotrophic and Eutrophic Lakes[a]

Plant Character	Oligotrophic	Eutrophic
Annual productivity (gm dry weight m^{-2} $year^{-1}$)	5–100	400–1000
Maximum photosynthetic rates (mg C gm^{-1} DW hr^{-1})	0.5–2.0	3–15
Bicarbonate uptake	None–little	May be substantial
CAM (dark CO_2 fixation)	Well developed in some species	Weak in some species
C_4 (light CO_2 fixation)	Weak in some species	Well developed in some species
Photorespiration	Present	Greatly reduced in some species
Carbon uptake from the sediment (% of total fixed)	40–100	<1–10

[a] Summarized from the literature cited in the text.

Bowes, 1980), it is quantitatively less important than in oligotrophic species. Fixation of carbon in the light via the C_4 enzyme PEP carboxylase is well developed in some eutrophic species (Bowes and Salvucci, 1984). Carbon uptake from the sediment is important in all oligotrophic species but quantitatively far less important in all eutrophic species (Boston *et al.*, 1987; Raven *et al.*, 1988).

IV. Structural Effects on Photosynthetic Performance

A useful model for evaluating how structure and function affect photosynthetic performance is an electrical circuit analogue of the resistances encountered by carbon dioxide along the path from the environment to the chloroplast (Nobel and Walker, 1985). For submerged aquatic macrophytes, stomatal resistance can be ignored, so the major limitations to carbon gain are the resistances of the boundary layer, cuticule, and mesophyll.

Because of the viscosity of water, boundary layers may be two to three orders of magnitude greater in submerged macrophytes than in terrestrial plants, creating a resistance that plays a dominant role in limiting carbon uptake (e.g., Black *et al.*, 1981; Raven, 1984). Although techniques are available for calculating boundary layers (e.g., Raven *et al.*, 1982), few data are available for comparing the oligotrophic and eutrophic species considered here. Several factors suggest, however, that boundary-layer effects are likely to play a greater role in oligotrophic species. These macrophytes have larger, bulkier leaves that are situated near the sediment. Eutrophic species possess thin, finely dissected foliage that is held nearer the water surface, where turbulence is greater. These considerations are complicated by differences in photosynthetic rate between oligotrophic and eutrophic species, which deplete carbon dioxide around the leaf to a different degree.

Cuticular resistance, which is closely correlated with cuticle thickness, is also greater in oligotrophic species. *Lobelia dortmanna*, for example, has a notably thick cuticle that effectively eliminates gas exchange across the leaves (Sand-Jensen and Prahl, 1982; Richardson *et al.*, 1984). *Isoetes* species and *Eriocaulon decangulare* L. also show substantial cuticular resistance (Raven *et al.*, 1988).

Several structural characteristics suggest that mesophyll resistance is also likely to be greater in oligotrophic macrophytes. The ratio of exposed mesophyll surface area to intercellular air space per unit leaf area (A^{mes}/A) is a determining factor in mesophyll resistance (Nobel, 1983). Comparative data on A^{mes}/A for aquatic macrophytes are lacking, but oligotrophic species have leaves with a lower surface : volume ratio and with relatively larger mesophyll cells than do eutrophic species such as *M. spicatum* and *P. pectinatus* (see Figures 1 and 2). One would therefore expect A^{mes}/A to be lower in the leaves of oligotrophic species, suggesting higher mesophyll resistance in these leaves relative to those of eutrophic species.

In addition, many eutrophic macrophytes have leaves with only two or three cell layers and with chloroplasts concentrated in the epidermal layer, factors that shorten the path of carbon uptake and greatly reduce mesophyll resistance. These considerations suggest that, on the whole, leaf resistance to carbon uptake is likely to be greater in oligotrophic species.

V. Interactive Stresses, Structure, Function, and Plant Productivity

Figure 3 presents a schematic model of how stresses in oligotrophic environments may interact in affecting the functional attributes of macrophytes. Low carbon availability, coupled with low nitrogen and phosphorus availability, contributes to low photosynthetic rates and, consequently, low primary production. Low photosynthetic rates generate less oxygen, and photorespiratory inhibition by oxygen is thus likely to be lower than in eutrophic environments. Low productivity will generate less shading and minimize limits imposed by light on growth.

Environments with severe carbon limitation might be expected to select

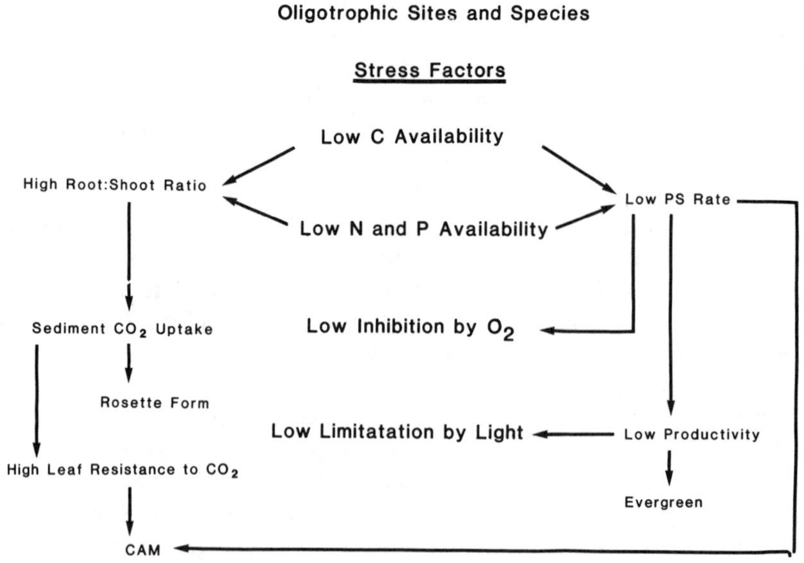

Figure 3 Interaction of stresses and plant structure, function, and productivity in oligotrophic lake environments. PS, photosynthesis.

for leaf structural attributes that reduce resistances to carbon uptake, a prediction not upheld for oligotrophic macrophytes (as discussed above). The explanation is simply that these species obtain much of their carbon from the sediment, a source much richer in carbon than the water column, despite its greater diffusive resistance to carbon dioxide (Raven *et al.*, 1988).

Other factors may also be involved. Bloom *et al.* (1985) suggest that "plants reduce disparity in supplies of carbon and nutrients by increasing their capacity to acquire the most limiting resource . . . [and] adjust allocation so that their growth is equally limited by all resources." In the face of severe nitrogen and phosphorus limitation, allocating more resources to root mass is to be expected in oligotrophic species. Indeed, some evidence indicates that these nutrients are more limiting to oligotrophic macrophytes than is carbon (Sand-Jensen and Sondergaard, 1979; Moeller, 1983; Boston and Adams, 1987). Such a commitment of resources to roots undoubtedly favors acquisition of carbon from the sediment. This, in turn, would select for other structural characteristics, such as the rosette growth form, which provides a short diffusion pathway from sediment to leaves, and extensive lacunal volume in both leaves and roots, which increases the diffusion of gases from roots to leaves. Under these conditions, one might expect the sediment to act as a source of carbon and the water as a sink; selection for higher cuticular resistance would thus be expected. CAM photosynthesis, which extends the diel period of carbon fixation, would be selected by low photosynthetic rates and, as leaf conductance decreases, may become more efficient, since leakage of carbon dioxide out of the leaves during decarboxylation would be minimized (Keeley *et al.*, 1983; Boston and Adams, 1987; Madsen, 1987). Characteristics such as thick leaves, large lacunal airspaces, and carbon uptake from the sediment would not be compatible with bicarbonate uptake. This incompatibility could explain why it may not be energetically advantageous for oligotrophic species to use this carbon source, even though bicarbonate can account for 25–90% of the total inorganic carbon pool in oligotrophic habitats (Hutchinson, 1975). Overall, the limitations to carbon gain in oligotropic lakes are substantial, resulting in low productivity. This factor may have been important in selection of the evergreen habit, which conserves nutrients and extends the annual period for photosynthetic production, even under winter ice cover (Boylen and Sheldon, 1976).

Submerged aquatic macrophytes indicative of eutrophic sites have responded quite differently to the largely biotic stresses of their environment (Fig. 4). Elevated levels of carbon and other inorganic nutrients allow for high photosynthetic rates. These high rates generate more oxygen and increase the potential for photorespiratory effects (Salvucci and Bowes, 1983; Holaday *et al.*, 1983). High photosynthetic rates also result in high

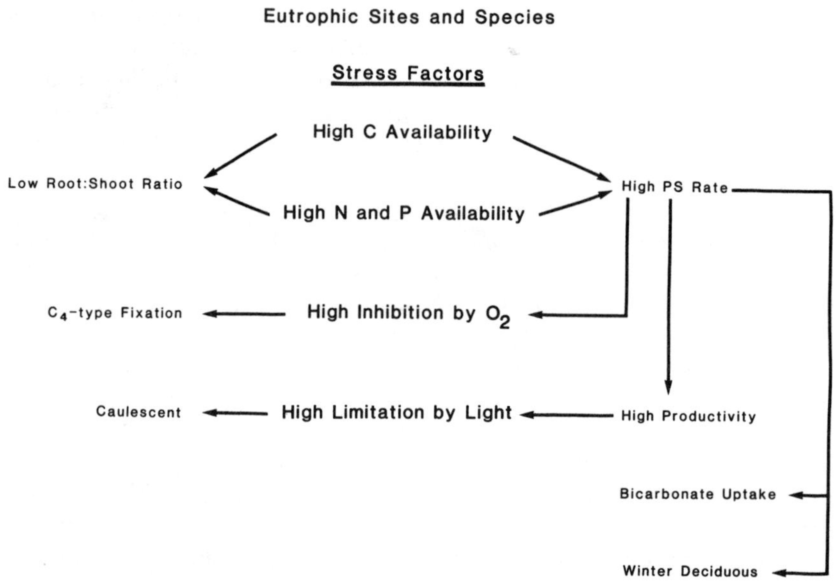

Eutrophic Sites and Species

Stress Factors

Figure 4 Interaction of stresses and plant structure, function, and productivity in eutrophic lake environments. PS, photosynthesis.

productivity. As biomass production (by macrophytes and microphytes) increases, the potential for shading increases, and selection for a caulescent growth form is expected (*e.g.,* Titus and Adams, 1979; Barko and Smart, 1981). High photosynthetic rates select for smaller, thinner leaves that minimize resistance to carbon uptake. Small leaves likewise reduce sedimentation and epiphyte establishment, thus reducing shading caused by those factors. High photosynthetic rates, coupled with the higher pH typical of eutrophic sites, put a premium on the ability to use alternative inorganic carbon species such as bicarbonate. Photosynthetic inhibition by high oxygen concentrations may be a major factor in selection for C_4 fixation of carbon dioxide in the light (e.g., *Hydrilla verticillata;* Holaday and Bowes, 1980); lack of this carboxylation pathway in *M. spicatum* (Salvucci and Bowes, 1983) and *P. pectinatus* (Winter, 1978), however, suggests there are other mechanisms for reducing photorespiratory effects. Low nutrient stress, coupled with selection for the caulescent habit to overcome limited light, has selected for low root:shoot ratios. High photosynthetic rates are commonly associated with high respiratory rates; this, coupled with cool temperatures and short days during winter, may have selected for the deciduous habit.

In aquatic habitats there are, of course, stress factors unrelated to productivity, such as turbulence from waves or currents and sediment

deposition. In general, the low-growing rosette of oligotrophic macrophytes may preadapt them for tolerating turbulence. Seddon (1965) noted
that *Isoetes echinospora* Dur. could avoid being outcompeted in eutrophic
environments only on sites where exposure to wave action prevented
growth of caulescent species. The rosette growth form, however, is mal-
adaptive on sites with rapid sedimentation (Spence, 1982).

VI. Perturbations and Predictions of Community Composition

The above considerations suggest that changes in site productivity should
be accompanied by predictable changes in species composition. Eutrophication has occurred over historical times in many aquatic systems, and
changes in macrophyte composition have always followed. Eutrophic enrichment with nitrogen and phosphate leads to luxurious growth of *Myriophyllum* and *Potamogeton* species and the disappearance of oligotrophic
species (Roelofs, 1983; Ozimek and Kowalczewski, 1984; Farmer and
Spence, 1986). An increase in one macrophyte species could potentially
change biotic stresses in the environment, and such changes may account
for the cyclic or wave pattern that invasions often follow, with one invasive
species replacing another (Carpenter 1980). As pollution intensifies and a
system becomes hypereutrophic, the availability of nutrients increases
selection for species with minimal resistance to carbon uptake; the resulting
phytoplankton "blooms" pose a severe light stress for rooted macrophytes.
Changes such as these (Fig. 5) have been documented for the natural

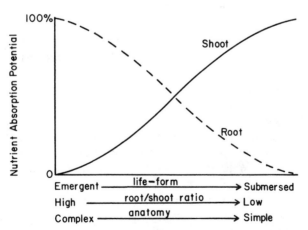

Figure 5 Hypothetical changes in the relative primary productivity of submerged, emer-
gent, epiphytic, and planktonic communities as nutrient enrichment increases. [Reprinted
from Davis and Brinson (1980).]

community succession that occurs in systems that are gradually becoming fertile (Wilson, 1935).

Acidification of lakes is now a significant factor affecting site productivity. The drop in pH caused by sulfuric-acid precipitation acidifies the water column and releases carbon dioxide from the water (Dillon *et al.*, 1987); these effects are most evident in poorly buffered oligotrophic habitats. The effect on the water column is to decrease the concentration of inorganic carbon. If the substrate is high in inorganic carbon, acidification of the overlying water column may increase the concentration of carbon dioxide in the sediment. Other effects include increased sulfate concentration; decreased nitrification, and thus higher ammonium concentration; and decreased microbial turnover of organic matter (Roelofs, 1983). These changes exacerbate the stress of low nutrient availability so much that isoetid species such as *Lobelia dortmanna* and *Isoetes* spp. disappear, resulting in an overall drop in macrophyte production and diversity (Roelofs, 1983; Schuurkes *et al.*, 1986).

The future effects of rising carbon dioxide levels in the atmospheric are open to speculation (Wetzel and Grace, 1983). In eutrophic habitats, the relative increase in carbon dioxide will be small, whereas in oligotrophic habitats, it will potentially be significant. Plants in these oligotrophic environments are likely to respond with increased photosynthetic rates, but this change would be transitory. For as primary production increases, other nutrients would become tied up in biomass, thus limiting productivity.

VII. Conclusions

Responses to stress described here for aquatic macrophytes have parallels for terrestrial systems. Abiotic stresses in oligotrophic aquatic environments have selected for a markedly different suite of plant traits from those selected for by the biotic stresses dominating eutrophic environments. Abiotically stressful terrestrial environments such as deserts are often dominated by species with high root : shoot ratios, a rosette growth form, and well-cutinized, evergreen leaves, and many such species have CAM photosynthesis (Chapter 14), all characteristics of abiotically stressful aquatic habitats as well. Highly productive terrestrial habitats are dominated by biotic stresses. Low root : shoot ratios, C_4 photosynthesis and the caulescent growth form are examples of characteristics held in common with aquatic species. Thus, the responses to stress are more closely tied to the form of the stress, biotic or abiotic, than whether it occurs on land or in the water.

Acknowledgments

This manuscript benefited from comments by John Raven, Harry Boston, John Titus, and the editors. Support was provided by NSF grant BSR 8705250.

References

Barko, J. W., and Smart, R. M. (1981). Comparative influences of light and temperature on the growth and metabolism of selected submerged freshwater macrophytes. *Ecol. Monogr.* 51: 219–235.

Black, M. A., Maberly, S. C., and Spence, D. H. N. (1981). Resistance to carbon dioxide fixation in four submerged freshwater macrophytes. *New Phytol.* 87: 557–568.

Bloom, A. J., Chapin, F. S., III, Mooney, H. A. (1985). Resource limitation in plants—an economic analogy. *Annu. Rev. Ecol. Syst.* 16: 363–392.

Boston, H. L. (1986). A discussion of the adaptations for carbon acquisition in relation to the growth strategy of aquatic isoetids. *Aquat. Bot.* 26: 259–270.

Boston, H. L., and Adams, M. S. (1987). Productivity, growth, and photosynthesis of two small 'isoetid' plants, *Littorella uniflora* and *Isoetes macrospora*. *J. Ecol.* 75: 333–350.

Boston, H. L., Adams, M. S., and Pienkowski, T. S. (1987). Utilization of sediment CO_2 by selected North American isoetids. *Ann. Bot.* 60: 485–494.

Bowes, G. (1985). Pathways of CO_2 fixation by aquatic organisms. *In* "Inorganic Carbon Uptake by Aquatic Photosynthetic Organisms" (W. J. Lucas and J. A. Berry, eds.), pp. 187–200. American Society of Plant Physiologists, Rockville, Maryland.

Bowes, G., and Salvucci, M. E. (1984). *Hydrilla:* Inducible C_4-type photosynthesis without Kranz anatomy. *In* "Advances in Photosynthesis Research" (C. Sybesma, ed.), Vol. III, pp. 829–832. Dr. W. Junk, The Hague.

Boylen, C. W., and Sheldon, R. R. (1976). Submerged macrophytes: Growth under winter ice cover. *Science* 182: 841–842.

Carpenter, S. R. (1980). The decline of *Myriophyllum spicatum* in a eutrophic Wisconsin Lake. *Can. J. Bot.* 58: 27–535.

Chambers, P. A. (1987). Light and nutrients in the control of aquatic plant community structure: In situ observations. *J. Ecol.* 75: 611–619.

Davis, G. J., and Brinson, M. M. (1980). Responses of submersed vascular plant communities to environmental change. U.S. Fish Wildl. Serv. FWS–OBS 79/33.

Dillon, P. J., Reid, R. A., and de Grosbois, E. (1987). The rate of acidification of aquatic ecosystems in Ontario, Canada. *Nature* 329: 45–48.

Farmer, A. M., and Spence, D. H. (1986). The growth strategies and distribution of isoetids in Scottish freshwater lochs. *Aquat. Bot.* 26: 247–258.

Hellquist, C. B. (1980). Correlation of alkalinity and the distribution of *Potamogeton* in New England. *Rhodora* 82: 331–334.

Holaday, A. S., and Bowes, G. (1980). C_4 acid metabolism and dark CO_2 fixation in a submersed aquatic macrophyte (*Hydrilla verticillata*). *Plant Physiol.* 65: 331–335.

Holaday, A. S., Salvucci, M. E., and Bowes, G. (1983). Variable photosynthesis/photorespiration ratios in *Hydrilla* and other aquatic macrophyte species. *Can. J. Bot.* 61: 229–236.

Hough, R. A. (1974). Photorespiration and productivity in submersed aquatic vascular plants. *Limnol. Oceanogr.* 19: 912–927.

Hutchinson, G. E. (1975). "A Treatise on Limnology," Vol. III, Limnological Botany. Wiley, New York.

Keeley, J. E., and Morton, B. A. (1982). Distribution of diurnal acid metabolism in submerged aquatic plants outside the genus *Isoetes*. *Photosynthetica* 16: 546–553.

Keeley, J. E., Walker, C. M., and Mathews, R. P. (1983). Crassulacean acid metabolism in *Isoetes bolanderi* in high-elevation oligotrophic lakes. *Oecologia* 58: 63–69.

Kirk, J. T. O. (1983). "Light and Photosynthesis in Aquatic Ecosystems." Cambridge Univ. Press, Cambridge.

Lucas, W. J., and Berry, J. A., eds. (1985). "Inorganic Carbon Uptake by Aquatic Photosynthetic Organisms." American Society of Plant Physiologists, Rockville, Maryland.

Madsen, T. V. (1987). Interactions between internal and external CO_2 pools in the photosynthesis of the aquatic CAM plants *Littorella uniflora* (L.) Aschers and *Isoetes lacustris* L. *New Phytol.* 106: 35–50.

Moyle, J. B. (1945). Some chemical factors influencing the distribution of aquatic plants in Minnesota. *Am. Midl. Nat.* 34: 402–420.

Moeller, R. E. (1983). Nutrient-enrichment of rhizosphere sediments: An experimental approach to the ecology of submersed macrovegetation. *In* "The Proceedings of the International Symposium on Aquatic Macrophytes." Nijmegen, The Netherlands.

Nobel, P. S. (1983). *Biophysical Plant Physiology and Ecology*. Freeman, San Francisco.

Nobel, P. S., and Walker, D. B. (1985). Structure of leaf photosynthetic tissue. *In* "Photosynthetic Mechanisms and the Environment" (J. Barber and N. R. Baker, eds.). Elsevier, New York.

Ozimek, T., and Kowalczewski, A. (1984). Long-term changes of the submerged macrophytes in eutrophic Lake Mikolajskie (North Poland). *Aquat. Bot.* 19: 1–11.

Raven, J. A. (1970). Exogenous inorganic carbon sources in plant photosynthesis. *Biol. Rev.* 45: 167–221.

Raven, J. A. (1981). Nutritional strategies of submerged brenthic plants: The acquisition of C, N, and P by rhizophytes and haptophytes. *New Phytol.* 88: 1–30.

Raven, J. A. (1984). "Energetics and Transport in Aquatic Plants." Alan R. Liss, New York.

Raven, J. A., Handley, L. L., Macfarlane, J. J., McInroy, S., McKenzie, L., Richards, J. H., and Samuelsson, G. (1988). The role of CO_2 uptake by roots and CAM in acquisition of inorganic C by plants of the isoetid life-form: A review, with new data on *Eriocaulon decangulare* L. *New Phytol.* 108: 125–148.

Richardson, K., Griffiths, H., Reed, M. L., Raven, J. A., and Griffiths, N. M. (1984). Inorganic carbon assimilation in the isoetids, *Isoetes lacustris* L. and *Lobelia dortmanna* L. *Oecologia* 61: 115–121.

Roelofs, J. G. M. (1983). Impact of acidification and eutrophication on macrophyte communities in soft waters in the Netherlands. I. Field observations. *Aquat. Bot.* 17: 139–155.

Salvucci, M. E., and Bowes, G. (1983). Two photosynthetic mechanisms mediating the low photorespiratory state in submersed aquatic angiosperms. *Plant Physiol.* 73: 488–496.

Samishi, Y. B. (1975). Oxygen build-up in photosynthesizing leaves and canopies is small. *Photosynthetica* 9: 372–375.

Sand-Jensen, K., and Prahl, C. (1982). Oxygen exchange with the lacunae across leaves and roots of the submerged vascular macrophytes. *New Phytol.* 91: 103–120.

Sand-Jensen, K., and Sondergaard, M. (1978). Growth and production of isoetids in oligotrophic Lake Kaalgaard, Denmark. *Verh. Int. Ver. Theor. Angew. Limnol.* 20: 659–666.

Sand-Jensen, K., and Sondergaard, M. (1979). Distribution and quantitative development of aquatic macrophytes in relation to sediment characteristics in oligotrophic Lake Kalgaard, Denmark. *Freshwater Biol.* 9: 1–11.

Schuurkes, J. A. A. R., Kok, C. J., and Den Hartog, C. (1986). Ammonium and nitrate uptake by aquatic plants from poorly buffered and acidified waters. *Aquat. Bot.* 24: 131–146.

Sculthorpe, C. D. (1967). "The Biology of Aquatic Vascular Plants." Edward Arnold, London.

Seddon, B. (1965). Occurrence of *Isoetes echinospora* in eutrophic lakes in Wales. *Ecology* 46: 747–748.

Seddon, B. (1972). Aquatic macrophytes as limnological indicators. *Freshwater Biol.* 2: 107–130.

Spence, D. H. N. (1982). The zonation of plants in freshwater lakes. *Adv. Ecol. Res.* 12: 37–125.

Titus, J. E., and Adams, M. S. (1979). Coexistence and the comparative light relations of the submersed macrophytes *Myriophyllum spicatum* L. and *Vallisneria americana* Michx. *Oecologia* 40: 273–286.

Van, T. K., Haller, W. T., and Bowes, G. (1976). Comparison of the photosynthetic characteristics of three submersed aquatic plants. *Plant Physiol.* 58: 761–768.

Westlake, D. F. (1967). Some effects of low velocity currents on the metabolism of aquatic macrophytes. *J. Exp. Bot.* 18: 187–205.

Wetzel, R. G., and Grace, J. B. (1983). Aquatic plant communities. *In* "CO_2 and Plants: The Response of Plants of Rising Levels of Atmospheric Carbon Dioxide" (E. R. Lemon, ed.), pp. 223–230. Westview Press, Boulder, Colorado.

Wilson, L. R. (1935). Lake development and plant succession in Vilas County, Wisconsin. I. The medium hard-water lakes. *Ecol. Monogr.* 5: 230–247.

Winter, K. (1978). Short-term fixation of [14]carbon by the submerged aquatic angiosperm *Potamogeton pectinatus*. *J. Exp. Bot.* 29: 1169–1172.

16

Shrub Life-Forms

Philip W. Rundel

I. Introduction

The shrub life-form is widely distributed, intensively studied, and ecologically important, but it remains a loosely defined entity. Ecologists commonly define shrubs as low woody plants with multiple stems, a definition that readily fits the characteristic 1–2 m shrubs and dwarf shrubs dominat-

Response of Plants to Multiple Stresses.

ing Mediterranean-climate and desert regions of the world, as well as the
low to prostrate dwarf and cushion shrubs of arctic and alpine tundras. In
other groups of low woody or semiwoody plants, however, the term becomes more difficult to apply. Multistemmed mallee eucalypts, for example, are often defined as shrubs (Box, 1981), although single-stemmed
mallot eucalypts of the same height and growing in the same environment
are defined as small trees. Are Krummholz trees at or above timberline best
classed as shrubs? How should one define succulent shrubby forms of
Cactaceae and Crassulaceae, which have both single and multiple stems?

Although a multiple-stemmed growth habit has been important in many
definitions of shrubs, this importance reflects a temperate bias. The majority of obligate understory shrubs, or *Schopfbaumchen* (Schimper, 1898), in
tropical rainforests are not multistemmed but rather have a single mono-
podial growth axis (Givnish, 1984; Rundel, unpublished data). In contrast,
most woody plants of low stature in temperate forests are multistemmed
and unequivocally classifiable as shrubs.

If shrubs are defined as low, woody, or semiwoody plants that retain
living tissues aboveground throughout the year, then they form a portion
of the phanerophytes and all of the chamaephytes as defined by Raunkiaer
(1934). Worldwide, this would include a significant percentage of the
earth's vascular plant flora. Among the 307 families of dicotyledons, 72%
include at least some shrubs, and 60% are characterized by shrub life-forms
(Cronquist, 1981). Even among the monocotyledons, where woody perennials are not characteristic, shrubs are remarkably common. Approximately
26% of the 65 families of monocotyledons include species that would be
considered shrubs, and 15% are characterized by shrubby life-forms.

No standard system of classifying or defining the shrub life-form has
found widespread acceptance. Ellenberg and Mueller-Dombois (1974) use
six different categories to encompass shrubs, whereas Box (1981) defines

Table I   Comparison of Three Systems of Shrub Classifications

Raunkiaer (1934)	Ellenberg and Mueller-Dombois (1974)	Box (1981)
Phanerophytes (in part)	Krummholz	Krummholz
	Shrubs	Shrubs
		Arborescents
	Tall succulents	Stem succulents (in part)
Chamaephytes	Woody dwarf shrubs	Dwarf shrubs (in part)
		Cushion shrubs (in part)
	Semiwoody dwarf shrubs	Dwarf shrubs (semishrubs)
		Cushion shrubs (in part)
	Low succulents	Stem succulents (in part)

Table II Shrubby Life-forms of World Vegetation[a]

Arborescents	
Evergreen	Mallee eucalypts
Rain-green thorn-scrub	*Acacia*
Summergreen	*Prosopis, Salix*
Leafless	*Haloxylon, Calligonum*
Krummholz	
Needle-leaved treeline	*Picea, Abies, Juniperus*
Shrubs	
Tropical broad evergreen	Rubiaceae
Temperate broad evergreen	
Mediterranean	*Protea, Quercus*
Typical temperate	*Ilex, Ligustrum*
Broad ericoid	*Rhododendron*
Hot-desert evergreen	*Larrea*
Leaf-succulent evergreen	*Crassula argentea*
Cold-winter xeromorphic	*Artemisia*
Summergreen broad-leaved	
Mesomorphic	*Rosa, Vaccinium*
Xeromorphic	Deciduous chaparral
Needle-leaved evergreen	*Juniperus*
Dwarf shrubs	
Mediterranean	*Thymus, Salvia*
Temperature evergreen	Ericaceae
Summergreen tundra	*Betula, Salix*
Xeric	*Ephedra, Retama*
Cushion Shrubs	
Perhumid evergreen	*Azorella*
Xeric	Puna hard cushions
Rosette shrubs	
Mesic	Understory palms
Xeric	*Agave, Aloe*
Stem succulents	
Arborescent	*Carnegia*
Typical	Unbranched cacti
Bush	Branched *Opuntia*

[a] Source: Box (1981).

seven different categories (Table I). Box further separates the seven shrubby life-forms into a lengthy series of subdivisions based largely on leaf morphology and phenology (Table II).

II. Evolution of the Shrub Life-Form

With a bias toward comparative stress physiology, physiological ecologists commonly assume that shrubs represent a highly derived life-form resulting from multiple stress interactions. This assumption is probably correct, yet it ignores the ancestral status of the shrub form among angiosperms. Stebbins (1965) argued that the earliest angiosperms were woody shrubs or subshrubs, rather than forest trees as other paleobotanists have suggested. He believes that the selective pressures that brought about angiospermous reproduction would have acted most strongly on smaller, short-lived plants inhabiting pioneer or transitional habitats, rather than on trees in stable mesic forests. Stebbins presents three lines of evidence to support this argument. First, lowland Jurassic forests show no indication of large gymnospermous trees, which should be ancestral to angiosperms. Second, in

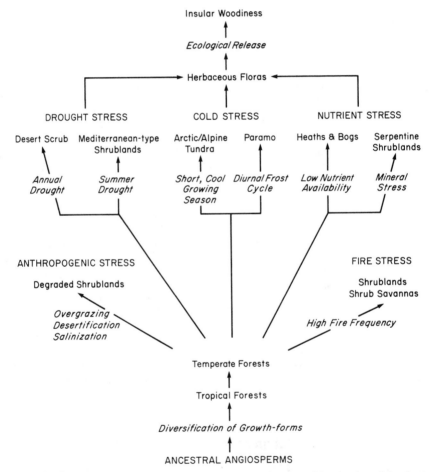

Figure 1 Evolutionary tree for the development of communities dominated by shrub life-forms under multiple environmental stresses.

tropical forests, little adaptive selection occurs for the reductionist trends in reproductive structures that led to angiospermy. Finally, angiosperms are thought to have evolved in semiarid upland habitats such as those dominated by shrubs today. This evolutionary trend from ancestral shrubs to trees in the angiosperms is reversible, and within specific lineages. Many examples exist of the evolution of shrubs from trees as well as from herbaceous progenitors.

An evolutionary tree for the ecological diversification of shrub growth forms (Fig. 1) would thus have the shrub ancestors of the angiosperms at its base. These ancestors quickly diversified in the late Mesozoic, forming extensive tropical and temperate forest communities dominated by angiosperms. The ecological dominance of shrubs in biological communities today can be seen in convergence toward the shrub life-form under selective pressures from a variety of environmental stresses. Single stress factors may predominate in each of the branches shown in Figure 1, but a complex of multiple stress interactions produces the evolutionary convergence in life-form. The common denominators in each line are multiple stress interactions that reduce and limit maximum growth rates in woody plants.

III. Shrub Distribution and Floristic Dominance

To a physiological plant ecologist interested in the mechanisms of plant tolerance to environmental stress, the shrub life-form generally brings to mind species in desert or Mediterranean-type ecosystems (MTES), or perhaps groups of chamaephytes in arctic or alpine ecosystems. Certainly, woody and semiwoody shrubs make up the ecologically dominant element of vegetation cover in these systems and have served as model species in developing an understanding of multiple-stress responses and the evolution of the shrub life-form. During the International Biological Program in the 1970s a variety of chaparral and matorral species in California and Chile served as models for convergent evolution in plant adaptations to MTES (Mooney and Dunn, 1970a,b; Mooney, 1977). In parallel, shrubby species of *Larrea* and other genera served as models for convergent evolution between the Sonoran Desert of Arizona and the Monte Desert of Argentina (Mabry *et al.*, 1977; Orians and Solbrig, 1977).

Shrubs are also widespread and often ecologically important in many world ecosystems where other life-forms dominate. In many tropical forests, for example, shrubs often constitute a significant percentage of the total flora. Shrubs make up 7.1% of the vascular plant flora of the evergreen rainforest of Barro Colorado Island, Panama (Croat, 1978), 10.8% of the kerangas forest flora of Sarawak (Brunig, 1974), and 23.3% of the dry tropical forest flora of Chamela, Mexico (Bullock, 1985). In contrast, shrubs form only 5.6% of the flora of the northern Mojave Desert (Beatley, 1976) and 13.0% of the Sonoran Desert flora at Deep Canyon, California (Zabriskie, 1979). Among the MTES of California, Chile, and the Mediterranean Basin, where shrubs dominate the vegetation, this life-form composes only 20–40% of the vascular plant flora (Rundel, 1991). Only in the fynbos and kwongan heaths of South Africa and western Australia and in serpentine scrublands of New Caledonia does the shrub life-form dominate floristically. Shrubs commonly make up 40% or more of the floras of mountain fynbos communities and up to 80% of some kwongan heathlands (Rundel, 1991). Jaffre (1980) reports that 86% of the flora of the serpentine scrub of the Massif du Koniambo in New Caledonia is woody, with only a few tree species present.

IV. Adaptations to Diverse Habitats

The shrub life-form represents an integrated evolutionary response to environmental stress. Although the architectural aspect of this adaptive response is loosely included in the definition of shrub, the architectural, phenological, morphological, and physiological characteristics of shrubs may vary greatly. This variation results from the differing forms of multiple stress interactions. In the following section, I will describe forms of adaptive responses in shrub vegetation types where the shrub life-form is ecologically dominant (see Fig. 1).

A. Desert Scrub

Warm-desert shrubs owe their dominance to a wide variety of form and functional adaptations to their arid habitats. Unlike other ecosystem types, where woody dominants appear to have largely evolved toward a common suite of adaptive traits, warm deserts have at least five shrub types that are widespread and ecologically important. These include drought-deciduous shrubs, evergreen shrubs, deciduous phreatophytes, succulents with crassulacean acid metabolism (CAM), and non-CAM stem succulents. These life-forms all represent successful modes of dealing morphologically and physiologically with extended periods of drought.

Drought-deciduous shrubs, the most characteristic desert group, rely heavily on morphological changes in the quality and quantity of their foliar biomass to remain metabolically active through much of the year. Typically, these species develop a relatively large canopy of mesomorphic leaves when water is available and thus maintain a relatively high rate of productivity. With increasing seasonal water stress, usually in summer, the total leaf surface area is reduced, and spring leaves are replaced by smaller and more xeromorphic summer leaves. Although these changes in total canopy leaf area and decreased leaf conductances associated with the xeromorphic (and often reflective) leaves combine to reduce productivity sharply, the associated increase in water use efficiency combines with adaptations in tissue water relations to allow photosynthetic activity through all but the most extreme water stress (Fig. 2). Good examples of this general pattern include *Encelia farinosa* in the Sonoran Desert (Ehleringer and Mooney, 1978), the cold-desert *Artemisia tridentata* (DePuit and Caldwell, 1975), and the C$_4$ *Hammada scoparia* from the Negev Desert (Kappen *et al.*, 1975).

Evergreen desert shrubs are not common, but some of the few examples, such as species of the genus *Larrea*, are remarkably successful in warm deserts throughout North and South America. These evergreen species appear to rely heavily on seasonal acclimation in the temperature response of photosynthesis rather than morphological adjustments to tolerate summer drought (Mooney *et al.*, 1977). The maintenance of positive turgor potentials at low soil water potential and effective stomatal regulation are also, as in the drought-deciduous shrubs, important components of their adaptive response.

Both drought- and winter-deciduous phreatophytes are important shrub dominants in desert environments where reliable groundwater resources exist. These species exhibit a variety of strategies in plant water relations, stomatal regulation, and phenology (Nilsen *et al.*, 1983, 1984).

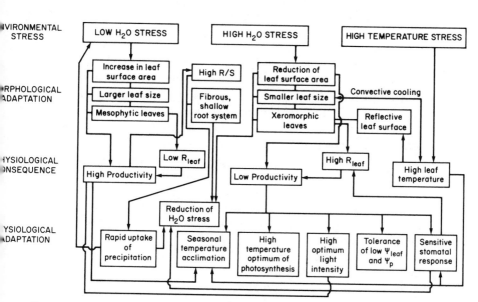

Figure 2 Multiple stress interactions in the evolution of drought-deciduous desert shrubs.

Cold desert environments present a particularly demanding set of multiple stresses for plant growth. Shrubs that dominate in these ecosystems must adapt to prolonged soil moisture stress; tremendous seasonal extremes in temperature, with physiological drought from frozen soils in winter; limited nitrogen availability; and widespread saline soils.

Because 60% of the annual precipitation in the Great Basin Desert of the United States from October to April falls as snow, water availability is greatest in early spring, and water stress increases steadily through the summer in most years. The dominant Great Basin shrubs appear well adapted to extract moisture from soils with low water potentials. *Artemisia tridentata* can take up water at -6 to -7 MPa (Campbell and Harris, 1977), and leaf water potentials of -11.5 MPa have been reported for *Atriplex confertifolia* and *Ceratoides lanata* (Moore *et al.*, 1972a). Several shrub species maintain positive rates of net photosynthesis at leaf water potentials below -5 MPa (Detling and Klikoff, 1973; White, 1976).

Summer temperatures may reach more than $40°$ C in cold desert ecosystems, and winter temperatures may drop to $-40°$ C. Thus, photosynthetic activity is largely precluded during the winter by low leaf temperatures and frozen soils. It seems surprising, therefore, that several prominent shrubs of the Great Basin are semievergreen. Overwintering leaves in *Artemisia tridentata* and *Atriplex confertifolia* form in mid- to late summer and persist until the following spring. When metabolic costs of leaf maintenance are considered, the value of these leaves is questionable, although they can take advantage of unpredictably favorable growth conditions in early spring. Despite extremely seasonal differences in thermal environment, little seasonal acclimation in the temperature response curve of photosynthesis has been found for cold-desert shrubs (DePuit and Caldwell, 1973, 1975).

Shrub species with C_4 metabolism are also important vegetation elements in cold desert ecosystems. Despite the prevailing cool temperatures characteristic of spring in the Great Basin, *Atriplex confertifolia* is active at the same time as C_3 shrubs (Caldwell *et al.*, 1977a). This contemporaneous timing of photosynthetic activity in C_4 and C_3 shrubs contrasts sharply with the offset seasonality of growth in C_4 and C_3 species of ephemerals and perennial herbs in warm desert ecosystems (Mulroy and Rundel, 1977). The evolutionary adaptation for photosynthesis at relatively low temperatures by *A. confertifolia* appears to be at a sacrifice of the high levels of photosynthetic capacity characteristic of herbaceous C_4 species. The maximum photosynthetic rate measured in *A. confertifolia* is only 14 μmol m^{-2}sec^{-1}, a rate lower than that present in C_3 shrubs from the same area (DePuit and Caldwell, 1975; Caldwell *et al.*, 1977b).

Much more than in warm deserts, cold desert regimes commonly have extensive areas of saline soils. Shrub species in the Great Basin demonstrate a wide range of tolerance to soil salinity and differential physiological mechanisms to deal with excess salt ions. *Artemisia tridentata* is only moderately salt tolerant, whereas *Allenrolfea occidentalis* is highly tolerant of salinity but restricted to moist habitats with available ground water (Fig. 3). *Ceratoides lanata*, a shrub with an intermediate level of tolerance, exudes excess salt ions at the site of water uptake in its root system. Species of *Atriplex*, also intermediate in tolerance, take up salts readily and excrete them into bladder hairs on the leaves (Breckle, 1975; Moore *et al.*, 1972b).

B. Mediterranean-Climate Shrublands

Drought stress is the primary factor influencing shrub dominance in both desert scrub and Mediterranean-climate shrublands. Under the unpredictable, but long, drought periods in MTES, trees are limited to sites where surface or ground water is readily available. An interactive model describing the evolution of evergreen, sclerophyllous-leaved shrubs was developed by Mooney and Dunn (1970b) to portray the multiple stress interactions found in MTES (Fig. 4). In this model, selective forces of climatic regime and fire lead to a series of physiological consequences for plant growth, resulting in convergent evolution toward an adaptive suite of morphological, phenological, and physiological characteristics.

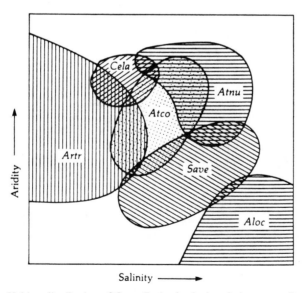

Figure 3 Habitat distribution of Great Basin shrubs in relation to gradients of salinity and aridity. Species abbreviations are as follows: *Aloc = Allenrolfea oxidentalis; Artr = Artemisia tridentata; Atco = Atriplex confertifolia; Atnu = Atriplex garderi; Cela = Ceratoides lanata; Save = Sarcobatus vermiculatus.* [From Breckle (1975).]

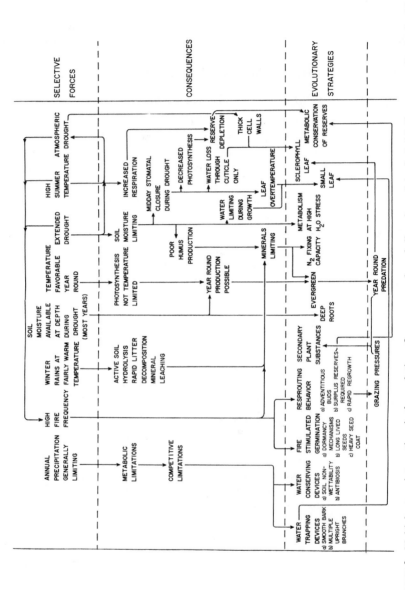

Figure 4 Multiple stress interactions in the evolution of evergreen, sclerophyllous-leaved shrubs in Mediterranean-climate shrublands. [Reprinted from Mooney and Dunn (1970b) with permission from *Evolution*.]

Studies of the ecological relations of Mediterranean-climate shrubs have largely focused on evergreen, sclerophyllous-leaved species. Evergreen and sclerophyllous leaves allow species to respond to dry summers and extreme seasonal unpredictability in precipitation by maintaining a canopy of drought-tolerant leaves throughout the year. These leaves have a relatively low photosynthetic capacity compared with malacophyllous, deciduous leaves, but they are more drought resistant and can thus amortize their cost of production over a longer period (Mooney and Dunn, 1970b; Harrison *et al.*, 1971; Mooney and Kummerow, 1971). Physiological models suggest that the duration of seasonal stress can explain how deciduous habits are selected for when either the dry season (desert and semidesert ecosystems) or winter cold (montane forest understory) becomes predictably long (Miller, 1979).

These models are useful, but they generally ignore the obvious biogeographical distribution of evergreen sclerophyllous species outside of Mediterranean regions. Well-developed chaparral communities occur in Arizona and along the upper and western margins of the Chihuahuan Desert in Mexico (Rundel and Vankat, 1989). Structurally similar to California chaparral, these communities inhabit areas with biseasonal or summer precipitation. Floristic links to California chaparral are apparent, and at least three species occur along entire precipitation gradients.

Climatic control of sclerophyllous leaf morphologies also appears to be poor in eastern Australia, where heathland communities extend northward out of Mediterranean-climate regions into areas of Queensland where annual precipitation levels exceed 1500 mm (Specht, 1981). There, low nutrient availability, rather than water availability, appears to be the controlling factor promoting the evolution of evergreen, sclerophyllous leaves.

Although the evolutionary convergence of evergreen shrubs in the five MTES of the world has long been recognized in recent years, increasing attention has been focused on the differences in diversity and leaf morphology between shrubs in the oligotrophic soils of the Cape Region of South Africa and western Australia and in the more nutrient-rich soils of the other three Mediterranean-climate regions (Kruger *et al.*, 1983). Low nutrient availability in South Africa and western Australia is clearly a factor in the heathlike form of many of the dominant ericoid shrubs and indicates a linkage to the evolution of shrub life-forms in nutrient-stressed soils of other ecosystems.

The dual root systems of Mediterranean-climate shrubs reflects the selective advantage of flexible belowground architecture to maximize use of surface precipitation (shallow roots) and semipermanent groundwater supplies (deep taproots) (Hellmers *et al.*, 1955). This simple model of rooting in Mediterranean sclerophyllous shrubs is still widely accepted even though root systems exhibit great diversity in morphology, root modifications, and rooting depth (see, for example, Kummerow [1981], Rundel, 1991). Clearly, patterns of root systems in sclerophyllous shrubs are much more complex than suggested by the simple dual root–system model. Much of this complexity almost certainly relates to differential selective factors associated with spatial and temporal patterns of belowground availability of water and nutrients and belowground storage of carbohydrates to facilitate postfire and postdrought growth (Rundel, 1991).

C. Arctic and Alpine Tundra

Since arctic and alpine ecosystems are largely defined by latitude, elevation, climate, and the absence of trees, the dominant vegetation elements may be highly variable. Low woody shrubs, herbaceous graminoids, mosses, or lichens may all be dominant. Environmental stresses thought to be most important in the low stature of tundra vegetation include winter desiccation and associated ice abrasion as well as a low annual carbon gain that is inadequate to build and support large woody structures (Billings and Mooney, 1968; Savile, 1972; Tranquillini, 1979).

Morphological adaptations of shrubs to wind desiccation and abrasion in these environments include prostrate growth, branch layering (rooting), leaves and branches arranged in a dense "cushion," and resinous scales or dead leaves that protect perennating buds. Shrubs taller than 1 m are generally restricted to protected microsites or locations where snow accumulation protects them from winter desiccation. Low cushion and rosette shrubs resist wind stress and thus characterize exposed rocky ridges and slopes where there is little winter snow and soils are well drained.

Although phenological patterns of growth in arctic and alpine shrubs do not differ substantively from those of temperate species, growth occurs at lower temperatures and is condensed into a shorter time. The low stature of these shrubs helps maximize microenvironmental temperatures of photosynthetic leaf tissues. The boundary layer between plant and bulk air masses commonly acts to increase leaf temperatures 3–8° C or more over ambient (Warren Wilson, 1957; Bliss, 1962; Molgaard, 1982). Prostrate *Salix* stems propped up above the ground surface to minimize boundary layer resistance produce 30–40% less biomass than control plants (Warren Wilson, 1959). The environmental and physiological constraints limiting the upper elevation of tree line are complex (Tranquillini, 1979), but metabolic energy balance is clearly a factor. Summer environmental conditions influence carbon fixation, shoot growth, and shoot maturation, all of which are important in determining whether tree growth is possible. The approximate position of the 10° isotherm for the mean temperature of the warmest month of the year has been found to be a fair predictor of the elevational and latitudinal limits of tree growth in the Northern Hemisphere (Wardle, 1974; Tranquillini, 1979).

Low growing-season temperatures are not the only environmental problem facing arctic and alpine shrubs. One important stress for many of these species, particularly those living in exposed and well-drained sites, is limited water availability, which affects plant distribution and productivity along local mesotopographical gradients. A number of studies have shown that species in sites with the lowest summer soil water potentials exhibit physiological drought adaptations allowing them to maintain positive turgor at relatively low leaf water potentials (Johnson and Caldwell, 1975; Teeri, 1973; Jackson and Bliss, 1984).

Availability of nutrients, particularly ammonium and phosphate, is also an important limiting factor in tundra habitats. Physiological studies with graminoids have shown that tundra species have evolved mechanisms of phosphate absorption that are less temperature sensitive than those in temperate species. Despite these compensations, nutrient absorption rates are still low because the concentrations of nutrients in the soil solution are very low. Nutrient absorption capacity in tundra shrubs appears to be related to leaf form. Deciduous shrubs with relatively high annual nutrient requirements have higher root absorption requirements than do evergreen shrubs (Chapin and Tryon, 1982).

The widespread distribution of both deciduous and evergreen shrubs in arctic and alpine tundras points out the breadth of adaptive strategies successful in meeting the multiple interactive stresses present in this type of environment. As in other temperate environments, evergreen leaves in tundra ecosystems have relatively low maximum rates of net photosynthesis relative to deciduous leaves (Fig. 5). Although most tundra communities contain a mixture of the two life-forms, deciduous shrubs characterize

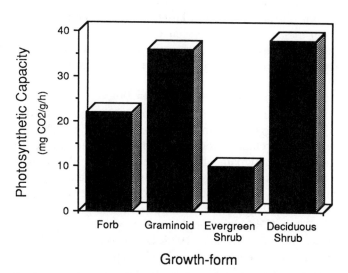

Figure 5 Comparative rates of maximum photosynthesis in leaves of arctic life-forms. [Data from Johnson and Tieszen (1976).]

more nutrient-rich or less water-stressed sites. Slower-growing evergreen shrubs generally occur on less fertile sites, with evergreen Ericaceae most common in areas of late-lying snow. Belowground biomass and productivity in both deciduous and evergreen shrubs is low compared with graminoids and other herbaceous perennials in the same habitats.

D. Tropical Alpine Fields

Tropical alpine environments pose special stresses because of the large amplitude of diurnal temperature changes. Whereas seasonal temperature changes are small, diurnal variations may be as great as the seasonal variations in typical temperate alpine ecosystems. Freezing is common and can occur on any night of the year (Hedberg, 1964; Troll, 1968). Although low daily air temperatures, soil frost heaving, and high radiant energy may be important environmental factors limiting plant growth, the most significant limiting factor is often low-temperature constraints on water availability (Meinzer and Goldstein, 1986). This is particularly true in drier paramo habitats with seasonal rainfall.

Low woody shrubs with a variety of drought adaptations occur in all the tropical alpine regions of the world, but the most conspicuous are giant rosette shrubs characterized by dense whorls of persistent evergreen leaves. This life-form has arisen in *Espeletia, Draba,* and *Lupinus* in the tropical Andes, in *Senecio* and *Lobelia* in the mountains of equatorial Africa, in *Argyroxiphium* on the high Hawaiian volcanoes, and in *Echium* in the Canary Islands. A variety of morphological and physiological adaptations has been shown to be critical in the convergent evolution toward this specialized life-form in such distant parts of the world. Marcescent leaves that protect the stem from diurnal freezing, a large volume of stem pith that buffers diurnal water loss, and thick pubescence that raises leaf temperatures over low ambient air temperatures during the day are all important adaptive traits in these giant rosette plants (Meinzer and Goldstein, 1986).

E. Heathlands and Bogs

Nutrient stress of many types can lead to dominance by shrub growth forms (see Fig. 1). Waterlogged soils and associated high acidity in heathlands and bogs, irrespective of climatic regime, strongly limit the availability of most macronutrients (Gimingham, 1972; Etherington, 1982; Larsen, 1982). Problems in nutrient availability largely result from anaerobic conditions, which promote slow decomposition of organic matter and relatively high humic and fulvic acid concentrations. The resulting acid soils commonly have a pH of 3 to 4, which strongly restricts microbial activity and limits nutrient availability (Collins *et al.,* 1978; Dickinson, 1983). Although nitrogen and phosphorus have been found to be especially limiting in bog soil, major cations (potassium, calcium, and magnesium) and micronutrients (copper, boron, and molybdenum) may also be limiting (Gore, 1961a,b;

Heilman, 1968; Small, 1972a,b; Etherington, 1982). These highly acid soils also increase the availability of potentially toxic elements such as aluminum, manganese, and iron (Bohn *et al.,* 1979; Etherington, 1982). Most modern vegetation studies of bogs and peatlands cite this complex of edaphic stresses as the primary factor influencing the dominance of woody shrubs. As bog soils become more acidic and nutrients less available, the stature of the dominant vegetation declines (Heinselman, 1970; Gore, 1983).

Waterlogged bog and heath soils produce a number of physiological stresses beyond these, however. Specialized metabolic strategies, for example, are necessary to mitigate biochemical problems associated with waterlogged conditions. These strategies include anaerobic respiration through the induction of enzymes that accelerate glycolysis; the excretion or translocation of toxic end products of glycolysis, or the production of alternative, nontoxic end products of glycolysis (Crawford, 1982, 1983).

A particularly interesting aspect of multiple stress effects on bog shrubs is the surprising occurrence of internal water deficits. The theory of physiological drought in bog plants was first suggested by Schimper (1898), and comparative studies of conspecific and congeneric shrubs in bog and adjacent non-bog sites have demonstrated greater water stress in the bog habitat (Marchand, 1975; Canfield, 1986). This effect can be seen in comparisons of $\delta^{13}C$ values for bog and adjacent bog populations of Hawaiian woody species (Table III). The bog populations average 3–4‰ higher (less negative) than the non-bog populations, suggesting higher levels of water-use efficiency. The relatively low water potentials and high water-use efficiencies in bog plants can be attributed to reduced root biomass in anaerobic soils and the resulting increase in hydraulic resistance to water flow through the soil–root–shoot continuum (Marchand, 1975; Bradbury and Grace, 1983).

The complex of multiple stress interactions affecting bog and heath plants has led to the dominance of evergreen leaves with xeromorphic characteristics. These leaf traits have been attributed to three specific factors: (1) increased nutrient use efficiencies with greater leaf longevity,

Table III Comparison of $\delta^{13}C$ Values of Leaf Tissue for Bog and Wet Forest Habitats in the Hawaiian Islands

	Island	Bog	Wet Forest	$\Delta\delta^{13}C$
Argyroxiphium grayanum	Maui	−25.82	−29.18	3.36
Dubautia laxa	Kauai	−25.36	−28.97	3.51
Dubautia plantaginea	Hawaii	−24.46	−27.89	3.43
Dubautia raillardioides	Kauai	−25.58	−27.05	1.57
Deschampsia nubigena	Kauai	−24.35	−29.14	4.79
Styphelia tameiameiae	Kauai	−26.45	−28.17	1.72
Metrosideros polymorpha	Kauai	−23.74	−27.77	4.03
Metrosideros polymorpha	Maui	−25.57	−27.71	2.14

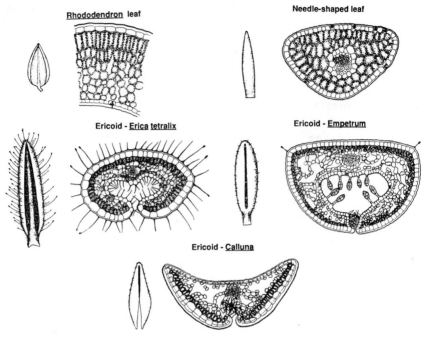

Figure 6 Comparative patterns of leaf morphology in the Ericales. [Adapted from Hagerup (1953).]

(2) decreased uptake of toxic reduced ions through reduced transpiration, and (3) controlled internal water deficits resulting from high root resistance through reduced transpiration (Small, 1972a,b, 1973; Grubb, 1977; Schulze, 1982; Crawford, 1983; Canfield, 1986).

Characteristic patterns of leaf morphology associated with habitat shifts from forest-edge shrublands or understory to heath or bog habitats have been documented in the Ericales (Hagerup, 1953). Groups of leaves of the *Rhododendron* type give rise to taxa with tiny needle-shaped leaves with stomata around the full circumference and, with increased evolutionary environmental stress, a variety of ericoid leaf morphologies with the stomata protected within a foliar cavity formed as the leaves curl back upon themselves (Fig. 6).

F. Serpentine Shrublands

Low nutrient availability is only one of many forms of mineral stress that can lead to the evolution of community dominance by shrubs in climatic regimes that would otherwise support forest vegetation. Serpentine shrublands provide a classic example of another form of nutrient stress. These shrublands may occur in temperate forest areas, such as the *Arctostaphylos-*

and *Ceanothus*-dominated serpentine formations in coniferous-forest areas of California (Kruckeberg, 1984), or in tropical regions, such as the extensive serpentine shrublands of New Caledonia (Jaffre, 1980; Brooks, 1987). The nutrient stresses acting in serpentine soils result from a complex of low cation availability caused by low soil calcium:magnesium ratios and potentially toxic concentrations of nickel, cobalt, and chromium. Comparable types of scrub vegetation may occur on other forms of nutrient-stressed soils.

G. Degraded Shrublands

Just as natural environmental stresses that reduce potential carbon gain lead to dominance of shrub life-forms over forest trees, so can chronic anthropogenic stresses. These may include air pollution, overgrazing, deforestation, pollution from toxic substances, soil erosion, trampling, and chronic gamma irradiation. These long-term effects of human activity ultimately lead to lower biomass and productivity, reduced soil fertility, and accelerated erosional loss of limiting nutrients. Such chronic degradation in habitat quality has been termed retrogression—in essence, a backward trajectory of succession (Whittaker and Woodwell, 1978).

Plant and community adaptation to chronic anthropogenic stresses is beyond the scope of this chapter, but the subject is reviewed broadly in many studies (Woodwell, 1968, 1970; Grime, 1979; Grubb and Hopkins, 1986; and Westman, 1986). Degradation of forest lands to shrublands has been best documented for the Mediterranean-climate regions of Europe (Tomaselli, 1977; Aschmann, 1973; Naveh and Dan, 1973). For MTES, the chronic stress of air pollution has reduced the diversity and stature of woody plants (Westman, 1979; Margaris *et al.*, 1986). Overgrazing or wood cutting can have much the same effect (Bahre, 1974; Balduzzi *et al.*, 1982; Arianoutsou-Faraggitaki, 1985), although overgrazing in arid grasslands can lead to dominance by woody shrubs over herbaceous graminoids (Humphrey, 1974).

Fire, considered below as a natural environmental stress, can also act as a chronic anthropogenic stress when human activities increase the frequency of natural fire cycles. When this happens, natural succession is changed as the timing of life-history events is upset by recurrent burning. Such changes in fire frequency can, for example, maintain early successional shrubs as dominants at the expense of longer-lived forest trees. In some areas, repeated annual burning is used as a management tool to replace shrublands with grass cover (Doman, 1968; Humphrey, 1974).

H. Shrub Savannas

Fire, as a natural ecosystem factor, can have profound effects on the structure and diversity of plant communities. Fire frequency, seasonality, and intensity may all alter the natural climatic patterns of life-form dominance (Rundel, 1981).

Shrubs and low trees become dominant because of fire in many tropical savannas, particularly those bordering tropical forests. In the extensively grazed savanna of both the Old and New World tropics, a major recurrent stress is the periodic or even annual fires that burn through the grass understory of these open communities in the dry season. Although many savannas are clearly climate-derived, others appear to be woodlands degraded by increased fire frequency. The dominant woody species in these savannas are seldom more than 3–8 m in height and single- to multi-stemmed, and they commonly have thick bark, which helps protect them from fire (Ferri, 1961). Furthermore, mature plants respond to fire damage by profuse resprouting from underground parts (Sarmiento and Monasterio, 1983), much like resprouting shrubs in Mediterranean ecosystems.

Fire frequency is not the only environmental stress facing woody plants in tropical savanna ecosystems. During the long dry season, high evaporative demand from the atmosphere and moisture depletion in the upper soil layers create the potential for plant water stress. Shrubs and small trees must rely on stored water reserves to maintain physiological activity through the dry season (Fig. 7). High rainfall in the wet season produces waterlogged soils and associated soil leaching, laterization, and foliar leaching, all of which restrict nutrient availability. The characteristic leaf of

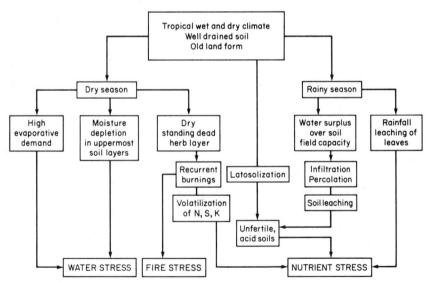

Figure 7 Multiple stress interactions in the development of evergreen shrubs and trees in tropical savanna ecosystems. [Adapted from Sarmiento *et al.* (1985).]

savanna dominants is broad, evergreen, and sclerophyllous, but deciduous species occur as well. Deep and extensive root systems aid greatly in maintaining metabolic activity during drought (Sarmiento *et al.*, 1985).

I. Oceanic Islands

The evolution of woody species from herbaceous progenitors in island ecosystems was first noted by Charles Darwin (1859), who wrote, "Islands often possess trees or bushes belonging to orders which elsewhere include only herbaceous species." Although a few authors have still held to a relict theory for the origin of insular woodiness, the collective evidence demonstrating an evolution from herbaceous species is very strong (Carlquist, 1974).

Because woody trees and shrubs from mainland habitats are generally less adapted for long-distance dispersal than are weedy herbaceous species, the former are less likely to colonize oceanic islands directly. Exceptions exist, but they are rare; the two most notable are *Metrosideros* (Myrtaceae) and *Weinmannia* (Cunoniaceae), with tiny stress-resistant seeds that have colonized high oceanic islands all across the Pacific.

Once herbaceous species have successfully colonized an island ecosystem with a relatively small number of established species and a diversity of ecological habitats, an "ecological release" occurs, which allows selection for a range of life-forms. On tropical or semitropical islands, release from climatic seasonality heightens this effect. One of the best examples of this phenomenon is the Hawaiian tarweeds (Fig. 8). This group has evolved from herbaceous tarweeds originating from the Pacific coast of North America into a complex of 28 species in three endemic genera (Carr *et al.*, 1990). The genus *Argyroxiphium*, with five species, and *Wilkesia*, with two species, both include monocarpic and polycarpic rosette shrubs. The genus *Dubautia*, with 21 species, includes cushion plants, mat-forming subshrubs, small shrubs, large woody shrubs, trees, and lianas. The collective range of tarweed habitats in the Hawaiian Islands goes from dry scrub with 400 mm precipitation per year to wet forests and bogs with more than 12,000 mm annual precipitation; Hawaiian tarweeds are distributed from 75 m to 3750 m in elevation. Despite this remarkable ecological and life-form diversity, as well as a wide range of species-specific leaf morphologies, the genetic cohesiveness of the Hawaiian tarweeds is readily apparent in the ease with which divergent species and genera hybridize, both naturally and artificially.

A second example of the plasticity possible in growth forms and leaf morphologies is the genus *Hebe* in New Zealand, a more temperate oceanic environment. From herbaceous ancestors in the Scrophulariaceae, *Hebe* has evolved approximately 100 species. Life-forms of *Hebe* range from low forest trees to broadleaved shrubs to cupressoid-leaved alpine cushion

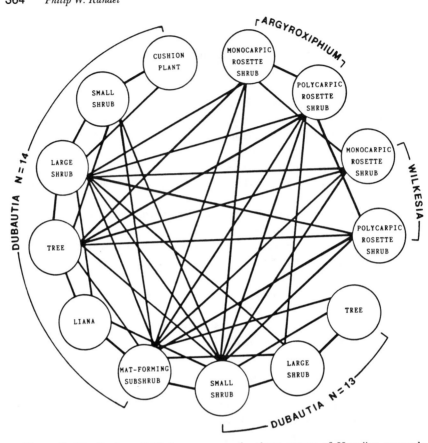

Figure 8 Distribution of life-forms among the three genera of Hawaiian tarweeds. Connecting lines indicate natural and artificial hybrids that have been produced between species and genera. [Reprinted from Carr *et al.* (1990) with permission from Annals of the Missouri Botanical Garden.]

plants (Fig. 9). The high degree of hybridization present in *Hebe* seems to provide strong evidence for recent evolutionary diversification without the formation of genetic barriers.

Although shrub species evolved from herbaceous progenitors are most characteristic of island ecosystems, examples of this direction of evolution also exist in mainland floras from Mediterranean and desert ecosystems. Two North American examples are the *Diplacus* group of shrubby species of *Mimulus* (Scrophulariceae) in California chaparral, and *Artemisia* (Compositae) in cold desert habitats of many northern temperate regions.

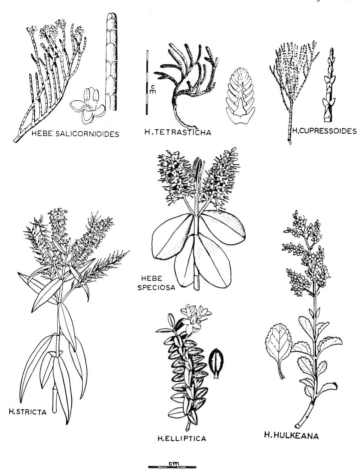

HEBE SALICORNIOIDES   H.TETRASTICHA   H.CUPRESSOIDES

HEBE SPECIOSA

H.STRICTA

H.ELLIPTICA   H.HULKEANA

cm

Figure 9 Range of life-forms and leaf morphologies in New Zealand species of *Hebe*. [Modified from Poole and Adams (1963).]

V. Conclusions

Shrubs, as a loosely defined life-form of low-growing woody plants, demonstrate remarkable phylogenetic diversity and ecological success. Much of what we know today about convergent evolution in response to multiple stress interactions has come from studies of shrubs. This growth form has evolved many times in many diverse ecosystems under environmental conditions that support woody plants but limit maximum growth. Duration and seasonality of drought, low- or high-temperature stress, macronutrient availability, trace element stress, fire frequency, and other chronic disturbances are all examples of significant environmental stresses that may interact to produce an evolutionary convergence toward a shrub life-form.

Although the shrub life-form is largely defined simplistically by aboveground architectural form, shrubs provide a wide range of divergent modes of belowground architecture, phenological patterns of leaf duration and growth, and physiological modes of adaptation. Despite the complexity of these traits, the shrub model remains an important basis for many of the important paradigms of plant adaptation to multiple environmental stresses, particularly in Mediterranean-climate, desert, and arctic and alpine ecosystems. These paradigms focus largely on problems of drought and temperature stress. With increasing knowledge of shrub life-forms in tropical and subtropical ecosystems, we will almost certainly be able to increase our understanding of the interactions of nutrient stresses as well.

References

Arianoutsou-Faraggitaki, M. (1985). Desertification by overgrazing in Greece: The case of Lesvos Island. *J. Arid Environ.* 9: 237–242.

Aschmann, H. (1973). Man's impact on the several regions with Mediterranean climates. *In* "Mediterranean-type Ecosystems: Origin and Structure" (F. di Castri and H. A. Mooney, eds.), pp. 363–371. Springer-Verlag, Berlin.

Bahre, C. (1974). Relationships between man and wild vegetation of the province of Coquimbo, Chile. Ph.D. dissertation, University of California, Riverside.

Balduzzi, A., Tomaselli, R., Serey, I., and Villasenor, R. (1982). Degradation of the Mediterranean-type vegetation in central Chile. *Ecol. Mediterr.* 8: 223–240.

Beatley, J. C. (1976). "Vascular Plants of the Nevada Test Site and Central–Southern Nevada." Energy Research and Development Administration, Washington, D.C.

Billings, W. D., and Mooney, H. A. (1968). The ecology of arctic and alpine plants. *Biol. Rev.* 43: 481–529.

Bliss, L. C. (1962). Adaptations of arctic and alpine plants to environmental conditions. *Arctic* 15: 117–144.

Bohn, H. L., McNeal, B. L., and O'Conner, G. A. (1979). "Soil Chemistry." Wiley, New York.

Box, E. D. (1981). "Macroclimate and Plant Forms." Dr. W. Junk, The Hague.

Bradbury, I. K., and Grace, J. (1983). Primary production in wetlands. *In* "Mires: Swamp, Bog, Fen, and Moor" (A. J. P. Gore, ed.), pp. 285–310. Elsevier, Amsterdam.

Breckle, S. W. (1975). Zur Okologie und zu den Mineralstoff-verhaltnissen absalzender und nichtabsalzender Xerohalophyten. Ph.D. dissertation, University of Bonn.

Brooks, R. R. (1987). "Serpentine and Its Vegetation: A Multidisciplinary Approach." Dioscordes Press, Portland, Oregon.

Brunig, E. F. (1974). "Ecological Studies in the Kerangas Forests of Sarawak and Brunei." Sarawak Forest Dept., Kuching, Malaysia.

Bullock, S. H. (1985). Breeding systems in the flora of a tropical deciduous forest in Mexico. *Biotropica* 17: 287–301.

Caldwell, M. M., Osmond, C. B., and Nott, D. L. (1977a). C_4 pathway photosynthesis at low temperature in cold-tolerant *Atriplex* species. *Plant Physiol.* 60: 157–164.

Caldwell, M. M., White, R. W., Moore, R. T., and Camp, L. B. (1977b). Carbon balance, productivity, and water use of cold-winter desert shrub communities dominated by C_3 and C_4 species. *Oecologia* 29: 275–300.

Campbell, G. S., and Harris, G. A. (1977). Water relations and water use patterns for *Artemisia tridentata* Nutt. in wet and dry years. *Ecology* 58: 652–659.

Canfield, J. E. (1986). The role of edaphic factors and plant water relations in plant distribution in the bog–wet forest complex of Alakai Swamp, Kauai, Hawaii. Ph.D. dissertation, University of Hawaii.

Carlquist, S. (1974). "Island Biology." Columbia Univ. Press, New York.

Carr, G. D., Robichaux, R. H., Witter, M. S., and Kyhos, D. W. (1990). Adaptive radiation of the Hawaiian silversword alliance (Compositae, Madinae): A comparison with Hawaiian picture-winged *Drosophila*. *Ann. Mo. Bot. Gard.* 77: 110–117.

Chapin, F. S., III, and Tryon, P. R. (1982). Phosphate absorption and root respiration of different plant growth forms from northern Alaska. *Holarct. Ecol.* 5: 164–171.

Collins, V. G., D'Sylva, B. T., and Latter, P. M. (1978). Microbial populations in peat. *In* "Production Ecology of British Moors and Montane Grasslands" (O. W. Heal and D. F. Perkin, eds.), pp. 94–112. Springer-Verlag, Berlin.

Crawford, R. M. M. (1982). Physiological responses to flooding. *In* "Encyclopedia of Plant Physiology, New Series" (O. L. Lange, P. S. Nobel, C. B. Osmond, and H. Ziegler, eds.), Vol. 12B, pp. 453–477. Springer-Verlag, Berlin.

Crawford, R. M. M. (1983). Root survival in flooded soils. *In* "Mires: Swamp, Bog, Fen, and Moor" (A. J. P. Gore, ed.), pp. 257–283. Elsevier, Amsterdam.

Croat, T. (1978). "Flora of Barro Colorado Island." Stanford Univ. Press, Stanford, California.

Cronquist, A. (1981). "An Integrated System of Classification of Flowering Plants." Columbia Univ. Press, New York.

Darwin, C. [1859] (1950). "On the Origin of Species by Means of Natural Selection." Watts, London.

DePuit, E. J., and Caldwell, M. M. (1973). Seasonal pattern of net photosynthesis of *Artemisia tridentata*. *Am. J. Bot.* 60: 426–435.

DePuit, E. J., and Caldwell, M. M. (1975). Gas exchange of three cool semi-desert species in relation to temperature and water stress. *J. Ecol.* 63: 835–858.

Detling, J. K., and Klikoff, L. G. (1973). Physiological response to moisture stress as a factor in halophyte distribution. *Am. Midl. Nat.* 90: 307–318.

Dickinson, C. H. (1983). Micro-organisms in peatlands. *In* "Mires: Swamp, Bog, Fen, and Moor" (A. J. P. Gore, ed.), pp. 225–245. Elsevier, Amsterdam.

Doman, E. R. (1968). Prescribed burning and brush-type conversion in California national forests. *Proc. Tall Timbers Fire Ecol. Conf.* 7: 225–233.

Ehleringer, J., and Mooney, H. A. (1978). Leaf hairs: Effects on physiological activity and adaptive value to a desert shrub. *Oecologia* 37: 183–200.

Ellenberg, H., and Mueller-Dombois, D. (1974). "Aims and Methods of Vegetation Ecology." Wiley, New York.

Etherington, J. P. (1982). "Environment and Plant Ecology." Wiley, New York.

Ferri, M. G. (1961). Problems of water relations of some Brazilian vegetation types, with special consideration of the concepts of scleromorphy and xerophytism. *In* "Plant Water Relations in Arid and Semiarid Conditions," pp. 191–197. UNESCO, Paris.

Gimingham, C. H. (1972). "Ecology of Heathlands." Chapman and Hall, London.

Givnish, T. J. (1984). Leaf and canopy adaptations in tropical forests. *In* "Physiological Ecology of Plants of the Wet Tropics." (E. Medina, H. A. Mooney, and C. Vazques-Yanes, eds.), pp. 51–84. Dr. W. Junk, The Hague.

Gore, A. J. P. (1961a). Factors limiting plant growth on high-level blanket peat. I. Calcium and phosphate. *J. Ecol.* 49: 399–402.

Gore, A. J. P. (1961b). Factors limiting plant growth on high-level blanket peat. II. Nitrogen and phosphate in the first year of growth. *J. Ecol.* 49: 605–616.

Gore, A. J. P. (1983). Introduction. *In* "Mires: Swamp, Bog, Fen, and Moor" (A. J. P. Gore, ed.), pp. 1–34. Elsevier, Amsterdam.

Grime, J. P. (1979). "Plant Strategies and Vegetation Processes." Wiley, New York.

Grubb, P. J. (1977). Control of forest growth and distribution on wet tropical mountains with special reference to mineral nutrition. *Ann. Rev. Ecol. Syst.* 8: 83–107.

Grubb, P. J., and Hopkins, A. J. M. (1986). Resilience at the level of the plant community. *In* "Resilience in Mediterranean-type Ecosystems" (B. Dell, A. J. M. Hopkins, and B. B. Lamont, eds.), pp. 21–38. Dr. W. Junk, Dordrecht.

Hagerup, O. (1953). The morphology and systematics of the leaves in Ericales. *Phytomorphology* 3: 459–464.

Harrison, A. T., Small, E., and Mooney, H. A. (1971). Drought relationships and distribution of two Mediterranean-climate California plant communities. *Ecology* 52: 869–875.

Hedberg, O. (1964). Features of Afroalpine plant ecology. *Acta Phytogeogr. Suec.* 49: 1–144.

Heilman, P. E. (1968). Relationship of availability of phosphorus and cations to forest succession and bog formation in interior Alaska. *Ecology* 49: 331–336.

Heinselman, M. C. (1970). Landscape evolution, peatland types, and the environment in the Lake Agassiz Peatlands Natural Area, Minnesota. *Ecol. Monogr.* 40: 235–261.

Hellmers, H., Horton, J. S., Juhren, G., and O'Keefe, J. (1955). Rootsystems of some chaparral plants in southern California. *Ecology* 36: 667–678.

Humphrey, R. R. (1974). Fire in the desert grassland of North America. *In* "Fire and Ecosystems" (T. T. Kozlowski and C. E. Ahlgren, eds.), pp. 365–400. Academic Press, New York.

Jackson, L. E., and Bliss, L. C. (1984). Phenology and water relations of three plant life-forms in a dry treeline meadow. *Ecology* 65: 1302–1314.

Jaffre, T. (1980). "Etude ecologique du peuplement végétal des sols dérivés de roches ultrabasiques en Nouvelle Caledonie." ORSTOM, Paris.

Johnson, D. A., and Caldwell, M. M. (1975). Gas exchange of four arctic and alpine tundra plant species in relation to atmospheric and soil moisture stress. *Oecologia* 21: 93–108.

Johnson, D. A., and Tieszen, L. L. (1976). Aboveground biomass allocation, leaf growth, and photosynthesis patterns in tundra plant forms in arctic Alaska. *Oecologia* 24: 159–173.

Kappen, L., Oertli, J. J., Lange, O. L., Schulze, E.-D., Evenari, M., and Buschbom, U. (1975). Seasonal and diurnal courses of water relations of the arido-active plant *Hammada scoparia* in the Negev Desert. *Oecologia* 21: 175–192.

Kruckeberg, A. R. (1984). "California Serpentines." Univ. of California Press, Berkeley.

Kruger, F. J., Mitchell, D. T., and Jarvis, J. U. M., eds. (1983). "Mediterranean-type Ecosystems: The Role of Nutrients." Springer-Verlag, Berlin.

Kummerow, J. (1981). Structure of roots and root systems. *In* "Mediterranean-type Shrublands" (F. di Castri, D. W. Goodall, and R. L. Specht, eds.), pp. 269–288. Elsevier, Amsterdam.

Larsen, J. A. (1982). "Ecology of the Northern Lowland Bogs and Conifer Forests." Academic Press, New York.

Mabry, T. J., Hunziker, J. H., and Difeo, D. R. (1977). "Creosote Bush." Dowden, Hutchinson and Ross, Stroudsburg, Pennsylvania.

Marchand, P. J. (1975). Apparent ecotypic differences in the water relations of some northern bog Ericaceae. *Rhodora* 77: 53–63.

Margaris, N. S., Arianoutsou-Faraggitaki, M., Iselas, S., and Loukas, L. (1986). Desertification due to air pollution in Attica. *In* "Desertification in Europe" (R. Fantechi and N. S. Margaris, eds.), pp. 166–169. D. Reidel, Dordrecht.

Miller, P. C. (1979). Quantitative plant ecology. *In* "Analysis of Ecosystems" (D. Horn, G. R. Stairs, and R. D. Mitchell, eds.), pp. 179–232. Ohio State University, Columbus.

Molgaard, P. (1982). Temperature observations in High Arctic plants in relation to microclimate in the vegetation of Peary Land, North Greenland. *Arct. Alp. Res.* 14: 105–115.

Meinzer, F. C., and Goldstein, G. (1986). Adaptations for water and thermal balance in Andean giant rosette plants. *In* "On the Economy of Plant Form and Function" (T. J. Givnish, ed.), pp. 381–411. Cambridge Univ. Press, Cambridge.

Mooney, H. A., ed. (1977). "Convergent Evolution in Chile and California Mediterranean-Climate Ecosystems." Dowden, Hutchinson and Ross, Stroudsburg, Pennsylvania.

Mooney, H. A., and Dunn, E. O. (1970a). Convergent evolution of Mediterranean sclerophyll shrubs. *Evolution* 24: 292–303.

Mooney, H. A., and Dunn, E. L. (1970b). Photosynthetic systems of Mediterranean-climate shrubs and trees of California and Chile. *Am. Nat.* 104: 447–453.

Mooney, H. A., and Kummerow, J. (1971). The comparative water economy of representative evergreen sclerophyll and drought-deciduous shrubs of Chile. *Bot. Gaz.* 132: 245–252.

Mooney, H. A., Bjorkman, O., and Collatz, G. J. (1977). Photosynthetic acclimation to temperature in the desert shrub *Larrea divaricata*. I. Carbon dioxide exchange characteristics of intact leaves. *Plant Physiol.* 61: 406–410.

Moore, R. T., Breckle, S. W., and Caldwell, M. M. (1972a). Mineral ion composition and osmotic relations of *Atriplex confertifolia* and *Eurotia lanata*. *Oecologia* 11: 67–78.

Moore, R. T., White, R. S., and Caldwell, M. M. (1972b). Transpiration of *Atriplex confertifolia* and *Eurotia lanata* in relation to soil, plant, and atmospheric moisture stresses. *Can. J. Bot.* 50: 2411–2418.

Mulroy, T. W., and Rundel, P. W. (1977). Annual plants: Adaptations to desert environments. *Bioscience* 27: 109–114.

Naveh, Z., and Dan, J. (1973). The human degradation of Mediterranean landscapes in Israel. *In* "Mediterranean-type Ecosystems: Origin and Structure" (F. di Castri and H. A. Mooney, eds.), pp. 373–390. Springer-Verlag, Berlin.

Nilsen, E. T., Sharifi, M. R., Rundel, P. W., Jarrell, W. M., and Virginia, R. A. (1983). Diurnal and seasonal water relations of the desert phreatophyte *Prosopis glandulosa* (honey mesquite) in the Sonoran Desert of California. *Ecology* 64: 1381–1393.

Nilsen, E. T., Sharifi, M. R., and Rundel, P. W. (1984). Comparative water relations of phreatophytes in the Sonoran Desert of California. *Ecology* 65: 767–778.

Orians, G. H., and Solbrig, O. T., eds. (1977). "Convergent Evolution in Warm Deserts." Dowden, Hutchinson and Ross, Stroudsburg, Pennsylvania.

Poole, A. L., and Adams, N. M. (1963). "Trees and Shrubs of New Zealand." R. E. Owen, Wellington.

Raunkiaer, C. (1934). "The Life Forms of Plants and Statistical Plant Geography." Clarendon, Oxford.

Rundel, P. W. (1981). Fire as an environmental factor. *In* "Encyclopedia of Plant Physiology, New Series" (O. L. Lange, P. S. Nobel, C. B. Osmond, and H. Ziegler, eds.), Vol. 12A, pp. 501–559. Springer-Verlag, Berlin.

Rundel, P. W. (1991). Adaptive significance of some morphological and physiological characteristics for water relations in Mediterranean-climate plants: Facts and fallacies. *In* "Time Scales of Water Stress Response of Mediterranean Biota" (J. Roy, J. Aronson, and F. di Castri, eds.). Springer-Verlag, Berlin (in press).

Rundel, P. W., and Vankat, J. L. (1989). Chaparral communities and ecosystems. *Nat. Hist. Mus. Los Angel. Cty. Sci. Ser.* 34: 127–139.

Sarmiento, G., and Monasterio, M. (1983). Life forms and phenology. *In* "Tropical Savannas" (F. Bourlière, ed.), pp. 79–108. Elsevier, Amsterdam.

Sarmiento, G., Goldstein, G., and Meinzer, F. (1985). Adaptive strategies of woody species in Neotropical savannas. *Biol. Rev.* 60: 315–355.

Savile, D. B. O. (1972). Arctic adaptations in plants. *Can. Dep. Agric. Res. Branch Monogr.* 6: 1–81.

Schimper, A. F. W. (1898). "Pflanzengeographie auf Physiologische Grundlage." Gustav Fischer, Jena, Germany.

Schulze, E.-D. (1982). Plant life forms and their carbon, water, and nutrient relations. *In* "Encyclopedia of Plant Physiology, New Series" (O. L. Lange, P. S. Nobel, C. B. Osmond, and H. Ziegler, eds.), Vol. 12B, pp. 615–676. Springer-Verlag, Berlin.

Small, E. (1972a). Water relations of plants in raised *Sphagnum* peat bogs. *Ecology* 53: 726–728.

Small, E. (1972b). Photosynthetic rates in relation to nitrogen recycling as an adaptation to nutrient deficiency in peat bog plants. *Can. J. Bot.* 50: 2227–2233.

Small, E. (1973). Xeromorphy in plants as a possible basis for migration between arid and nutritionally deficient environments. *Bot. Not.* 126: 534–539.

Specht, R. L. (1981). Heathlands. *In* "Australian Vegetation" (R. H. Groves, ed.), pp. 253–275. Cambridge Univ. Press, Cambridge.

Stebbins, G. L. (1965). The probable growth habit of the earliest flowering plants. *Ann. Mo. Bot. Gard.* 52: 457–468.

Teeri, J. (1973). Polar desert adaptations of a High Arctic plant species. *Science* 179: 496–497.

Tomaselli, R. (1977). The degradation of European maquis. *Ambio* 6: 356–362.

Tranquillini, W. (1979). "Physiological Ecology of the Alpine Timberline." Springer-Verlag, Berlin.

Troll, C., ed. (1968). "Geo-ecology of the Mountain Regions of the Tropical Americas." UNESCO, New York.

Wardle, P. (1974). Alpine timberlines. *In* "Arctic and Alpine Environments" (J. D. Ives and R. G. Barry, eds.), pp. 371–402. Methuen, London.

Warren Wilson, J. (1957). Observations on the temperatures of arctic plants and their environment. *J. Ecol.* 45: 499–531.

Warren Wilson, J. (1959). Notes on wind and its effects in arctic-alpine vegetation. *J. Ecol.* 47: 415–427.

Westman, W. E. (1979). Oxidant effects on California coastal sage scrub. *Science* 205: 1001–1003.

Westman, W. E. (1986). Resilience: Concepts and measures. *In* "Resilience in Mediterranean-type Ecosystems" (B. Dell, A. J. M. Hopkins, and B. B. Lamont, eds.), pp. 5–19. Dr. W. Junk, Dordrecht.

White, R. S. (1976). Seasonal patterns of photosynthesis and respiration in *Atriplex confertifolia* and *Ceratoides lanata*. Ph.D. dissertation, Utah State University, Logan.

Whittaker, R. H., and Woodwell, G. M. (1978). Retrogression and coenocline distance. *In* "Ordination of Plant Communities" (R. H. Whittaker, ed.), pp. 51–70. Dr. W. Junk, The Hague.

Woodwell, G. M. (1968). Effects of chronic gamma irradiation on plant communities. *Q. Rev. Biol.* 43: 42–55.

Woodwell, G. M. (1970). Effects of pollution on the structure and physiology of ecosystems. *Science* 168: 429–433.

Zabriskie, J. G. (1979). "Plants of Deep Canyon." Philip L. Boyd Deep Canyon Research Center, Riverside, California.

17

Responses of Evergreen Trees to Multiple Stresses

R. H. Waring

I. Introduction

Trees that maintain evergreen leaves for more than one year are widely distributed throughout the world in forests, savannas, and at timberline. Evergreen gymnosperms and angiosperms often dominate the vegetation in the subalpine and boreal forests, in Mediterranean woodlands, and in tropical and subtropical areas where rainfall is highly variable or the soils infertile. In temperate regions, evergreens usually do not dominate except after disturbances or where the soil is infertile; the giant conifers characteristic of the temperate forests of the Pacific Northwest are an exception. Their dominance is associated with unusually dry summers with cool nights and mild, wet winters (Waring and Franklin, 1979). The ancient evergreen gymnosperm and angiosperm trees native to New Zealand may survive more for reasons of evolutionary isolation than special adaptations (Benecke and Nordmeyer, 1982; Francey *et al.*, 1985).

Evergreens assimilate carbon dioxide throughout the year when conditions are favorable. Their photosynthetic organs must therefore be able to withstand the full range of environmental hazards to which they are exposed. Moreover, evergreens that maintain multiyear foliage have a limited potential for rapidly replacing more than a fraction of their total canopy in one year. Evergreens thus make major internal and structural adjustments before shedding foliage prematurely. These adjustments are measurable and may serve to integrate the combined effects of environmental stresses on plant physiology and growth.

To appreciate the evolutionary advantages and constraints of evergreen characteristics, we must first understand how a balance among photosynthesis, respiration, and structural partitioning is normally maintained. Then deviations from normal can be interpreted as expressions of chronic or periodic stress. The kinds of deviations observed provide a clue to the types of stress imposed. In presenting these ideas, I will emphasize three key features of both broadleaf and coniferous evergreen trees: (1) low photosynthetic capacity, (2) large amounts of respiring tissue in sapwood, and (3) a low requirement for nutrients.

II. Photosynthetic Activity

Evergreens almost exclusively use the C_3 pathway of photosynthesis. Even the rare C_4 evergreen *Euphorbia forbesii*, which grows in Hawaiian rainforests, photosynthesizes and grows at rates similar to surrounding C_3 trees (Pearcy and Robichaux, 1985). Wherever evergreens grow, from the tiaga to the tropics, their maximum photosynthetic capacities are usually half to less than a quarter that of co-occurring deciduous species (Mooney and Dunn, 1970; Larcher, 1980; Medina and Kinge, 1983; Oechel and Lawrence, 1985; Matyssek, 1986); this is true even within a single genus that includes both deciduous and evergreen species. As evergreen foliage ages, photosynthetic capacity decreases, in some cases to about 5% of the initial capacity (Lassoie *et al.*, 1985). Still, because the older foliage often makes up 50% to more than 80% of the total leaf area, it may assimilate far more in total than the current year's foliage (Schulze *et al.*, 1983, Caldwell *et al.*, 1986).

The difference in photosynthetic capacity between evergreen and deciduous trees is related to variations in the levels of carboxylating enzymes and, to a lesser extent, chlorophyll (Berry and Downton, 1982). Older leaves, as a result of their subservient position, adapt to a more shaded environment by continually adjusting pigments, enzymes, and phloem (Björkman, 1981; Ewers, 1982; Hollinger, 1989).

In association with low photosynthetic capacity and a relatively high weight per unit of surface area, evergreen foliage contains few nutrients. Nitrogen is often as low as 0.6% of dry weight, and phosphorus may be below 0.02% (Vitousek and Sanford, 1986). Yet expressed in terms of leaf area, nutrient concentrations in evergreens may equal those reported in deciduous leaves. If evergreens accumulate high levels of nitrogen, it is evidently not synthesized into enzymes or pigments essential for photosynthesis (Fig. 1). If mineral imbalances occur, photosynthetic capacity may be reduced, regardless of the absolute concentrations of nitrogen or other elements (Zech *et al.*, 1969; Kreutzer, 1972; Linder and Rook, 1984; Sheriff *et al.*, 1986).

Although evergreens generally have a low photosynthetic capacity per unit area of foliage, in many environments they can develop a canopy of leaves dense enough to absorb nearly all the visible (400–700 nm) solar

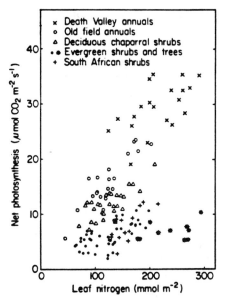

Figure 1 In a survey of 137 species, evergreen shrubs and trees exhibited the lowest photosynthetic rates and the least change in response to increasing foliar nitrogen. [Reprinted from Field and Mooney (1986) with permission from Cambridge Univ. Press.]

radiation (Jarvis and Leverenz, 1983). Moreover, evergreen foliage may photosynthesize while neighboring deciduous trees lack foliage. Evergreens in some regions are thus able to grow up through competing deciduous vegetation, putting deciduous plants at a disadvantage once the evergreens dominate.

To calculate the relative effectiveness of evergreen foliage in absorbing photosynthetically active radiation (PAR), we may apply a simple Beers-law analysis, with knowledge of an extinction coefficient *(K)* that averages about 0.5 for many evergreen trees (Table I). When the product of L and K equals 3.0, 95% of the visible light reaching the canopy is absorbed (Fig. 2). Many evergreen forests, particularly those growing in maritime or subalpine environments, support leaf areas twice that required to absorb 95% of all the incoming radiation (Jarvis and Leverenz, 1983; Waring, 1983). For example, Table I shows that *Pseudotsuga menziesii* can have an L of 11 but only needs an L of 6 to absorb 95% of the visible radiation. This extra leaf area is relatively inconsequential photosynthetically; its development and maintenance costs are low enough that each branch is metabolically self-supporting. Ecologically, the extra leaf area serves to capture more nutrients and light, thus reducing competition from other types of vegetation. Pruning and thinning experiments reported in the forestry literature support the conclusion that excessive leaf areas may be reduced, often by half, without demonstrably reducing tree or stand growth (Waring *et al.*, 1981; Waring, 1983; Larsson *et al.*, 1983; Ericcson *et al.*, 1985).

Table I Leaf Area Index (L_{95}) Required to Intercept 95% of PAR by Selected Evergreen Canopies[a]

Species	Leaf area index (L)	Extinction coefficient (K)	L_{95}
Coniferous			
Pinus resinosa	2.6	0.40	7.5
Pinus radiata	8.3	0.51	5.9
Pinus sylvestris	2.8	0.62	4.8
Picea sitchensis	9.8	0.53	5.7
Pseudotsuga menziesii	11.0	0.48	6.2
Tsuga mertensiana	5.6	0.50	6.0[b]
Broadleaf			
Eucalyptus maculata	2.8	0.57	5.3
Castanopsis cuspidata	7.0	0.50	6.0
Quercus coccifera	5.4	0.55	5.4[c]

[a] After Jarvis and Leverenz (1983).
[b] Source: Waring *et al.* (1987).
[c] Source: Caldwell *et al.* (1986).

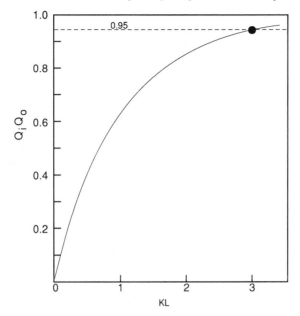

Figure 2 The relationship between canopy interceptance (light below Q_i to light above Q_o) and the product of leaf area index L and the extinction coefficient K indicates that at a value of 3, 95% of the incoming PAR has been absorbed by the canopy. [Reprinted from Jarvis and Leverenz (1983) with permission from Springer-Verlag.]

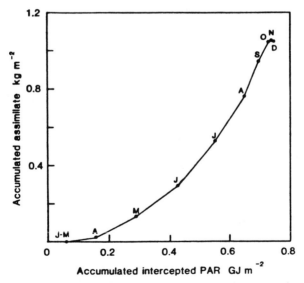

Figure 3 Accumulated photosynthate in a young Scots pine forest in relation to accumulated intercepted PAR over a year. [Reprinted from Jarvis and Leverenz (1983) with permission from Springer-Verlag.]

It has long been recognized in agriculture that net photosynthesis over a growing season is often closely related to a canopy's ability to absorb PAR (Baker, 1965; Lemon, 1967; Denmead, 1976; Acock *et al.*, 1978; Baldocchi *et al.*, 1981), but only recently has the relationship been evaluated for evergreen forests (Fig. 3). Taking into account seasonal changes in solar radiation and the light-absorbing properties of the canopy in an evergreen oak forest, Caldwell *et al.* (1986) also showed that the measured annual carbon uptake is closely related to the absorbed PAR integrated over the year. Global satellite coverage offers the potential of remotely monitoring seasonal changes in canopy spectral absorption and relating these changes to photosynthetic activity and net primary production (Goward *et al.*, 1985; Tucker *et al.*, 1986; Goward *et al.*, 1987).

III. Respiration

A certain amount of the photosynthate assimilated by a leafy canopy is respired as carbon dioxide during the synthesis of new tissues and maintenance of living cells. Biochemical synthesis of a given kind and amount of tissue produces the same amount of carbon dioxide, regardless of the temperature at which synthesis occurs (Penning de Vries, 1974; Williams *et al.*, 1987). In contrast, maintenance respiration of living cells increases exponentially with temperatures between 0 to about 40° C, and linearly with the quantity of enzymes present (Penning de Vries, 1975; Jones *et al.*, 1978; Penning de Vries and van Laar, 1982).

Synthesis costs vary considerably with the biochemical composition of tissue. For example, lignin and cellulose, major components in the cell walls of all evergreen leaves, cost in glucose equivalents about a third of what it costs to synthesize lipids and half what it costs for proteins (Penning de Vries, 1974). This means that evergreen foliage, although high in calories, may be less costly to synthesize than leaves with higher photosynthetic capacities and more protein content.

Per unit weight the maintenance respiration of evergreen foliage is also likely to be less than for deciduous foliage. First, the large proportion of cell wall material, once formed, costs nothing to maintain. Second, the relatively low photosynthetic capacity of evergreen foliage is associated with lower enzyme content, reducing maintenance respiration proportionately.

A surplus of nitrogen may pose special hazards to many evergreens because maintenance respiration increases without a parallel increase in photosynthetic capacity. All inorganic nitrogen, once absorbed by roots, must eventually be reduced into organic forms to be useful. This requires specific enzymes, some of which turnover unusually rapidly and thus increase maintenance costs (Zielke and Filner, 1971). Many evergreens have modest capacities to reduce nitrogen in the form of nitrate (Smirnoff *et al.*, 1984; Smirnoff and Stewart, 1985). Excess nitrogen in dense evergreen forests may ultimately result in a negative carbon balance in some leaves and branches, forcing premature shedding (Aber *et al.*, 1989). Even if excluded from root uptake, excess nitrogen in the soil may reduce availability of other nutrients and lead to nutritional imbalances in trees (Schulze, 1989).

Although deciduous and evergreen trees may grow a similar amount of foliage annually, a more extensive vascular system is required to support a multiaged evergreen canopy. The water conducting sapwood contains from 5% to 30% living parenchyma, depending on the taxonomic group (Panshin *et al.*, 1964), and these live cells require carbohydrates that otherwise might go to the synthesis of new tissues.

Although the fraction of living cells in sapwood may be low, parenchyma cell weight may match that of the foliage in a 40-year-old pine forest (Waring *et al.*, 1979). Because the enzyme content of parenchyma cells is less than in leaf tissue, the actual cost of maintenance is not directly proportionate to weight. Nevertheless, maintenance costs of tree boles, branches, and large-diameter roots may be considerable as trees enlarge (Fig. 4; Waring, 1987; Ryan, 1990). One adaptation of trees that counterbalances this trend is the formation of heartwood, which provides support but contains no living cells. The formation of heartwood is irrevers-

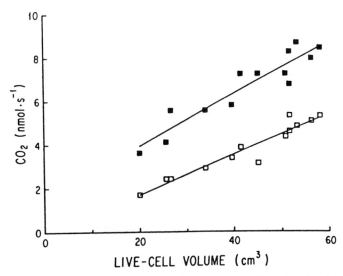

Figure 4 Maintenance respiration of lodgepole pine is linearly related to sapwood volume at a temperature of 0.5° C (open boxes) and also at 13° C (closed boxes). [From Ryan (1990).]

ible, however, so when evergreen trees lose 20–40% of their foliage in one year, the sapwood supporting the original leaves usually remains alive and continues to consume carbohydrates (Margolis *et al.*, 1988). Trees with large amounts of respiring nonphotosynthetic tissue are therefore more vulnerable to excessive foliage loss than trees with lower maintenance costs.

It may be possible to rank the relative sensitivity of various species to foliage loss from a knowledge of their sapwood–to–leaf area relationships. Within a particular forest region (e.g., the western U.S.), species that characteristically occupy harsher environments, such as lodgepole pine and juniper, support only a small leaf area for a given cross-sectional area of sapwood compared with species adapted to more moderate or maritime climates (Table II).

The relative importance of maintenance respiration by tree boles and branches can be roughly assessed by comparing how rapidly the ratio of sapwood volume increases in relation to leaf area as trees grow. Three subalpine conifers show significantly different trends in this regard, differences that reflect their relative abilities to sustain growth rates on a particular site as they increase in size (Fig. 5). Although initially fast-growing, the growth rate of pine decreases two to three times faster than that of spruce or fir of similar diameter (Kaufmann and Ryan, 1986). Not surprisingly, without disturbance, pine is eventually replaced by the slower-growing, more shade-tolerant conifers with lower maintenance respirations in their branches, stems, and large roots.

How a loss of foliage will affect trees of similar size can be addressed by examining differences in photosynthetic activity. The pine holds fewer age

Table II   Ratios of Projected Leaf Area to Sapwood
Cross-Sectional Area for Selected Conifer Species
Found along a West–East Transect[a]

Species[b]	Environment	Ratio $(m^2\ cm^{-2})$
Sitka spruce	Maritime	0.44
Douglas fir	Moderate	0.32
Noble fir	Humid subalpine	0.27
Lodgepole pine	Continental	0.15
Mountain hemlock	Continental	0.16
Ponderosa pine	Semiarid	0.17
Western juniper	Arid	0.07

[a] Transect runs from the maritime Pacific Coast over two mountain ranges to an arid interior plateau in Oregon (lat. 44°N). [After Waring (1980).]

[b] The sapwood of these conifers contains 5–8% living cells by volume.

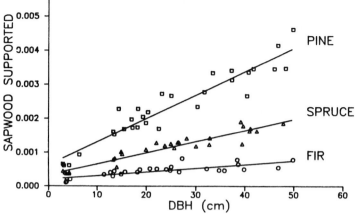

Figure 5 In the subalpine zone, the ratio of sapwood volume to a given leaf area changes much more rapidly for lodgepole pine than for its successional replacements, Engelmann spruce or subalpine fir. The rate of change reflects the relative importance of sapwood respiration as trees grow in size. [From Ryan (1988); see also Ryan (1989).]

classes of foliage than spruce or fir. Its needles are better adapted to photosynthesize under high evaporative demand, with two to three times the water-conducting tissue per unit of leaf area. With an opening in the canopy associated with partial defoliation, both radiation and evaporative demand increase in the remaining canopy. In this case, a large loss of foliage is likely to be less detrimental to early successional species, such as pine, because they have relatively high amounts of sapwood per unit of foliage and fewer age classes of foliage to replace than more shade-tolerant species, such as spruce or fir (Whitehead and Jarvis, 1981).

IV. Partitioning

Assimilate is translocated within a plant to produce leaves, stems, branches, roots, and reproductive organs, and it helps support fungal and bacterial symbionts. Assimilate may also be synthesized into stored reserves, such as starch or oils, and a variety of defensive compounds. The overall pattern of assimilate distribution includes partitioning to organs, allocation into various biochemical constituents, and various respiratory losses (see Chapters 5, 6, 12). Changes in allocation and partitioning of assimilate indicate the kind and degree of stress plants experience (Table III).

A. Shade

As a tree grows, its crown normally expands and begins to shade other trees. In a forest, wood production per unit of foliage therefore decreases

Table III   Changes in Carbon Partitioning and Allocation Associated
with Various Stresses on Trees[a]

Stress	Root growth	Stem taper	Foliage growth	Other
Shade	Reduced	Reduced	Increased	Umbrella-shaped crown, decreased reserves
Drought	Increased	Increased	Reduced	Loss of older foliage
Low temperature	Increased	Reduced	Reduced	Increased reserves
Mechanical	Increased	Increased	Reduced	Asymmetrical shape of bole, branches, and roots
Nitrogen deficiency	Increased	Increased	Reduced	Higher C reserves, less amino acids
Nitrogen surplus	Reduced	Reduced	Increased	Decreased reserves, more amino acids

[a] Summarized from literature cited in the text.

as the total canopy leaf area increases (Fig. 6). The rate and extent of that decrease depend not only on competition among trees for light but also on competition with understory plants for water and nutrients (Oren *et al.*, 1987).

Once a forest reaches maximal canopy, leaves are distributed on fewer and fewer trees as stand growth continues. For reasons associated with disproportionate growth in sapwood volume, the surviving trees' requirements for photosynthate may increase more rapidly than their canopies can sustain (see Fig. 5). Because maintenance costs increase so dramatically with tree size, small saplings have a better chance of surviving under a dense canopy than do larger, pole-sized individuals. Trees adapted to extreme shade often have umbrella-shaped crowns, thin leaves, little conducting tissue, very little taper in their stems from the base to the top, and greatly reduced root growth and seed production (Waring and Schlesinger, 1985).

In shade-stressed trees, new foliage is favored over other structures, and carbohydrate reserves are always low. Such suppressed trees have shapes that make them susceptible to damage by wind and snow. If light conditions suddenly improve, they do poorly in competition with more normally shaped trees, in particular with deciduous species, for reasons mentioned previously.

B. Drought

Partial drought creates conditions where photosynthesis still continues, but shoot elongation is inhibited. Leaves produced under such conditions will be smaller than normal but may contain more stored carbohydrates and

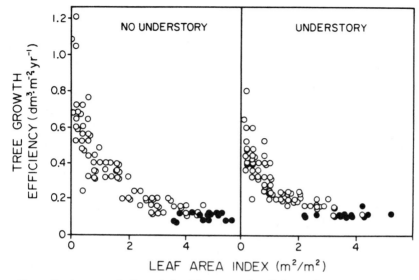

Figure 6 As canopy leaf area index L increases, the efficiency of stemwood production decreases. Competition with understory vegetation for water and nutrients further reduces growth efficiency until L is sufficient to shade out the understory. [Reprinted from Oren *et al.* (1987) with permission from the Society of American Foresters, Washington, D.C.]

defensive compounds. With reduced shoot growth, sugar concentrations increase in the phloem transport system. The flux of carbon to roots is disproportionately increased, although the velocity of solute flow may decrease (Magnuson *et al.*, 1979). Combined with changes in nutrient availability, root growth is favored over the elongation of shoots (Vessey and Layzell, 1987).

C. Temperature

Temperature changes may activate or deactivate certain enzyme pathways so that enzyme kinetics affects carbon partitioning and allocation. Lowering the temperature particularly influences the ability of roots to take up water and nutrients (Kramer, 1983). Even a small drop in stem temperature may halt phloem transport of carbohydrates to roots in some plants, as shown with short-lived isotopes of carbon (Goeschl *et al.*, 1984).

The general effect of cooler temperatures is to reduce shoot growth, which permits roots to capture adequate nutrients and water. Reduced growth and cooler-than-normal temperatures increase carbohydrate reserves because of a decrease in synthesis and maintenance respiration. Critically low temperatures, of course, limit the growing season, immobilize nutrients, and adversely affect water relations and photosynthetic activity.

D. Nutrition

Assimilate and other resources appear to be shifted in a manner that helps maintain adequate photosynthetic capacity. These shifts may be internal reallocations or involve structural changes. If resources cannot be sequestered for the established leaves, branches, roots, and seed-bearing organs when needed, then growth is likely to cease, and some organs may be shed.

To maintain even moderate photosynthetic capacity, nitrogen and phosphorus are required in relatively large amounts compared with other nutrients. Nitrogen, of course, is critical for the synthesis of enzymes and pigments essential for photosynthesis, and phosphorus is required in ATP, an energy source for biochemical reactions. Other nutrients are also required but often may be obtained through translocation from older, less active leaves and other tissues at the cost of a reduction in overall growth (Schulze, 1989). Deficiencies in nitrogen, and to a lesser extent phosphorus, favor growth of small-diameter roots over that of the shoot (Ingestad, 1979; Axelsson and Axelsson, 1986).

Vessey and Layzell (1987) clarify why excess nitrogen results in a relative reduction in root growth. Once sufficient nitrogen is available to meet the plant's ability to maximize photosynthesis in a given light environment, the excess accumulates in other forms. Nitrogen in the form of amino acids (never nitrate) is translocated back down the phloem to the roots. Under normal conditions, the roots sequester most of this soluble nitrogen. When their requirements are exceeded, however, proportionately less nitrogen is used, and the excess is shunted to the vascular system, where it is transported by mass flow back to the leaves. The greater the flux of nitrogen to foliage, the greater the opportunity for foliage growth. Drought, which forces plants to extract water from deeper, less fertile soil horizons where nutrient concentrations are reduced, in turn also reduces the flux of nitrogen and other minerals to the foliage. In this way drought can lead to a shift in the ratio of mobile carbohydrates to nitrogen similar to that experienced by plants with nitrogen deficiencies, with parallel shifts in partitioning observed.

Experimental studies in Sweden support the general theories of nutrient flux and relative growth rate developed by Ingestad (1987) and Vessey and Layzell (1987). There, following addition of a balanced nutrient solution to a pine forest throughout the growing season, nearly a three-fold increase in leaf area and stem growth was observed with only a slight improvement in the photosynthetic capacity of the foliage (Linder and Rook, 1984). A parallel reduction in the production of small-diameter roots occurred as foliage and stemwood production increased (Axelsson and Axelsson, 1986).

Carbohydrate reserves are low in trees well supplied with nitrogen because growth starts earlier and is greater, and the tissues formed require more carbohydrates for maintenance than in nitrogen-stressed trees (Margolis and Waring, 1986a; Birk and Matson, 1986). At the end of the growing season, however, large carbohydrate reserves may accumulate (Margolis and Waring, 1986b; Birk and Matson, 1986). Whether rapid leaf growth and expenditure of stored reserves are good or bad depends on what stresses occur later in the growing season.

E. Combined Stresses

Usually no one environmental factor limits the growth of evergreen trees. For example, on an open site when tree seedlings first become established, frost may limit photosynthesis. As the forest canopy closes, light and water become more limiting. Normally, climate controls the rates of nutrient release from organic matter. Insect defoliation, thinning, fire, and other disturbances tend to increase the cycling of essential nutrients, whereas the absence of disturbance tends to limit the rates of nutrient release. Thus, evergreens, as well as other vegetation, often grow better when essential nutrients are supplied somewhat in excess of that available through normal decomposition of organic matter, weathering, and atmospheric inputs, assuming there is growing space for an expanded canopy.

Air pollutants such as ozone may act like frost in reducing photosynthetic capacity, but older foliage rather than new foliage is more likely to be prematurely shed. Other pollutants, such as sulfur dioxide and oxides of nitrogen, appear to alter nutrient balances and reduce carbon transport to roots (Whitmore, 1985; Vogels *et al.*, 1986; Kuhn and Beck, 1987). This may help account for observations of more wood being added to the upper bole and less to the base of spruce trees in European forests showing general symptoms of forest decline (Schutt and Cowling, 1985).

Three principles may be deduced from a study of carbon allocation and partitioning in trees under stress. First, the photosynthetic process is generally a high priority for evergreen trees. The maintenance and replacement of photosynthetic machinery is essential, as indicated by the observation that conducting sapwood was not reduced despite a temporary removal of 20–40% of the foliage (Margolis *et al.* 1988).

Second, to support leaves, a minimum phloem and water-conducting system must be maintained. When roots do not conduct sufficient water to leaves, the current photosynthate and some mobilized reserves may go to build more roots. In time, a tree's transport system adjusts to reflect consistent carbohydrate demands. The phloem cross-sectional area per unit of leaf area (or sapwood area) has been shown to indicate such changes (Gifford and Evans, 1981).

Third, when nitrogen is in relative excess to other nutrients, evergreens appear to restrict growth of their small-diameter roots. This response may be an evolutionary adaptation because most environments where ever-

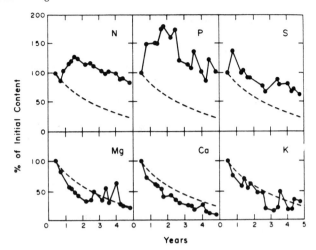

Figure 7 Loss of nutrients from Scots pine needles in a Swedish forest shows that N, P, and S are selectively accumulated by microbes using the litter as substrate, whereas Mg, Ca, and K are released at rates greater than the general weight loss (---). [After Staaf and Berg (1982). Reprinted from Waring and Schlesinger (1985) with permission from Academic Press.]

greens thrive release other nutrients more rapidly during litter decomposition (Fig. 7; Vitousek and Melillo, 1979, Vitousek *et al.* 1982). Deciduous trees, in contrast to evergreens, are more adapted to variation in nutrient availability—particularly to large flushes of nutrients—owing to their abilities to adjust leaf area more quickly and their generally higher growth and photosynthetic potentials.

V. Susceptibility to Symbionts, Pathogens, and Herbivores

Most evergreen trees have survived because of dependency on various symbionts and resistance to native pathogens and herbivores. Evergreens require photosynthate to produce protective organs, such as thorns or bark, or to synthesize defensive compounds, such as tannins, resins, and terpenes. Few evergreens produce defensive compounds like alkaloids, with nitrogen in their molecular structure (Swain, 1977). Thus, any environmental stress that greatly reduces the supply of assimilate available for producing carbon-based defensive structures or compounds, or for maintaining a symbiotic association, creates a situation favorable for many pathogens and herbivores.

A reduction in the availability of carbohydrates is easily induced by increasing the availability of nitrogen or provoking any general imbalance in nutrition. The excess nitrogen is usually converted to amino acids, and an excess of amino acids provides improved substrate for many pathogens (Fig. 8) and herbivores, particularly if combined with a reduction in carbon-based defensive compounds (Waring and Schlesinger, 1985).

The size of a plant is not directly correlated with its susceptibility to herbivory. When colonizing a disturbed site, tree seedlings may find abundant resources; later, as saplings, they may face harsh competition for resources. Large trees, on the other hand, eventually slow their height growth, allowing competing trees to encroach. When this happens, the lower branches of the large trees contribute little to the total assimilate, and the trees become increasingly vulnerable to attack from a variety of organisms (Cole and Amman, 1969; Waring and Pitman, 1985; Christiansen *et al.*, 1987). Likewise, in harsh environments, large areas of mature trees can die when infected by native pathogens. Younger trees of the same species can become established again in forest openings, although pathogens still reside on the dead roots of the original trees (McCauley and Cook, 1980; Entry *et al.*, 1986; Waring *et al.*, 1987).

Many fungi form symbiotic relationships with the roots of a wide variety of plants. Some types of mycorrhizal fungi are efficient at using dead organic material to help sustain themselves (Trojanowski *et al.*, 1984). There is no evidence, however, that mycorrhizal fungi can establish themselves on roots without access to current photosynthate (Harley and Smith, 1983: p. 146–151; Cairney *et al.*, 1989). This means that any stress that reduces the export of carbohydrates to roots is also likely to inhibit coloni-

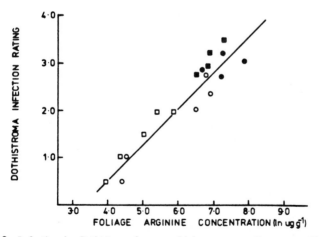

Figure 8 Infection by *Dothistroma* fungus, which causes needle cast on *Pinus radiata*, increases with the concentration of the amino acid arginine (□ untreated; ○ sulfur added; ● moderate nitrogen added; ■ high amounts of nitrogen added). [Reprinted from Turner and Lambert (1986) with permission from Annual Reviews, Inc., Palo Alto, California.]

zation of symbionts and adversely affect water and nutrient uptake (Bowen, 1982, 1984; France and Reid, 1983).

VI. Conclusions

Sustained stress on evergreen trees results in premature shedding of foliage and a loss of competitive status relative to many deciduous trees. The remaining evergreen foliage cannot easily adapt to a more extreme environment, and the original conducting tissue still requires carbohydrates for maintenance. Nutrient imbalances, particularly those associated with relative excess of nitrogen, are difficult for evergreen trees to accommodate, particularly in forests where dense canopies limit photosynthesis and expansion of leaf area.

Various kinds of environmental stresses induce predictable changes in how carbohydrates are partitioned to leaves, stem, roots, and storage reserves. Chronic stress on evergreens often prevents accumulation of storage carbohydrates in twigs, stems, and large roots; reduces stem growth per unit of leaf area; changes the normal ratios of nutrients in foliage and other tissues; and alters demands for carbohydrates by various organs. Sustained changes in how carbohydrates are partitioned among different organs may induce parallel changes in the relative conducting area of phloem and xylem tissue.

References

Aber, J. D., Nadelhoffer, K. J., Steudler, P., and Melillo, J. M. (1989). Nitrogen saturation in northern forest ecosystems. *Bioscience* 39: 378–386.

Acock, B., Charles-Edwards, D. A., Fitter, D. J., Hand, D. W., Ludwig, J., Warren, J. W., and Withers, A. C. (1978). The contribution of leaves from different levels within a tomato crop to canopy net photosynthesis: An experimental examination of two canopy models. *J. Exp. Bot.* 29: 815–827.

Axelsson, E., and Axelsson, B. (1986). Changes in carbon allocation patterns in spruce and pine trees following irrigation and fertilization. *Tree Physiol.* 2: 189–204.

Baker, D. N. (1965). Effects of certain environmental factors on net assimilation in cotton. *Crop Sci.* 5: 53–56.

Baldocchi, D. D., Verma, S. B., and Roseberg, N. J. (1981). Mass and energy exchanges of a soybean canopy under various environmental regimes. *Agron. J.* 73: 706–710.

Benecke, U., and Nordmeyer, A. H. (1982). Carbon uptake and allocation by *Nothofagus solandri* var. *cliffortioides* (Hook. f.) and *Pinus contorta* at montane and subalpine altitudes. *In* "Carbon Uptake and Allocation in Subalpine Ecosystems as a Key to Management" (R. H. Waring, ed.), pp. 9–21. Forest Research Lab., Oregon State University, Corvallis, Oregon.

Berry, J. A., and Downton, W. H. S. (1982). Environmental regulation of photosynthesis. *In* "Photosynthesis: Development, Carbon Metabolism, and Plant Productivity" (Govindjee, ed.), Vol. 2, pp. 263–343. Academic Press, New York.

Birk, E. M., and Matson, P. A. (1986). Site fertility affects seasonal carbon reserves in loblolly pine. *Tree Physiol.* 2: 17–27.

Björkman, O. (1981). Responses to different quantum flux densities. *In* "Encyclopedia of Plant Physiology, New Series" (O. L. Lange, P. S. Nobel, C. B. Osmond, and H. Ziegler eds.), Vol. 12A, pp. 57–107. Springer-Verlag, Berlin.

Bowen, G. D. (1982). The root–microorganism ecosystem. *In* "Biological and Chemical Interactions in the Rhizophere" (T. Rosswall, ed.), pp. 3–42. Swedish Natural Sciences Research Council, Stockholm.

Bowen, G. D. (1984). Tree roots and the use of soil nutrients. *In* "Nutrition of Plantation Forests" (G. D. Bowen and E. K. S. Nambiar, eds.), pp. 147–180. Academic Press, Orlando, Florida.

Cairney, J. W. G., Ashford, A. E., and Allaway, W. G. (1989). Distribution of photosynthetically fixed carbon within root systems of *Eucalyptus pilularis* plants ectomycorrhizal with *Pisolithus tinctorius*. *New Phytol.* 112: 495–500.

Caldwell, M. M., Meister, H.-P., Tenhunen, J. D., and Lange, O. L. (1986). Canopy structure, light microclimate, and leaf gas exchange of *Quercus coccifera* L. in a Portuguese macchia: Measurements in different canopy layers and simulations with a canopy model. *Trees* 1: 25–41.

Cole, W. E., and Amman, G. D. (1969). Mountain pine beetle infestations in relation to lodgepole pine diameters. U.S. For. Serv. Res. Note INT 95.

Christiansen, E., Waring, R. H., and Berryman, A. A. (1987). Resistance of conifers to bark beetle attack: Searching for general relationships. *For. Ecol. Manage.* 22: 89–106.

Denmead, O. T. (1976). Temperate cereals. *In* "Vegetation and the Atmosphere" (J. L. Monteith, ed.), pp. 1–33. Academic Press, New York.

Ericcson, A., Hellkvist, C., Lanstrom, B., Larsson, S., and Tenow, O. (1985). Effects on growth of simulated and induced pruning by *Tomicus piniperda* is related to carbohydrate and nitrogen dynamics in Scots pine. *J. Appl. Ecol.* 22: 105–124.

Entry, J. R., Martin, N. E., Cromack, K., Jr., and Stafford, S. G. (1986). Light and nutrient limitations in *Pinus monticola:* Seedling susceptibility to *Armillaria* infection. *For. Ecol. Manage.* 17: 189–198.

Ewers, F. W. (1982). Secondary growth in needle leaves of *Pinus longaeva* (bristlecone pine) and other conifers: Quantitative data. *Am. J. Bot.* 69: 1552–1559.

Field, C., and Mooney, H. A. (1986). The photosynthesis–nitrogen relationship in wild plants. *In* "On the Economy of Plant Form and Function" (T. J. Givnish, ed.), pp. 25–55. Cambridge Univ. Press, London.

France, R. C., and Reid, C. P. P. (1983). Interactions of nitrogen and carbon in the physiology of ectomycorrhizae. *Can. J. Bot.* 61: 964–984.

Francey, R. J., Gifford, R. M., Sharkey, T. D., and Weir, B. (1985). Physiological influence on carbon isotope discrimination in huon pine *(Lagarostrobos franklinii)*. *Oecologia* 66: 211–218.

Gifford, R. M., and Evans, L. T. (1981). Photosynthesis, carbon partitioning, and yield. *Annu. Rev. Plant Physiol.* 32: 485–509.

Goeschl, J. D., Magnuson, C. E., Fzres, Y., Jaeger, C. H., Nelson, C. E., and Strain, B. R. (1984). Spontaneous and induced blocking and unblocking of phloem transport. *Plant, Cell Environ.* 7: 607–613.

Goward, S. N., Tucker, C. J., and Dye, D. G. (1985). North American vegetation patterns observed with the NOAA-7 advanced very high resolution radiometer. *Vegetatio* 64: 3–14.

Goward, S. N., Dye, D., Kerber, A., and Kalb, V. (1987). Comparison of North and South American biomes from AVHRR observations. *Geocarto Int.* 2: 27–39.

Harley, J. L., and Smith, S. E. (1983). "Mycorrhizal Symbiosis." Academic Press, London.

Hollinger, D. Y. (1989). Canopy organization and foliage photosynthetic capacity in a broadleaved evergreen montane forest. *Funct. Ecol.* 3: 53–62.

Ingestad, T. (1979). Nitrogen stress in birth seedlings. II. N, K, P, Ca, and Mg nutrition. *Physiol. Plant.* 45: 149–157.

Ingestad, T. (1987). New concepts on soil fertility and plant nutrition as illustrated by research on forest trees and stands. *Geoderma* 40: 237–252.

Jarvis, P. G., and Laverenz, J. W. (1983). Productivity of temperate, deciduous, and evergreen forests. *In* "Encyclopedia of Plant Physiology, New Series" (O. L. Lange, P. S. Nobel, C. B. Osmond, and H. Ziegler, eds.), Vol. 12D, pp. 233–280. Springer-Verlag, Berlin.

Jones, M. B., Leafe, E. L., Stiles, W., and Collett, B. (1978). Pattern of respiration of a perennial ryegrass crop in the field. *Ann. Bot.* 42: 693–703.

Kaufmann, M. R., and Ryan, M. G. (1986). Physiographic, stand, and environmental effects on individual tree growth and growth efficiency in subalpine forests. *Tree Physiol.* 2: 47–59.

Kramer, P. J. (1983). "Water Relations of Plants." Academic Press, New York.

Kreutzer, K. (1972). Die Wirkung des Manganmangels auf die Farbe, die Pigmente, und den Gaswechsel von Fichtennadeln (*Picea abies* Karst.). *Forstwiss. Centralbl.* 91: 80–98.

Kuhn, U., and Beck, E. (1987). Conductance of needle and twig axis phloem of damaged and intact Norway spruce (*Picea abies* [L.] Karst.) as investigated by application of ^{14}C in situ. *Trees* 1: 207–214.

Larcher, W. (1980). "Physiological Plant Ecology," 2nd ed. Springer-Verlag, Berlin.

Larsson, S., Oren, R., Waring, R. H., and Barrett, J. W. (1983). Attacks of mountain pine beetle as related to tree vigor of ponderosa pine. *For. Sci.* 29: 395–402.

Lassoie, J. P., Hinckley, T. M., and Grier, C. C. (1985). Coniferous forests of the Pacific Northwest. *In* "Physiological Ecology of North American Plant Communities" (B. F. Chabot and H. A. Mooney, eds.), pp. 127–161. Chapman and Hall, New York.

Lemon, E. R. (1967). Aerodynamic studies of CO_2 exchange between atmosphere and the plant. *In* "Harvesting the Sun" (A. San Petro, F. A. Greer, and T. J. Army, eds.), pp. 117–137. Academic Press, New York.

Linder, S., and Rook, D. A. (1984). Effects of mineral nutrition on carbon dioxide exchange and partitioning of carbon in trees. *In* "Nutrition of Plantation Forests" (G. D. Bowen and E. K. S. Nambiar, eds.), pp. 211–236. Academic Press, Orlando, Florida.

McCauley, K. J., and Cook, S. A. (1980). *Phillinus weirii* infestation of two mountain hemlock forests in the Oregon Cascades. *For. Sci.* 26: 23–29.

Magnuson, C. E., Goeschl, J. D., Sharpe, P. J. H., and Demichele, D. W. (1979). Consequences of insufficient equations in models of the Munch hypothesis of phloem transport. *Plant, Cell Environ.* 2: 181–188.

Margolis, H. A., and Waring, R. H. (1986a). Carbon and nitrogen allocation patterns of Douglas-fir seedlings fertilized with nitrogen in autumn. I. Overwinter metabolism. *Can. J. For. Res.* 16: 897–902.

Margolis, H. A., and Waring, R. H. (1986b). Carbon and nitrogen allocation patterns of Douglas-fir seedlings fertilized with nitrogen in autumn. II. Field performance. *Can. J. For. Res.* 16: 903–909.

Margolis, H. A., Gagnon, R. R., Pothier, D., and Pineau, M. (1988). The adjustment of growth, sapwood area, heartwood area, and sapwood-saturated permeability of balsam fir after different intensities of pruning. *Can. J. For. Res.* 18: 723–727.

Matyssek, R. (1986). Carbon, water, and nitrogen relations in evergreen and deciduous conifers. *Tree Physiol.* 2: 177–187.

Medina, E., and Kinge, H. (1983). Tropical forests and tropical woodlands. *In* "Encyclopedia of Plant Physiology, New Series" (O. L. Lange, P. S. Nobel, C. B. Osmond, and H. Ziegler, eds.), Vol. 12D, pp. 281–303. Springer-Verlag, Berlin.

Mooney, H. A., and Dunn, E. L. (1970). Convergent evolution of Mediterranean-climate evergreen sclerophyll shrubs. *Evolution* 24: 292–303.

Oechel, W. C., and Lawrence, W. T. (1985). Tiaga. *In* "Physiological Ecology of North

American Plant Communities" (B. F. Chabot and H. A. Mooney, eds.), pp. 66–94. Chapman and Hall, New York.

Oren, R., Waring, R. H., Stafford, S. G., and Barrett, J. W. (1987). Twenty-four years of ponderosa pine growth in relation to canopy leaf area and understory competition. *For. Sci.* 33: 538–547.

Panshin, A. J., de Zeeuw, C., and Brown, H. P. (1964). "Textbook of Wood Technology," 2nd ed. McGraw-Hill, New York.

Pearcy, R. W., and Robichaux, R. H. (1985). Tropical and subtropical forests. *In* "Physiological Ecology of North American Plant Communities" (B. F. Chabot and H. A. Mooney, eds.), pp. 278–295. Chapman and Hall, New York.

Penning de Vries, F. W. T. (1974). Substrate utilization and respiration in relation to growth and maintenance in higher plants. *Neth. J. Agr. Sci.* 22: 40–44.

Penning de Vries, F. W. T. (1975). The cost of maintenance processes in plant cells. *Ann. Bot.* 39: 77–92.

Penning de Vries, F. W. T., and van Laar, H. H., eds. (1982). "Simulation of Plant Growth and Crop Production." Simulation Monographs. Pudoc, Wageningen.

Ryan, M. G. (1988). The importance of maintenance respiration by living cells in sapwood of subalpine conifers. Ph.D. dissertation, Oregon State University, Corvallis, Oregon.

Ryan, M. G. (1989). Sapwood volume for three subalpine conifers: Predictive equations and ecological implications. *Can. J. For. Res.* 19: 1397–1401.

Ryan, M. G. (1990). Growth and maintenance respiration in stems of *Pinus contorta* and *Picea engelmannii*. *Can. J. For. Res.* 20: 48–57.

Schulze, E.-D. (1989). Air pollution and forest decline in a spruce *(Picea abies)* forest. *Science* 244: 776–783.

Schulze, E.-D., Kuppers, M., and Matyssek, R. (1983). The role of carbon balance and branching pattern in the growth of woody species. *In* "On the Economy of Plant Form and Function" (T. J. Givnish, ed.), pp. 585–602. Cambridge Univ. Press, Cambridge.

Schutt, P., and Cowling, E. B. (1985). *Waldsterben,* a general decline of forests in central Europe: Symptoms, development, and possible causes. *Plant Dis.* 69: 548–558.

Sheriff, D. W., Nambiar, E. K. S., and Fife, D. N. (1986). Relationships between nutrient status, carbon assimilation, and water use efficiency in *Pinus radiata* (D. Don) needles. *Tree Physiol.* 2: 73–88.

Smirnoff, N., and Stewart, G. R. (1985). Nitrate assimilation and translocation by higher plants: Comparative physiology and ecological consequences. *Physiol. Plant.* 64: 133–140.

Smirnoff, N., Todd, P., and Stewart, G. R. (1984). The occurrence of nitrate reduction in the leaves of woody plants. *Ann. Bot.* 54: 363–374.

Staaf, H., and Berg, B. (1982). Accumulation and release of plant nutrients in decomposing Scots pine needle litter. II. Long-term decomposition in a Scots pine forest. *Can. J. Bot.* 60: 1561–1568.

Swain, T. (1977). Secondary compounds as protective agents. *Annu. Rev. Plant Physiol.* 28: 479–501.

Trojanowski, J., Haider, K., and Huttermann, A. (1984). Decomposition of ^{14}C-labelled lignin, holocellulose, and lignocellulose by mycorrhizal fungi. *Arch. Microbiol.* 139: 202–206.

Tucker, C. J., Fung, I. Y., Keeling, C. D., and Gammon, R. H. (1986). Relationship between CO_2 variations and a satellite-derived vegetation index. *Nature* 319: 195–199.

Turner, J., and Lambert, M. J. (1986). Nutrition and nutritional relationships of *Pinus radiata*. *Annu. Rev. Ecol. Syst.* 17: 325–350.

Vessey, J. K., and Layzell, D. B. (1987). Regulation of assimilate partitioning in soybean. *Plant Physiol.* 83: 341–348.

Vitousek, P. M., and Melillo, J. M. (1979). Nitrate losses from disturbed ecosystems: Patterns and mechanisms. *For. Sci.* 25: 605–619.

Vitousek, P. M., and Sanford, R. L., Jr. (1986). Nutrient cycling in moist tropical forests. *Annu. Rev. Ecol. Syst.* 17: 137–167.

Vitousek, P. M., Gosz, J. R., Grier, C. C., Melillo, J. M., and Reiners, W. A. (1982). A comparative analysis of potential nitrification and nitrate mobility in forest ecosystems. *Ecol. Monogr.* 52: 155–177.

Vogels, K., Guderian, R., and Masuch, G. (1986). Studies on Norway spruce (*Picea abies* Karst.) in damaged stands and in climatic chamber experiments. *In* "Acidification and Its Policy Implications" (T. Scheider, ed.), pp. 171–186. Elsevier, Amsterdam.

Waring, R. H. (1980). Site, leaf area, and phytomass production in trees. *In* "Mountain Environments and Subalpine Tree Growth" (U. Benecke, ed.), pp. 125–135. *N. Z. For. Serv. For. Res. Inst. Tech. Pap.* 70.

Waring, R. H. (1983). Estimating forest growth and efficiency in relation to canopy leaf area. *Adv. Ecol. Res.* 13: 327–354.

Waring, R. H. (1987). Characteristics of trees predisposed to die. *Bioscience* 37: 569–574.

Waring, R. H., and Franklin, J. F. (1979). Evergreen coniferous forests of the Pacific Northwest. *Science* 204: 1380–1386.

Waring, R. H., and Pitman, G. B. (1985). Modifying lodgepole pine stands to change susceptibility to mountain pine beetle attack. *Ecology* 66: 889–897.

Waring, R. H., and Schlesinger, W. H. (1985). "Forest Ecosystems: Concepts and Management." Academic Press, Orlando, Florida.

Waring, R. H., Whitehead, D., and Jarvis, P. G. (1979). The contribution of stored water to transpiration in Scots pine. *Plant, Cell Environ.* 2: 309–317.

Waring, R. H., Newman, K., and Bell, J. (1981). Efficiency of tree crowns and stemwood production at different canopy leaf densities. *Forestry* 54: 15–23.

Waring, R. H., Cromack, K., Jr., Matson, P. A., Boone, R. D., and Stafford, S. G. (1987). Responses to pathogen-induced disturbance: Decomposition, nutrient availability, and tree vigor. *Forestry* 60: 219–227.

Whitehead, D., and Jarvis, P. G. (1981). Coniferous forests and plantations. *In* "Water Deficits and Plant Growth" (T. T. Kozlowski, ed.), Vol. 6, pp. 49–152. Academic Press, New York.

Whitmore, M. E. (1985). Effects of SO_2 and NO_x on plant growth. *In* "Sulfur Dioxide and Vegetation" (W. E. Winner, H. A. Mooney, and R. A. Goldstein, eds.), pp. 281–295. Stanford Univ. Press, Stanford, California.

Williams, K., Percival, F., Merino, J., and Mooney, H. A. (1987). Estimation of tissue construction cost from heat of combustion and organic nitrogen content. *Plant, Cell Environ.* 10: 725–734.

Zech, W., Koch, W., and Franz, F. (1969). Nettoassimilation und Transpiration von Kiefernzweigen in Abhängigkeit von Kaliumversorgung und Lichtintensität. *Forstwiss. Centralbl.* 88: 372–378.

Zielke, H. R., and Filner, P. (1971). Synthesis and turnover of nitrate reductase induced by nitrate in cultured tobacco cells. *J. Biol. Chem.* 246: 1772–1779.

18

Effects of Environmental Stresses on Deciduous Trees

T. T. Kozlowski

I. Introduction

Deciduous trees are subject to many adverse abiotic and biotic stresses of varying intensity and duration (Kozlowski, 1979, 1985a). Whereas some stresses affect trees more or less continually, others do so cyclically. Because of shading by dominant trees in a forest stand, for example, understory trees continually undergo shading stress. Furthermore, because

Response of Plants to Multiple Stresses. Copyright © 1991 by Academic Press, Inc. All rights of reproduction in any form reserved.

leaves overlap, even the crown interiors of dominant trees encounter persistent shading. Hence, the growth of trees represents an integrated response to the influences of different continual and periodic environmental stresses. The importance of individual stress factors, however, may vary appreciably over the short or long term. In Wisconsin, for example, the correlation between temperature and cambial growth in *Quercus ellipsoidalis* decreases toward late summer as soil moisture is progressively depleted, and growth is limited by internal water deficits (Kozlowski *et al.*, 1962). Sudden imposition of a severe stress (e.g., fire or disease) can predominate over the other factors that previously regulated growth. A major difficulty with correlating growth and changes in individual environmental factors is that such correlations may suggest that the factor studied is influencing growth-controlling processes. Yet this factor may only be correlated with some other factor or factors that are more important in regulating growth but were not even included in the analysis (Kozlowski *et al.*, 1991).

With the foregoing considerations in mind, I will characterize in this chapter some unique features of deciduous trees and relate them to growth and survival in stressful environments. Stress will be considered to be any combination of abiotic factors, biotic factors, or both, that reduces growth below the maximum attainable amount.

II. Abscission

Environmental stress has had a decisive influence on the evolution of deciduousness. Deciduous broadleaf trees first appeared at lower middle latitudes in the Northern Hemisphere during the early Cretaceous, associated with broadleaf evergreens in areas of warm temperate climates. Deciduous broadleaf trees then migrated to higher latitudes. Evolution of the deciduous habit exemplifies both preadaptation and zonal adaptation (Axelrod, 1966).

In cool temperate forests, the loss of leaves from deciduous trees is almost entirely an autumn phenomenon (Kozlowski, 1973). Individual trees shed most of their leaves within a few days or weeks. Deciduous trees may nevertheless lose leaves at any time during the growing season in response to injury or unfavorable environmental conditions.

Autumnal leaf shedding, which involves formation of an abscission layer at the base of the petiole, is preceded by leaf senescence, during which there is a general trend from anabolism to catabolism. Both photosynthetic and respiratory capacities of leaves decline. Mineral nutrients are lost by translocation into branches and by increased leaching from senescing cells. The leaves of some species turn yellow, reflecting loss of chlorophyll and unmasking of carotenoid pigments. In other species the leaves turn various shades of red as they synthesize anthocyanin pigments (Kramer and Kozlowski, 1979).

As the time for leaf abscission approaches, several changes occur in the vicinity of the abscission zone, including development of tyloses, blocking of vessels, increase in protoplasm density, and starch deposition. The cell walls swell, and digestion of pectic and cellulosic compounds takes place. Finally, the leaf is shed after one or more cell wall layers dissolve (Kozlowski, 1973).

Several environmental factors, including light intensity, day length, temperature, drought, flooding, mineral supply, and air pollution, either accelerate or retard abscission (Table I; Addicott and Lyon, 1973). Whereas moderate or light frosts promote abscission, heavy frosts retard it by injuring the abscission zone and thereby impeding the normal processes of

Table I   Effects of Some Environmental Factors on Leaf
Abscission[a]

Factors	Promotion → Retardation ←
Temperature	
Moderate	→
Light frost	→
Extremes: heat or frost	←
Light	
Photosynthetic: moderate	→
deficiency or excess	←
Photoperiodic: long days	←
short days	→
Water	
Drought or flooding	→
High humidity	→
Gases	
Oxygen	→
< 20% oxygen	←
Carbon dioxide	⇄
Ethylene	→
NH_3	→
Mineral and soil factors	
Nitrogen	←
Deficiencies of N, Zn, Ca, S, Mg, K, B, Fe	→
Excessive Zn, Fe, Cl, I	→
Salinity and alkalinity	⇄
Biotic factors	
Insect or fungus injury to leaf blade	→

[a] Reprinted from Addicott (1968) with permission from the American Society of Plant Physiologists.

abscission. The interactions among environmental factors are well illustrated by an overriding effect of drought on photoperiodic regulation of abscission (Addicott, 1982).

The effects of environmental stresses on abscission are mediated by hormonal and enzymatic changes (see Chapter 9). Major regulatory roles are played by auxins, which prevent abscission, and ethylene, which stimulates it. As a leaf senesces in the autumn, the flow of auxin across the abscission zone declines, and ethylene synthesis increases. A variety of abiotic stresses, as well as insect attack and disease, stimulates ethylene production and often accelerates premature abscission in deciduous trees. According to Osborne (1973), other growth hormones are involved in modifying the auxin–ethylene control system. For example, abscisic acid stimulates ethylene production.

Environmental stresses often influence the growth of various deciduous species differently because of variations among them in abscission patterns in response to stress. In many tropical species, the amount of leaf shedding is correlated with periods of drought. In fact, water supply determines whether some species are classified as deciduous or evergreen. For example, *Gossampinus malabarica* and *Tectona grandis* are deciduous in regions with alternating wet and dry seasons but nondeciduous in areas that are constantly wet (Merrill, 1945). In Singapore, *Hevea brasiliensis* retains its leaves for 13.3 months; in North Malaysia for 12 months; and in Ceara, Brazil, and South Malaysia for 10 months (Addicott, 1982). Some species do not abscise all of their leaves except when droughts are unusually severe. Such species may retain their leaves in wet years and shed them in dry years (Merrill, 1945).

A number of deciduous trees also abscise branches by a process similar to that involved in leaf abscission. Such shedding (cladoptosis) has been most extensively documented in species of *Populus*, but it also occurs in *Salix*, *Juglans*, *Fraxinus*, *Prunus*, and *Quercus* (Millington and Chaney, 1973).

III. Growth Patterns

Annual cambial growth in deciduous trees generally stops earlier in the summer than it does in evergreen trees. In Wisconsin, for example, seasonal diameter growth of several species of deciduous trees stops earlier than in *Tsuga canadensis* (Winget and Kozlowski, 1965). Seasonal cambial growth of pines in the southeastern United States usually continues for several weeks after it stops in deciduous trees (Kozlowski, 1971).

Whereas the dry weight of deciduous trees decreases during the winter, in evergreens it increases late into the autumn and sometimes throughout the winter. In southern Scotland, for example, the dry weight of *Picea sitchensis* seedlings approximately doubled between October and April. Most of the increase occurred during October, late March, and April, but some took place in midwinter (Bradbury and Malcolm, 1978). Winter increases in dry weight of *Pinus radiata* seedlings in Wales have also been demonstrated (Pollard and Wareing, 1968). In deciduous trees, the winter decreases in dry weight following autumnal leaf shedding are traceable largely to respiratory losses, whereas the increases in dry weights of evergreens are associated with higher rates of photosynthesis.

IV. Carbohydrate Relations

Because deciduous trees lack foliage during certain periods when some of their tissues continue to grow, their growth depends more on mobilization and partitioning of carbohydrate reserves than does that of evergreens (Kozlowski and Keller, 1966). In addition to their role in maintaining respiration, reserve carbohydrates are mobilized for early-season growth of shoots, stems, and some reproductive structures.

The annual carbohydrate cycles of deciduous trees and evergreens differ appreciably. In the temperate zone, evergreen species accumulate carbohydrates much later into the winter, and seasonal variations in carbohydrate reserves are much smaller than in deciduous trees. In California, carbohydrate reserves vary much more in the drought-deciduous *Aesculus californica* than in the evergreen *Quercus agrifolia*. During autumn fruit production, the stems of *Aesculus californica* are leafless, and their carbohydrate content decreases from 17 to 10% (Fig. 1). As new leaves emerge in February, reserve carbohydrates are further depleted to a low value of 5%. During March and April, after the trees have leafed out, reserve carbohydrates in the branches are restored to a seasonal high value. In contrast to the instability of carbohydrate reserves in *Aesculus californica* twigs, carbohydrate contents of *Quercus agrifolia* are quite stable, varying only from about 3.5 to 5% (Mooney and Hays, 1973).

The large amplitude in reserve carbohydrates of deciduous trees reflects a high rate of depletion by respiration and growth early in the growing season, followed by rapid leaf expansion and efficient and early carbohydrate production and export by expanded leaves to growing and storage tissues.

During the early stages of shoot elongation, deciduous trees depend heavily on carbohydrate reserves accumulated during the previous season. This dependence is shown by rapid depletion of stored carbohydrates in twigs as buds open and shoots expand (Gibbs, 1940; Quinlan, 1969; Hansen, 1971; Hansen and Grauslund, 1973). Hansen (1971) estimates that stored reserves supply one-half to two-thirds of the carbohydrates used

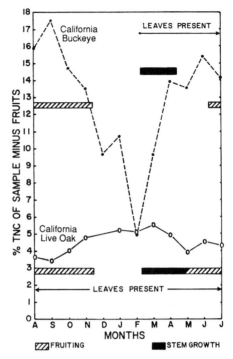

Figure 1 Seasonal cycles of total nonstructual carbohydrates (TNC) in twigs of the drought-deciduous California buckeye *(Aesculus californica)* and the evergreen California live oak *(Quercus agrifolia).* [Reprinted from Mooney and Hays (1973) with permission from Gustav Fischer Verlag.]

for the growth of shoots and flowers in *Malus* trees very early in the growing season. In Florida, carbohydrate reserves in roots of *Quercus laevis* and *Q. incana* are lowest between late April and early May. The period of greatest depletion coincides with the time the new leaves are approaching full size. Total carbohydrate reserves at this time are only half as great as during the dormant season (Woods *et al.,* 1959). As early as 1903, Schimper reported that a bud of a species of *Brownia* in the tropics expanded for several days at a rate of 2.6 cm per day while the leaves were still rolled and photosynthetically inactive.

Young, growing leaves do not export photosynthate to storage tissues. Once leaves mature, however, export of photosynthetic products predomi-nates. The individual leaves of deciduous trees generally expand in a few days to a few weeks. The early leaves of *Betula* and *Populus,* for example, expand within two to three weeks after buds open (Kozlowski and Clausen, 1966). The rapid growth of leaves in deciduous species contrasts with the relatively slow growth of leaves of evergreens. In *Pinus resinosa,* current-year needles do not elongate appreciably until after internode elongation ceases (Larson, 1964).

Early export of carbohydrates by mature leaves to storage tissues has been shown for deciduous forest trees and fruit trees. Examples are *Populus deltoides* and *Malus* (Hansen, 1967). In *Malus* large amounts of carbohydrates accumulate during the last 30 to 45 days before leaf fall, especially in roots (Hansen, 1967; Hansen and Grauslund, 1973).

Sprouting after the top of a tree has died from disturbance or physiological disorders is much more common in deciduous trees than in evergreens (Kramer and Kozlowski, 1979). Sprouts that originate from the root collar and the lower part of the stem arise from previously dormant buds. When the bark over a dormant bud becomes very thick, the bud may not break through it and develop into a sprout; hence, the sprouting capacity of deciduous trees declines with tree age (Smith, 1986).

Following disturbance of tree stands, several species of deciduous trees regenerate by root suckers originating from adventitious buds on the roots. Reproduction by root suckers is best known in *Populus tremuloides* and *P. grandidentata*, but it also occurs in *Liquidambar, Fagus, Ailanthus, Robinia, Sassafras,* and *Nyssa*. During light seed years, regeneration of *Fagus americana* depends on survival of seedlings and root sprouts under closed stands (Jones and Raynal, 1987).

Differences in growth of stump sprouts and root suckers have been related to carbohydrate levels in roots of parent trees at the time the trees were harvested (Roth and Hepting, 1943; Schier and Zasada, 1973). In Missouri, *Quercus marilandica* trees that were girdled in June, when carbohydrate reserves were minimal, had fewer sprouts per tree, smaller sprout clumps, and higher mortality of sprouts than trees girdled during any other month. In addition, practically all sprouting occurred during the first two seasons after girdling (Fig. 2).

In India, survival of sprouts of deciduous trees was higher at the forest periphery than under dense canopy, suggesting the importance of light in accumulation of carbohydrates (Khan *et al.,* 1986). Levels of reserve carbohydrates in roots and stumps of *Acer saccharum* affected sprouting, but the response was modified by changing photoperiod and light intensity. Low carbohydrate reserves shortly after buds opened from mid-May to mid-June restricted sprout growth, even though photoperiod and light intensity were favorable for a time. After trees were cut in late June and July, shorter days and decreasing light intensity resulted in reduced growth of sprouts, despite median levels of reserve carbohydrates (Wargo, 1971, 1979; MacDonald and Powell, 1985). These observations emphasize that if one wants a cut tree stand to regenerate by sprouting, cutting during the dormant season will produce a stand of maximum density and vigor. If trees are cut during the growing season, when carbohydrate reserves are low, sprouting

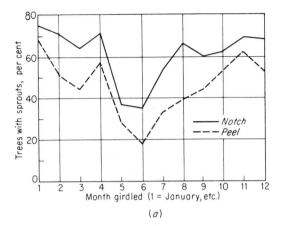

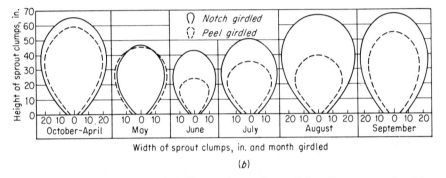

Figure 2 Effect of season of girdling on the number and size of sprouts produced by *Quercus marilandica* trees. Observations were made six seasons after girdling. (a) Number of trees producing sprouts after girdling in various months; (b) height and width of clumps of sprouts produced after girdling in various months. [From Clark and Liming (1953).]

will be reduced. When trees are cut very late in the growing season, by which time reserve carbohydrates have been replenished, the abundant sprouts that follow often fail to harden adequately and may be killed by frost (Kramer and Kozlowski, 1979). The shortest rotation at which sucker production in *Populus tremuloides* and *P. grandidentata* stands could be physiologically sustained was more than ten years. Declining yields from short rotations were attributed to starvation of rootstocks by frequent removal of the photosynthesizing tops (Stiell and Berry, 1986).

V. Responses to Environmental Stresses

In addition to leaf shedding and accumulation of large amounts of reserve carbohydrates, certain deciduous trees have evolved various other adaptations to environmental stresses. Species and genotypes vary greatly in their capacity to tolerate stress.

A. Aridity

Many deciduous trees tolerate drought by avoiding desiccation or postponing protoplasmic dehydration (desiccation tolerance). In addition to shedding leaves, desiccation-avoiding trees exhibit such water-conserving characteristics as extensive rooting, stomatal control of water loss, and a heavily cutinized epidermis. Desiccation-tolerant deciduous trees can actually endure dehydration of tissues, but most drought-tolerant trees are desiccation-avoiding plants (Kozlowski, 1982a).

Early shedding of leaves during droughts often prevents lethal desiccation in tropical and temperate-zone woody plants (Kozlowski, 1976a,b,c, 1982a). In the caatinga and cerrado of Brazil, for example, some species shed all their leaves early in the dry season and use already accumulated carbohydrates to leaf out when the rains resume (Alvim, 1964; Eiten, 1972). The 12-month period following June 1933 was one of the driest recorded in the midwestern United States (Kincer, 1934), leading to extensive early defoliation of deciduous trees (Albertson and Weaver, 1945). In California, *Aesculus californica* shed its leaves three to four weeks earlier than usual in the drought years of 1972 and 1977 (Addicott, 1982). Early defoliation by drought also has been reported for *Malus* (Landsberg and Jones, 1981) and *Prunus persica* (Proebsting and Middleton, 1980).

The capacity for leaf shedding during drought varies appreciably among species. In woodlands of the Sierra Nevada foothills, *Quercus douglasii* occupies much drier sites and has lower predawn water potentials than does *Quercus agrifolia*. When late summer or autumn soil water supplies are low, *Q. douglasii* drops most of its leaves, thereby avoiding desiccation. Because *Q. agrifolia* lacks such leaf shedding capacity, it does not dominate consistently dry woodlands (Griffin, 1973).

Generalizations about the effects of environmental stresses on different species of deciduous trees are complicated by wide intraspecific variations in response. Because water deficits frequently limit growth and survival of trees, selective pressure for adaptation to drought is often high (Pallardy, 1981). Genetically related differences in survival during droughts have been reported for *Quercus rubra* (Kriebel *et al.*, 1976) and *Acer saccharum* (Kriebel, 1957).

Many drought-tolerant deciduous trees avoid injury from dehydration by restricting water loss. Trees originating from seed collected from the more arid parts of a species' range or from xeric habitats often have slower rates of shoot growth than those from wetter areas or more mesic habitats (e.g., *Acer rubrum;* Townsend and Roberts, 1973). In addition, leaves of *Populus deltoides* clones from mesic provenances are larger than those from xeric portions of the species' range (Ying and Bagley, 1976). Small leaves also characterize *Betula alleghaniensis* and *Cercis canadensis* from xeric sites (Dancik and Barnes, 1975).

Stomatal adaptations for drought tolerance include reductions in stomatal frequency, size, or both; sunken stomata; and changes in stomatal response to environmental or internal water deficits. Variations in stomatal size and frequency have been found among *Populus* clones (Siwecki and Kozlowski, 1973; Ceulemans *et al.*, 1978a; Pallardy and Kozlowski, 1979a) and provenances of *Juglans nigra* (Carpenter, 1974).

The stomatal apertures of *Populus* clones differ in response to the drying of soil in a constant aerial environment. Stomata of a clone of *Populus trichocarpa* closed at much lower soil water potentials than did those of two *Populus trichocarpa* × *P. deltoides* clones and a clone of *P. robusta* (Ceulemans *et al.*, 1978b). Responses to changes in vapor-pressure deficit (VPD) varied appreciably in two *Populus* clones (Pallardy, 1981). When VPD was low, leaf resistance differed little between clones of *Populus candicans* × *P. berolinensis* and *Populus deltoides*. However, when evaporative demand was high, stomatal closure, particularly of the adaxial stomata, was much more pronounced in *Populus candicans* × *P. berolinensis*. In another study, two *Populus* clones exhibited differences in stomatal responses to two levels of light and VPD (Table II). Total leaf diffusive resistance of a clone of *Populus candicans* × *P. berolinensis* was more responsive to changing VPD and less responsive to a change in light intensity than was that of a clone of *Populus deltoides* × *P. caudina*. After the simple effects of low light and high VPD had been removed, *Populus candicans* × *P. berolinensis* showed a greater interactive increase in total leaf resistance in response to a combination of low light and high VPD than did *Populus deltoides* × *P. caudina* (Pallardy and Kozlowski, 1979b).

Some xeric species of deciduous trees exhibit osmotic adaptation to drought. Species of *Quercus* tend to have low osmotic potentials and show osmotic adjustment in response to drought. More mesic species, such as *Acer saccharum*, have relatively high osmotic potentials and do not show osmotic adjustment. Leaves of drought-tolerant species of *Quercus* also show greater dehydration tolerance than do mesic species (Pallardy *et al.*, 1983).

B. Flooding

Flood tolerance of deciduous trees varies greatly with species and genotype, with some species surviving inundation of the soil for at least two growing seasons while others are killed by less than four weeks of flooding (Hall and Smith, 1955). Whereas *Nyssa aquatica*, *N. sylvatica*, *Salix nigra*, and *Fraxinus pennsylvanica* are considered flood tolerant, *Betula papyrifira*, *Cornus florida*, and *Liquidambar styraciflua* are flood intolerant (Kramer and Kozlowski, 1979).

Table II Response of Adaxial, Abaxial, and Total Leaf Resistance of Two *Populus* Clones to Two Levels of Photon Flux Density and Vapor-Pressure Deficit[a]

Photon flux density (μE m^{-2} sec^{-1})	VPD (kPa)	*Populus candicans* × *P. berolinensis*			*Populus deltoides* × *P. caudina*		
		r_{AB}	r_{AD}	r_T	r_{AB}	r_{AD}	r_T
520	0.78	1.68 ± 0.04	3.71 ± 0.10	1.14 ± 0.02	2.27 ± 0.06	3.28 ± 0.13	1.32 ± 0.04
520	3.16	2.81 ± 0.14	10.97 ± 0.52	2.19 ± 0.09	3.31 ± 0.09	5.79 ± 0.25	2.05 ± 0.05
65	0.78	3.15 ± 0.22	8.92 ± 0.89	2.22 ± 0.14	7.20 ± 0.67	17.13 ± 2.41	4.68 ± 0.48
65	3.16	8.04 ± 0.86	51.08 ± 4.09	6.40 ± 0.61	13.02 ± 0.92	29.42 ± 2.88	8.27 ± 0.54

[a] r_{AD}, Adaxial resistance; r_{AB}, abaxial resistance; r_T, total resistance; VPD, vapor-pressure deficit. Resistance in sec cm^{-1}. Reprinted from Pallardy and Kozlowski (1979b) with permission of the American Society of Plant Physiologists.

Adaptations of flood-tolerant deciduous trees include a capacity for absorbing oxygen into aerial tissues, transport of oxygen through the stem to the roots, diffusion of oxygen to the rhizosphere, and production of adventitious roots on submerged stems and roots (Kozlowski 1982b, 1984a,b, 1986).

Oxidation of the rhizosphere is associated with oxidation of reduced soil compounds such as ferrous and manganous ions, which are toxic to roots (Opik, 1980). Entry of oxygen through leaves has been shown for species of *Salix* and *Populus* (Chirkova, 1968; Armstrong, 1968).

Several flood-tolerant deciduous species produce lenticels on the submerged portions of stems and roots. Such lenticels may facilitate aeration of the stem as well as the release of potentially toxic compounds produced by flooded plants. Flood-induced lenticels have been demonstrated in *Salix nigra, Populus deltoides, Fraxinus pennsylvanica,* and *Platanus occidentalis* (Hook, 1984; Angeles *et al.,* 1986).

Some species of deciduous trees produce adventitious roots that contribute to flood tolerance by compensating physiologically for any loss of absorbing capacity caused by decay of the original root system (Pereira and Kozlowski, 1977; Tang and Kozlowski, 1982, 1984; Fig. 3). Water absorp-

Figure 3 Extensive development of adventitious roots on the previously submerged stem of a *Platanus occidentalis* seedling. The horizontal line indicates the height of flooding. [Reprinted from Kozlowski (1984b) with permission of the American Institute of Biological Sciences.]

tion can be almost twice as high in seedlings of *Fraxinus pennsylvanica* with adventitious roots on submerged portions of stems as in seedlings from which such roots have been excised (Sena Gomes and Kozlowski, 1980). Early excision of flood-induced adventitious roots from submerged stems reduces subsequent growth of *Platanus occidentalis* seedlings (Tsukahara and Kozlowski, 1985). Furthermore, flood-induced adventitious roots of *Nyssa aquatica* are important in oxidizing and detoxifying soil toxins, whereas the original roots are not (Hook *et al.*, 1979; Hook, 1984).

C. Light Intensity

Both the structure and response to shading of sun-grown and shade-grown leaves differ appreciably. This applies to shaded leaves within a tree crown, compared with those in the crown periphery, as well as leaves of entire trees grown in the shade or full sun.

Sun-grown leaves are usually smaller and thicker and have more palisade parenchyma and more chlorophyll per unit of leaf area than shade leaves do. Whereas shade-grown leaves of *Ficus benjamina* have a single, poorly developed palisade layer with large chloroplasts dispersed throughout the palisade cells, sun-grown leaves have one or two layers of well-developed palisade cells with chloroplasts aligned primarily along the radial walls (Fails *et al.*, 1982). Sun leaves also have more stomata per unit of leaf area and larger interveinal areas (Kramer and Kozlowski, 1979). Because sun leaves have lower stomatal and mesophyll resistances to carbon dioxide diffusion (Boardman, 1977), they exhibit higher rates of photosynthesis per unit of leaf area and become light saturated at a higher light intensity than shade leaves do. At low light intensities, however, sun leaves have lower rates of photosynthesis than shade leaves (Kramer and Kozlowski, 1979).

Differences among deciduous species in response to shading have important implications in plant competition and succession. Whereas *Fagus grandifolia*, *Acer saccharum*, and *Cornus florida* tolerate shade and thrive in the understory of a forest stand, *Populus tremuloides*, *Betula papyrifera*, and *Liriodendron tulipifera* require high light intensities for maximum growth (Kramer and Kozlowski, 1979).

Shade-tolerant species such as *Acer saccharum* and *Fagus grandifolia*, with a long, suppressed seedling stage, grow vigorously in small gaps in forests, but can also grow rapidly in larger gaps. Shade-intolerant species, such as *Liriodendron tulipifera*, however, can grow rapidly in large gaps but not in small ones (Runkle, 1985).

In the Piedmont plateau of the southeastern United States, the natural course of succession is from pine forests to hardwood forests. Pine stands are readily established on open fields. The understory of such young stands consists largely of deciduous trees, and the overstory pines eventually are thinned out by the time they reach overmaturity. An important factor in the success of the deciduous trees is that they achieve maximum photosynthetic rates at one-third or less of full sunlight. By comparison, the pines require full sunlight for maximum photosynthesis (Kramer and Decker, 1944).

D. Low Temperature

Deciduous species and genotypes native to warm regions cannot be moved to very cold regions because they do not develop enough cold hardiness. Whereas northern species such as *Populus tremuloides* and *Betula papyrifera* can resist freezing to $-80°$ C, *Magnolia grandiflora* survives temperatures only down to $-15°$ C (Sakai and Weiser, 1973). *Acer saccharum* and *Quercus rubra* seedlings from northern provenances resist freezing better than those from southern sources (Kriebel, 1957; Flint, 1972). Acclimation of northern provenances to freezing often is related to phenology, with seed sources that set buds early showing less injury from early frosts (Sakai and Larcher, 1987).

Young leaves of even the most hardy deciduous trees are very sensitive to frost. When young leaves are killed by early spring freezes, a new crop of leaves emerges from previously dormant buds. This emergence is possible because of the large amounts of stored carbohydrates in buds and twigs.

An important difference between deciduous and evergreen trees is that many evergreens undergo extensive winter desiccation injury, whereas deciduous trees do not. Such injury to evergreens, characterized by leaf scorching and abscission (and sometimes tree death), occurs because the rate of transpirational water loss exceeds the rate of water absorption when the air temperature rises above freezing and steepens the vapor-pressure gradient between the leaves and air (Kozlowski, 1976b, 1982a).

E. Soil Fertility

Because of their less efficient use of mineral nutrients and lower carbon gain per unit of nutrient acquired, deciduous trees do not thrive on the more infertile sites. This is shown by domination by evergreens of infertile soils and ridgetops and by growth of deciduous trees on adjacent, more fertile soils (Monk, 1966).

Recovery of mineral nutrients from senescing leaves is of approximately the same order in deciduous and evergreen trees. About half of the nitrogen and one-third of the phosphorus is recovered by resorption from the leaves to other parts of the plant before the leaves are shed, regardless of leaf lifespan (Chabot and Hicks, 1982). The timing of recovery, however, varies greatly. In deciduous trees, mineral nutrients are resorbed shortly before the leaves are shed. In *Betula papyrifera,* for example, about half of the nitrogen and phosphorus is resorbed during the few days of autumnal color change just before leaf abscission (Chapin and Kedrowski, 1983). In evergreens, mineral nutrients are stored in old leaves during the winter and translocated to new, growing leaves in the subsequent growing season. Thus, withdrawal of nutrients from the foliage of evergreens occurs over a period of years, rather than a few weeks as in deciduous trees (Chabot and Hicks, 1982). Deciduous trees lose more nutrients by litterfall than evergreens do, reflecting the higher mineral content of the former. Monk (1971) estimated that deciduous trees, depending on species, lost about 85% of annual nutrient uptake in litterfall, whereas evergreens lost approximately 10 to 25%. The less efficient internal cycling of mineral nutrients in deciduous trees results in short internal turnover times for nutrients and lower carbon gain per unit of nutrient turned over. For example, deciduous trees often gain less than half as much carbon per gram of nitrogen turned over as evergreens do (Small, 1972; Schlesinger and Chabot, 1977).

Through gradual leaf shedding and subsequent decay, small amounts of nutrients may become available to evergreens throughout the year. In deciduous stands on sandy soils with moderate to heavy rainfall, soil nutrients tend to be depleted by the return of large amounts during a short time, a more wasteful process than the gradual nutrient return throughout the year by evergreens (Monk, 1966). Nevertheless, some deciduous species can adjust to environmental stresses by increased resorption of mineral nutrients from the leaves to branches just before abscission. For example, proportional resorption of nitrogen and phosphorus from leaves of *Quercus alba* and *Acer rubrum* trees was greater when they were grown on an infertile soil than on a very fertile one (Boerner, 1984).

F. Air Pollution

Air pollutants, alone and in combination, often reduce growth and injure deciduous trees. They also alter plant community structure (Kozlowski, 1980, 1985b; Kozlowski and Constantinidou, 1986a). Combined air pollutants are often more toxic than single pollutants (Constantinidou and Kozlowski, 1979a,b; Freer-Smith, 1984; Fig. 4).

Interactions of environmental influences on deciduous trees have been well illustrated in studies of air pollution. Much interest has been shown in determining threshold dosages of environmental pollutants that injure different species of trees. The effect of a specific dosage of a pollutant is influenced, however, by prevailing environmental regimes as well as those occurring before and after a pollution episode (Kozlowski and Constantinidou, 1986b). The effects of prevailing light intensity, humidity, temperature, and water supply are particularly important. Leaves absorb more gaseous air pollutants and are injured more in the daytime, when the stomata are open, than at night, when stomatal conductance is lower. They also absorb more pollutants when the soil is wet than when it is dry because

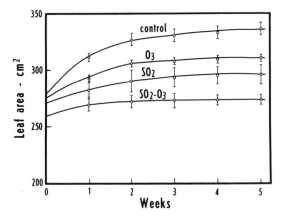

Figure 4   Effects of sulfur dioxide (SO₂), ozone (O₃), and SO₂–O₃ mixtures on leaf areas of *Ulmus americana* seedlings. [Reprinted from Constantinidou and Kozlowski (1979a) with permission from the National Research Council of Canada.]

leaf turgor is higher and stomata are more open in trees growing in wet soil (Kozlowski and Constantinidou, 1986b).

Response to a given dosage of an air pollutant may vary appreciably with changes in air humidity. Stomata of some species act as humidity sensors by opening when air humidity is high and closing when it is low. Such changes often occur independently of the overall hydration of the leaf (Sena Gomes *et al.*, 1987). For example, stomata of *Betula papyrifera* seedlings absorbed more sulfur dioxide at high than at low humidity because stomata were more open at high humidity (Norby and Kozlowski, 1982).

Prevailing temperature regimes also influence responses of deciduous trees to air pollutants. Seedlings of *Betula papyrifera* and *Fraxinus pennsylvanica* fumigated with sulphur dioxide at 30° C had more open stomata and absorbed more of the pollutant than seedlings fumigated at 12° C (Norby and Kozlowski, 1981a).

Environmental conditions following a pollution episode may also influence responses of deciduous trees to air pollutants. Such effects are usually mediated by changes in plant metabolism (Norby and Kozlowski, 1981b). An extensive literature documents wide intraspecific variation in tolerance to air pollution. Differences in responses of *Populus* clones to sulfur dioxide and ozone were reported by Karnosky (1976, 1977). Variations in tolerance to ozone were confirmed for families from different regions, for example, in *Acer rubrum, Fraxinus americana,* and *F. pennsylvanica* (Townsend and Dochinger, 1974; Steiner and Davis, 1979). Variations in pollution tolerance of different genotypes often are closely related to stomatal conductance (Kozlowski and Constantinidou, 1986b).

VI. Conclusions

Control of growth in deciduous trees by environmental stresses is very complex and differs from that of evergreens. Correlations between single environmental factors and plant responses generally are too simplistic to account for success or failure of a given species or genotype in an ecosystem. As emphasized by Osmond *et al.* (1987), we understand fairly well some of the cellular mechanisms of stress physiology. To predict their ecological impact on deciduous trees, however, we need to better understand the temporal and spatial variations of specific stresses, genetic variations in response to stress, the capacity of plants to acclimate to stress, and the nature of the interactive effects of various stresses on plant responses. The influences of environmental stresses on accumulation of carbohydrate reserves are particularly important.

References

Addicott, F. T. (1968). Environmental factors in the physiology of abscission. *Plant Physiol.* 43: 1471–1479.

Addicott, F. T. (1982). "Abscission." Univ. of California Press, Berkeley.

Addicott, F. T., and Lyon, J. L. (1973). The physiological ecology of abscission. *In* "Shedding of Plant Parts" (T. T. Kozlowski, ed.), pp. 475–524. Academic Press, New York.

Albertson, F. W., and Weaver, J. E. (1945). Injury and death or recovery of trees in prairie climate. *Ecol. Monogr.* 15: 393–433.

Angeles, G., Evert, R. F., and Kozlowski, T. T. (1986). Development of lenticels and adventitious roots in flooded *Ulmus americana* seedlings. *Can. J. For. Res.* 16: 585–590.

Alvim, P. de T. (1964). Tree growth and periodicity in tropical climates. *In* "The Formation of Wood in Forest Trees" (M. H. Zimmermann, ed.), pp. 479–495. Academic Press, New York.

Armstrong, W. (1968). Oxygen diffusion from the roots of woody species. *Physiol. Plant.* 21: 539–543.

Axelrod, D. I. (1966). Origin of deciduous and evergreen habits in temperate forests. *Evolution* 20: 1–15.

Boardman, N. K. (1977). Comparative photosynthesis of sun and shade plants. *Annu. Rev. Plant Physiol.* 28: 355–377.

Boerner, R. E. J. (1984). Foliar nutrient dynamics and nutrient use efficiency of four deciduous species in relation to site fertility. *J. Appl. Ecol.* 21: 1029–1040.

Bradbury, I. K., and Malcolm, D. C. (1978). Dry matter accumulation by *Picea sitchensis* seedlings during winter. *Can. J. For. Res.* 8: 207–213.

Carpenter, S. B. (1974). Variation in leaf morphology in black walnut (*Juglans nigra* L.) and its possible role in photosynthetic efficiency. *In* "Proc. Central States Forest Tree Improvement Conf., 8th, 1972," pp. 24–27.

Ceulemans, I., Impens, I., Lemeur, R., Moermans, R., and Samsuddin, Z. (1978a). Water movement in the soil–poplar–atmosphere system. I. Comparative study of stomatal morphology and anatomy and the influence of stomatal density and dimensions on the leaf diffusion characteristics of different poplar clones. *Oecol. Plant.* 13: 1–12.

Ceulemans, I., Impens, I., Lemeur, R., Moermans, R., and Samsuddin, Z. (1978b). Water movement in the soil–poplar–atmosphere system. II. Comparative study of transpiration regulation during water stress situations in four different poplar clones. *Oecol. Plant.* 13: 139–146.

Chabot, B. F., and Hicks, D. J. (1982). The ecology of leaf life spans. *Annu. Rev. Ecol. Syst.* 13: 229–259.

Chapin, F. S., III, and Kedrowski, R. A. (1983). Seasonal changes in nitrogen and phosphorus fractions and autumnal retranslocation in evergreen deciduous taiga trees. *Ecology* 64: 376–391.

Chirkova, T. V. (1968). Features of the O_2 supply of roots of certain woody plants in anaerobic conditions. *Fiziol. Rast. (Mosc.)* 15: 565–568.

Clark, F. B., and Liming, F. G. (1953). Sprouting of blackjack oak in the Missouri Ozarks. *U.S. For. Serv. Cent. States For. Exp. Stn. Tech. Pap.* 137.

Constantinidou, H. A., and Kozlowski, T. T. (1979a). Effects of sulfur dioxide and ozone on *Ulmus americana* seedlings. I. Visible injury and growth. *Can. J. Bot.* 57: 170–175.

Constantinidou, H. A., and Kozlowski, T. T. (1979b). Effects of sulfur dioxide and ozone on *Ulmus americana* seedlings. II. Carbohydrates, proteins, and lipids. *Can. J. Bot.* 57: 176–184.

Dancik, P. B., and Barnes, B. V. (1975). Leaf variability in yellow birch *(Betula alleghaniensis)* in relation to environment. *Can. J. For. Res.* 5: 149–159.

Eiten, G. (1972). The cerrado vegetation of Brazil. *Bot. Rev.* 38: 201–341.

Fails, B. S., Lewis, A. J., and Barden, J. A. (1982). Light acclimatization potential of *Ficus benjamina*. *J. Am. Soc. Hortic. Sci.* 197: 762–766.

Flint, H. L. (1972). Cold hardiness of twigs of *Quercus rubra* L. as a function of geographic origin. *Ecology* 53: 1163–1170.

Freer-Smith, P. H. (1984). The responses of six broadleaved trees during long-term exposure to SO_2 and NO_2. *New Phytol.* 97: 49–61.

Gibbs, R. D. (1940). Studies in tree physiology. II. Seasonal changes in the food reserves of field birch *(Betula populifolia* Marsh.). *Can. J. Res.* 18: 1–9.

Griffin, J. R. (1973). Xylem sap tension in three woodland oaks of central California. *Ecology* 54: 152–159.

Hall, T. F., and Smith, G. E. (1955). Effects of flooding on woody plants: West Sandy dewatering project, Kentucky Reservoir. *J. For.* 53: 281–285.

Hansen, P. (1967). ^{14}C-studies on apple trees. III. The influence of season on storage and mobilization of labelled compounds. *Physiol. Plant.* 20: 1103–1111.

Hansen, P. (1971). ^{14}C-studies on apple trees. VII. The early seasonal growth in leaves, flowers, and shoots as dependent upon current photosynthates and existing reserves. *Physiol. Plant.* 25: 469–473.

Hansen, P., and Grauslund, J. (1973). ^{14}C-studies on apple trees. VIII. The seasonal variation and nature of reserves. *Physiol. Plant.* 28: 24–32.

Hook, D. D. (1984). Adaptations to flooding with fresh water. *In* "Flooding and Plant Growth" (T. T. Kozlowski, ed.), pp. 265–294. Academic Press, New York.

Hook, D. D., Brown, C. L., and Kormanik, P. P. (1970). Lenticels and water root development of swamp tupelo under various flooding conditions. *Bot. Gaz.* 131: 217–224.

Jones, R. H., and Raynal, D. J. (1987). Root sprouting in American beech: Production, survival, and the effect of parent tree vigor. *Can. J. For. Res.* 17: 539–544.

Karnosky, D. F. (1976). Threshold levels for foliar injury to *Populus tremuloides* by sulphur dioxide and ozone. *Can. J. For. Res.* 6: 166–169.

Karnosky, D. F. (1977). Evidence for genetic control of response to sulphur dioxide and ozone in *Populus tremuloides*. *Can. J. For. Res.* 7: 437–440.

Khan, M. L., Rai, J. P. N., and Tripathi, R. S. (1986). Regeneration and survival of tree seedlings and sprouts in tropical deciduous and sub-tropical forests of Meghalaya, India. *For. Ecol. Manage.* 14: 293–304.

Kincer, J. B. (1934). Data on the drought. *Science* 80: 179.

Kozlowski, T. T. (1971). "Growth and Development of Trees," Vol. II. Academic Press, New York.

Kozlowski, T. T. (1973). Extent and significance of shedding of plant parts. *In* "Shedding of Plant Parts" (T. T. Kozlowski, ed.), pp. 1–44. Academic Press, New York.

Kozlowski, T. T. (1976a). Drought resistance and transplantability of shade trees. *U.S. For. Serv. Gen. Tech. Rep. NE* 22: 77–90.

Kozlowski, T. T. (1976b). Water relations and tree improvement. *In* "Tree Physiology and Yield Improvement" (M. Cannell and F. T. Last, eds.), pp. 307–327. Academic Press, London.

Kozlowski, T. T. (1976c). Water supply and leaf shedding. *In* "Water Deficits and Plant Growth" (T. T. Kozlowski, ed.), Vol. IV, pp. 191–231. Academic Press, New York.

Kozlowski, T. T. (1979). "Tree Growth and Environmental Stresses." Univ. of Washington Press, Seattle.

Kozlowski, T. T. (1980). Impacts of air pollution on forest ecosystems. *Bioscience* 30: 88–93.

Kozlowski, T. T. (1982a). Water supply and tree growth. Part I. Water deficits. *For. Abstr.* 43: 57–95.

Kozlowski, T. T. (1982b). Water supply and tree growth. Part II. Flooding. *For. Abstr.* 43: 145–161.

Kozlowski, T. T. (1984a). Responses of woody plants to flooding. *In* "Flooding and Plant Growth" (T. T. Kozlowski, ed.), pp. 129–164. Academic Press, New York.

Kozlowski, T. T. (1984b). Plant responses to flooding of soil. *BioScience* 34: 162–167.

Kozlowski, T. T. (1985a). Tree growth in response to environmental stresses. *J. Arboric.* 11: 97–111.

Kozlowski, T. T. (1985b). Effects of SO_2 on plant community structure. *In* "Sulfur Dioxide and Vegetation." (W. E. Winner, H. A. Mooney, and R. Goldstein, eds.), pp. 431–451. Stanford Univ. Press, Stanford, California.

Kozlowski, T. T. (1986). Soil aeration and growth of forest trees. *Scand. J. For. Res.* 1: 113–123.

Kozlowski, T. T., and Clausen, J. J. (1966). Shoot growth characteristics of heterophyllous woody plants. *Can. J. Bot.* 44: 827–843.

Kozlowski, T. T., and Constantinidou, H. A. (1986a). Responses of woody plants to environmental pollution. Part I. Sources, types of pollutants, and plant responses. *For. Abstr.* 47: 5–51.

Kozlowski, T. T., and Constantinidou, H. A. (1986b). Responses of woody plants to environmental pollution. Part II. Factors affecting responses to pollution. *For. Abstr.* 47: 105–132.

Kozlowski, T. T., and Keller, T. (1966). Food relations of woody plants. *Bot. Rev.* 32: 293–382.

Kozlowski, T. T., Winget, C. H., and Torrie, J. H. (1962). Daily growth of oak in relation to maximum and minimum temperature. *Bot. Gaz.* 124: 9–17.

Kozlowski, T. T., Kramer, P. J., and Pallardy, S. G. (1991). "The Physiological Ecology of Woody Plants." Academic Press, San Diego.

Kramer, P. J., and Decker, J. P. (1944). Relation between light intensity and rate of photosynthesis of loblolly pine and certain hardwoods. *Plant Physiol.* 19: 350–358.

Kramer, P. J., and Kozlowski, T. T. (1979). "Physiology of Woody Plants." Academic Press, New York.

Kriebel, H. B. (1957). Patterns of genetic variation in sugar maple. *Ohio Agric. Exp. Stn. Res. Bull.* 791.

Kriebel, H. B., Bagley, W. T., Deneke, F. J., Funsch, R. W., Roth, P., Jokela, J. J., Merritt, C.,

Wright, J. W., and Williams, R. D. (1976). Geographic variation in *Quercus rubra* in north central United States plantations. *Silvae Genet.* 25: 118–122.

Landsberg, J. J., and Jones, H. G. (1981). Apple orchards. *In* "Water Deficits and Plant Growth" (T. T. Kozlowski, ed.), Vol. VI, pp. 419–469. Academic Press, New York.

Larson, P. R. (1964). Contribution of different-aged needles to growth and wood formation of young red pines. *For. Sci.* 10: 224–238.

MacDonald, J. E., and Powell, G. R. (1985). First growth period development of *Acer saccharum* stump sprouts arising after different dates of cut. *Can. J. Bot.* 63: 819–825.

Merrill, E. D. (1945). "Plant Life of the Pacific World." MacMillan, New York.

Millington, W. F., and Chaney, W. R. (1973). Shedding of shoots and branches. *In* "Shedding of Plant Parts" (T. T. Kozlowski, ed.), pp. 149–204. Academic Press, New York.

Monk, C. D. (1966). An ecological significance of greenness. *Ecology* 47: 504–505.

Monk, C. D. (1971). Leaf decomposition and loss of ^{45}Ca from deciduous and evergreen trees. *Am. Midl. Nat.* 86: 370–384.

Mooney, H. A., and Hays, R. I. (1973). Carbohydrate storage cycles in two Californian Mediterranean-climate trees. *Flora* 162: 295–304.

Norby, R. J., and Kozlowski, T. T. (1981a). Relative sensitivity of three species of woody plants to SO_2 at high or low exposure temperature. *Oecologia* 51: 33–36.

Norby, R. J., and Kozlowski, T. T. (1981b). Interaction of SO_2 concentration and post-fumigation temperature on growth of woody plants. *Environ. Pollut.* 25: 27–39.

Norby, R. J., and Kozlowski, T. T. (1982). The role of stomata in sensitivity of *Betula papyrifera* Marsh. seedlings to SO_2 at different humidities. *Oecologia* 53: 34–39.

Opik, H. (1980). "The Respiration of Higher Plants." Edward Arnold, London.

Osborne, D. J. (1973). Internal factors regulating abscission. *In* "Shedding of Plant Parts" (T. T. Kozlowski, ed.), pp. 125–147. Academic Press, New York.

Osmond, C. B., Austin, M. P., Berry, J. A., Billings, W. D., Boyer, J. S., Dacey, J. W. H., Nobel, P. S., Smith, S. D., and Winner, W. E. (1987). Stress physiology and the distribution of plants. *Bioscience* 37: 38–47.

Pallardy, S. G. (1981). Closely related woody plants. *In* "Water Deficits and Plant Growth" (T. T. Kozlowski, ed.), Vol. VI, pp. 511–548. Academic Press, New York.

Pallardy, S. G., and Kozlowski, T. T. (1979a). Frequency and length of stomata of 21 *Populus* clones. *Can. J. Bot.* 57: 2519–2523.

Pallardy, S. G., and Kozlowski, T. T. (1979b). Stomatal responses of *Populus* clones to light intensity and vapor pressure deficit. *Plant Physiol.* 64: 112–114.

Pallardy, S. G., Parker, W. C., Whitehouse, D. L., Hinckley, T. M., and Teskey, R. O. (1983). Physiological responses to drought and drought adaptations in woody species. *In* "Current Topics in Plant Biochemistry–Plant Physiology" (D. D. Randall, D. Blevins, and R. Larson, eds.), Vol. 2, pp. 185–189. Univ. of Missouri Press, Columbia.

Pereira, J. S., and Kozlowski, T. T. (1977). Variations among woody angiosperms in response to flooding. *Physiol. Plant.* 41: 184–192.

Pollard, D. F. W., and Wareing, P. F. (1968). Rates of dry matter production in forest tree seedlings. *Ann. Bot.* 32: 573–591.

Proebsting, E. L., Jr., and Middleton, J. E. (1980). The behavior of peach and pear trees under extreme drought stress. *J. Am. Soc. Hortic. Sci.* 105: 380–385.

Quinlan, J. D. (1969). Mobilization of ^{14}C in the spring following autumn assimilation of $^{14}CO_2$ by apple rootstock. *J. Hortic. Sci.* 44: 107–110.

Roth, E. R., and Hepting, G. H. (1943). Origin and development of the oak stump sprouts as affecting their likelihood to decay. *J. For.* 41: 27–36.

Runkle, J. R. (1985). Disturbance regimes in temperate forests. *In* "The Ecology of Natural Disturbance and Patch Dynamics" (S. T. A. Pickett and P. S. White, eds.), pp. 17–33. Academic Press, New York.

Sakai, A., and Larcher, W. (1987). "Frost Survival of Plants." Springer-Verlag, Berlin.

Sakai, A., and Weiser, C. J. (1973). Freezing resistance of trees in North America with reference to tree regions. *Ecology* 54: 118–126.

Schier, G. A., and Zasada, J. C. (1973). Role of carbohydrate reserves in the development of root suckers in *Populus tremuloides*. *Can. J. For. Res.* 3: 243–250.

Schimper, F. W. (1903). "Plant Geography upon a Physiological Basis" English translation. Oxford Univ. Press (Clarendon), London.

Schlesinger, W. H., and Chabot, B. F. (1977). The use of water and minerals by evergreen and deciduous shrubs in Okefenokee Swamp. *Bot. Gaz.* 138: 490–497.

Sena Gomes, A. R., and Kozlowski, T. T. (1980). Growth responses and adaptations of *Fraxinus pennsylvanica* seedlings to flooding. *Physiol. Plant.* 49: 373–377.

Sena Gomes, A. R., Kozlowski, T. T., and Reich, P. B. (1987). Some physiological responses of *Theobroma cacao* var. *Catongo* seedlings to air humidity. *New Phytol.* 107: 591–602.

Siwecki, R., and Kozlowski, T. T. (1973). Leaf anatomy and water relations of excised leaves of six *Populus* clones. *Arbor. Kornickie* 8: 83–105.

Small, E. (1972). Photosynthetic rates in relation to nitrogen recycling as an adaptation to nutrient deficiency in peat bog plants. *Can. J. Bot.* 50: 2227–2233.

Smith, D. M. (1986). "The Practice of Silviculture." Wiley, New York.

Steiner, K. C., and Davis, D. D. (1979). Variation among *Fraxinus* families in foliar response to ozone. *Can. J. For. Res.* 9: 106–109.

Stiell, W. M., and Berry, A. B. (1986). Productivity of short-rotation aspen stands. *For. Chron.* 62: 10–15.

Tang, Z. C., and Kozlowski, T. T. (1982). Physiological, morphological, and growth responses of *Platanus occidentalis* seedlings to flooding. *Plant Soil* 66: 243–255.

Tang, Z. C., and Kozlowski, T. T. (1984). Water relations, ethylene production, and morphological adaptations of *Fraxinus pennsylvanica* seedlings to flooding. *Plant Soil* 77: 183–192.

Townsend, A. M., and Dochinger, L. S. (1974). Relationship of seed sources and developmental stage to the ozone tolerance of *Acer rubrum* seedlings. *Atmos. Environ.* 8: 957–964.

Townsend, A. M., and Roberts, B. R. (1973). Effect of moisture stress on red maple seedlings from different seed sources. *Can. J. Bot.* 51: 1989–1995.

Tsukahara, H., and Kozlowski, T. T. (1985). Importance of adventitious roots to growth of flooded *Platanus occidentalis* seedlings. *Plant Soil* 88: 123–132.

Wargo, P. M. (1971). Seasonal changes in carbohydrate levels in roots of sugar maple. *U.S. For. Serv. Res. Pap. NE* 213.

Wargo, P. M. (1979). Starch storage and radial growth in woody roots of sugar maple. *Can. J. For. Res.* 9: 49–56.

Winget, C. H., and Kozlowski, T. T. (1965). Seasonal basal area growth as an expression of competition in northern hardwoods. *Ecology* 46: 786–793.

Woods, F. W., Harris, H. C., and Caldwell, R. E. (1959). Monthly variations of carbohydrates and nitrogen in roots of sandhill oaks and wiregrass. *Ecology* 40: 292–295.

Ying, C. C., and Bagley, W. T. (1976). Genetic variation of eastern cottonwood in an eastern Nebraska provenance study. *Silvae Genet.* 25: 67–73.

Index

Physiological Ecology

A Series of Monographs, Texts, and Treatises

Continued from page ii

CODY, M. L. Habitat Selection in Birds, 1985

HAYNES, R. J., CAMERON, K. C., GOH, K. M., and SHERLOCK, R. R. Mineral Nitrogen in the Plant–Soil System, 1986

KOZLOWSKI, T. T., KRAMER, P. J., and PALLARDY, S. G. The Physiological Ecology of Woody Plants, 1991

DATE DUE
